Introduction to Applied Statistical Signal Analysis

The Aksen Associates Series in Electrical and Computer Engineering

Principles of Applied Optics
Partha P. Banerjee
Syracuse University
Ting-Chung Poon
Virginia Polytechnic Institute

Software Engineering
Stephen R. Schach
Vanderbilt University

Introduction to Applied Statistical Signal Analysis
Richard Shiavi
Vanderbilt University

Communication Networks: A First Course
Jean Walrand
University of California
Berkeley

Advisory Editors

Jacob A. Abraham
University of Texas at Austin

Leonard A. Gould
Massachusetts Institute of Technology

Frederic J. Mowle
Purdue University

James D. Plummer
Stanford University

Stuart C. Schwartz
Princeton University

Introduction to Applied Statistical Signal Analysis

Richard Shiavi
Vanderbilt University

Aksen
Associates
Incorporated
Publishers

IRWIN
Homewood, IL 60430
Boston, MA 02116

© Richard D. Irwin, Inc., and Aksen Associates, Inc., 1991

All rights reserved. No part of this publication may be reproduced, stored in a retrieval system, or transmitted, in any form or by any means, electronic, mechanical, photocopying, recording, or otherwise, without the prior written permission of the publisher.

Cover and text designer: *Harold Pattek*
Compositor: *Science Typographers, Inc.*
Typeface: $10\frac{1}{2}/12\frac{1}{2}$ Century Schoolbook
Printer: *R. R. Donnelley & Sons Company*

Library of Congress Cataloging-in-Publication Data

Shiavi, Richard.
Introduction to applied statistical signal analysis/Richard Shiavi.
p. cm.—(The Aksen Associates series in electrical and computer engineering)
Includes bibliographical references and index.
ISBN 0-256-08862-4
1. Signal processing—Statistical methods. I. Title.
II. Series.
TK5102.5.S474 1991
621.382′2—dc20

90–1164
CIP

Printed in the United States of America
1 2 3 4 5 6 7 8 9 0 DO 7 6 5 4 3 2 1 0

About the Author

Richard Shiavi is professor of biomedical and electrical engineering at Vanderbilt University in Nashville, Tennessee. Dr. Shiavi received the B.E. degree in electrical engineering at Villanova University and the M.S. and Ph.D. degrees in biomedical engineering from Drexel University. Since 1972, Dr. Shiavi has been actively engaged in teaching and research at Vanderbilt University and the Veterans Administration Medical Center. His primary professional interests are in signal processing and his main research interests are in electromyographic signal processing and knee and gait kinematics. His published research appears in the biomedical and biomechanics literature and conference proceedings, and he has written chapters for the *Handbook of Pattern Recognition and Image Processing* and for *Gait and Rehabilitation*. Dr. Shiavi is an active member of the Engineering in Medicine and Biology Society of the Institute of Electrical and Electronics Engineers and of the International Society for Electrophysiologic Kinesiology.

Table of Contents

	Preface	**xiii**
	Acknowledgments and Dedications	**xvii**
	List of Symbols	**xix**
Chapter 1	**Introduction and Terminology**	**1**
1.1	Introduction	1
1.2	Signal Terminology	4
	1.2.1 Domain Types	4
	1.2.2 Amplitude Types	5
	1.2.3 Basic Signal Forms	7
	1.2.4 The Transformed Domain, The Frequency Domain	10
	1.2.5 General Amplitude Properties	11
1.3	Analog-to-Digital Conversion	13
1.4	Measures of Signal Properties	14
	1.4.1 Time Domain	14
	1.4.2 Frequency Domain	15
	References	16
Chapter 2	**Empirical Modeling and Approximation**	**18**
2.1	Introduction	18
2.2	Model Development	21
2.3	Generalized Least Squares	26
2.4	Generalities	28

2.5	Models from Linearization	30
2.6	Orthogonal Polynomials	34
2.7	Interpolation and Extrapolation	40
	2.7.1 Lagrange Polynomials	43
	2.7.2 Spline Interpolation	48
2.8	Overview	55
	References	56
	Exercises	57
	Appendix	65
	2.1 Properties of Orthogonal Function Sets	65
Chapter 3	**Fourier Analysis**	**68**
3.1	Introduction	68
3.2	Overview of Fourier Transform Relationships	71
	3.2.1 Continuous versus Discrete Frequency	71
	3.2.2 Continuous versus Discrete Time	73
	3.2.3 Discrete Time and Frequency	75
3.3	Discrete Fourier Transform	77
	3.3.1 Definition Continued	77
	3.3.2 Partial Summary of DFT Properties	79
3.4	Fourier Analysis	82
	3.4.1 Frequency Range and Scaling	83
	3.4.2 The Effect of Discretizing Frequency	86
	3.4.3 The Effect of Truncation	87
	3.4.4 Windowing	93
	3.4.5 Resolution	94
	3.4.6 Detrending	98
3.5	Procedural Summary	100
3.6	Selected Applications	100
	References	104
	Exercises	106
	Appendices	109
	3.1 DFT of Ionosphere Data	109
	3.2 Brightness Data of a Variable Star	110
	3.3 International Airline Passenger Data	115
	3.4 Review of Fourier Series and Transform	115
	3.5 Data and Spectral Windows	124
Chapter 4	**Concepts of Probability and Estimation**	**126**
4.1	Introduction	126
4.2	Introduction to Random Variables	127
	4.2.1 Probability Descriptors	128

4.2.2 Moments of Random Variables 134
4.2.3 Gaussian Random Variable 136
4.3 Joint Probability 139
4.3.1 Bivariate Distributions 139
4.3.2 Moments of Bivariate Distributions 141
4.4 Concept of Sampling and Estimation 145
4.4.1 Sample Moments 145
4.4.2 Significance of the Estimate 148
4.5 Estimate of Correlation 151
4.6 Density Function Estimation 155
4.6.1 General Principle 155
4.6.2 Detailed Procedure 157
4.7 General Properties of Estimators 162
4.7.1 Convergence 162
4.7.2 Recursion 164
References 164
Exercises 166
Appendices 170
4.1 Plots and Formulae for Five Probability Density Functions 170
4.2 Values of the Error Function 172
4.3 Chart of River Flows 173
4.4 Student's *t* Distribution 173
4.5 Chi-Square Distribution 174
4.6 Time Intervals between Heartbeats 176

Chapter 5 Introduction to Random Processes and Time-Domain Description 178
5.1 Introduction 178
5.2 Definition of Stationarity 180
5.3 Definition of Moment Functions 183
5.3.1 General Definitions 183
5.3.2 Moments of Stationary Processes 184
5.4 Time Averages and Ergodicity 188
5.5 Estimating Correlation Functions 192
5.5.1 Estimator Definition 192
5.5.2 Estimator Bias 195
5.5.3 Consistency and Ergodicity 195
5.5.4 Sampling Properties 197
5.5.5 Asymptotic Distributions 198
5.6 Random Numbers and Random Process Simulation 201
5.6.1 Random Number Generation 201

5.6.2 Change of Mean and Variance 205
5.6.3 Density Shaping 205
5.6.4 Correlation and Structure 210
5.7 Assessing Stationarity of Signals 217
References 220
Exercises 222
Appendices 226
5.1 Yields from a Distillation Process 226
5.2 Wolfer's Sunspot Numbers 227
5.3 Variance of Autocovariance Estimate 228
5.4 Parametric Stationarity Tests 230

Chapter 6 Random Signals, Linear Systems, and Power Spectra 232
6.1 Introduction 232
6.2 Power Spectra 232
6.3 System Definition Review 235
6.3.1 Basic Definitions 235
6.3.2 Relationships between Input and Output 238
6.4 Systems and Signal Structure 241
6.4.1 Moving Average Process 241
6.4.2 Structure with Autoregressive Systems 243
6.4.3 Higher-Order AR Systems 248
6.5 Time-Series Models for Spectral Density 254
References 261
Exercises 262

Chapter 7 Spectral Analysis for Random Signals—Classical Estimation 266
7.1 Spectral Estimation Concepts 266
7.1.1 Developing Procedures 270
7.1.2 Sampling Moments of Estimators 272
7.2 Sampling Distribution for Spectral Estimates 278
7.2.1 Spectral Estimate for White Noise 278
7.2.2 Sampling Properties for General Random Processes 283
7.3 Consistent Estimators—Direct Methods 288
7.3.1 Spectral Averaging 288
7.3.2 Confidence Limits 292
7.3.3 Summary of Procedure for Spectral Averaging 302
7.3.4 Welch Method 303

7.3.5 Spectral Smoothing 304
7.3.6 Additional Applications 308
7.4 Consistent Estimators—Indirect Methods 310
7.4.1 Spectral and Lag Windows 310
7.4.2 Important Details for Using FFT Algorithms 316
7.4.3 Statistical Characteristics of BT Approach 316
7.5 Autocorrelation Estimation 326
References 326
Exercises 328
Appendices 332
7.1 Variance of Periodogram 332
7.2 Proof of Variance of BT Spectral Smoothing 336
7.3 Window Characteristics 338
7.4 Lag Window Functions 338
7.5 Spectral Estimates from Smoothing 339
7.6 Hospital Census Data 340

Chapter 8 Random Signal Modeling and Modern Spectral Estimation 341
8.1 Introduction 341
8.2 Model Development 345
8.3 Random Data Modeling Approach 348
8.3.1 Basic Concepts 348
8.3.2 Solution of the General Model 353
8.3.3 Model Order 356
8.3.4 Levinson–Durbin Algorithm 363
8.3.5 Burg Method 369
8.3.6 Summary of Signal Modeling 373
8.4 Spectral Density Function Estimation 374
8.4.1 Definition and Properties 374
8.4.2 Statistical Properties 381
8.4.3 Other Spectral Estimation Methods 384
8.4.4 Comparison of Modern and Classical Methods 384
References 386
Exercises 388
Appendices 391
8.1 Mean Squared Error for Two-Term Linear Prediction 391

8.2 Matrix Form of Levinson–Durbin Recursion 392
8.3 Surface Roughness for a Grinding Wheel 397
8.4 Average Annual Rainfall in the Eastern United States 398
8.5 Vibration in a Mechanical System 398

Chapter 9 Theory and Application of Cross Correlation and Coherence 399

9.1 Introduction 399
9.2 Properties of Cross Correlation Functions 402
9.2.1 Theoretical Function 402
9.2.2 Estimators 403
9.3 Detection of Time-Limited Signals 410
9.3.1 Basic Concepts 411
9.3.2 Application of Pulse Detection 413
9.3.3 Random Signals 415
9.3.4 Time Difference of Arrival 416
9.3.5 Marine Seismic Signal Analysis 419
9.3.6 Procedure for Estimation 419
9.4 Cross Spectral Density Functions 421
9.4.1 Definition and Properties 421
9.4.2 Properties of Cross-Spectral Estimators 424
9.5 Applications 427
9.6 Tests for Correlation between Time Series 431
9.6.1 Coherence Estimators 431
9.6.2 Statistical Properties of Estimators 432
9.6.3 Confidence Limits 434
9.6.4 Procedure for Estimation 437
9.6.5 Application 438
References 441
Exercises 443
Appendices 445
9.1 Bivariate Gas Furnace Data 445
9.2 Bivariate Temperature and Birth Data 449

Index 451

Preface

This book presents a practical introduction to signal analysis techniques that are commonly used in broad range of engineering areas, such as speech, biomedical signals, geophysics, acoustics, communications, pattern recognition, and so forth. In order to emphasize the analytic approaches, a certain background is necessary. The book is designed for an individual who has a basic background in mathematics, science, and computer programming that is required in an undergraduate engineering curriculum. In addition, one *needs* to have an *introductory-level* background in probability and statistics and discrete-time systems.

The sequence of material begins with a discussion of techniques for modeling and representing discrete data measurements, interpolation, and a definition of time series. Then the classical and some modern techniques for analyzing signals are treated. These topics include Fourier spectra, statistical properties of time series, correlation functions, spectral density functions, and time-series modeling. Integrated into this treatment of these techniques are the methodologies for estimating them properly. The presentation style is designed for the individual who wants a theoretical introduction to the basic principles and then the knowledge necessary to implement them practically. The mode of presentation is to define a theoretical concept, to show areas of engineering in which these concepts are useful, to define the algorithms and assumptions needed to implement them, and then to present detailed examples that have been implemented on a computer. The exposure to engineering applications hopefully will develop an appreciation for the utility and necessity of signal processing methodologies.

The exercises at the end of each chapter are designed with several goals. Some focus directly on the material presented and some extend the material for applications that are less often encountered. The degree of difficulty ranges from simple pencil-and-paper problems to computer implementation of algorithms. All computer-oriented examples and problems are directly implementable on personal computers. One can use either signal processing environments such as MATLAB and ILS or standard computing approach with a FORTRAN or C compiler integrated with suitable graphics hardware and software. There are many software subroutine packages that are useful. For an introductory course, the software recommended are those that are not so overly sophisticated and complex that the student cannot comprehend the code. The sources that have been used in the examples and exercises are listed at the end of the Preface. This software has been proven to be reliable and easily implementable. The libraries are sold with textbooks that explain how to implement them and briefly review the basic theory. An additional benefit with the software libraries from IEEE Press and Wiley is that test programs are available in order to be able to check the correctness of the software.

When used as a course textbook, most of the material in this book can be studied in one semester in a senior undergraduate or first-year graduate course. The topic selection is obviously the instructor's choice. A short review of Fourier transforms and probability and statistics is provided in Chapters 3 and 4.

Software Libraries

Digital Signal Processing Committee (1979). *Programs for Digital Signal Processing*. IEEE Press, New York.*

Peerless Engineering Service (1984). *Scientific Subroutine Library*. John Wiley and Sons, New York.†

Press, W., B. Flannery, S. Teukolsky, and W. Vetterling (1986). *Numerical Recipes—The Art of Scientific Computing*. Cambridge University Press, New York.†

Stearns, S. and R. David (1988). *Signal Processing Algorithms*. Prentice-Hall, Englewood Cliffs, NJ.*

Signal Processing Environments

MATLAB—The MathWorks, Inc. Sherborn, MA.

ILS, Interactive Laboratory System—Signal Technology, Inc. Goleta, CA.

ISP, The Interactive Signal Processor—Bedford Research. Bedford, MA.

SIG, A General Purpose Signal Processing Package—Techni-Soft; P.O. Box 2525, Livermore, CA.

TIMESLAB—Newton, J. (1988). *Timeslab: A Time Series Analysis Laboratory*. Wadsworth and Brooks/Cole, Pacific Grove, CA.

*FORTRAN libraries are available.
†FORTRAN and C libraries are available.

Dedications

This book is dedicated to my wife, Gloria, and to my parents who encouraged me and gave me the opportunity to be where I am today.

Acknowledgments

The author of a textbook is usually helped significantly by the institution by which he is employed and through surrounding circumstances. In particular I am indebted to the Department of Biomedical Engineering and the School of Engineering at Vanderbilt University for giving me some released time and for placing a high priority on writing this book for academic purposes. The reviewers have been very helpful in their constructive criticism toward developing a presentation with balance of theory and application. In particular, I would like to thank, in alphabetical order, K. S. Arun—University of Illinois, Edward Delp—Purdue University, O. K. Esroy—Purdue University, Alfred Hero—University of Michigan, and David Munson—University of Illinois for their detailed remarks and scholarly review of the manuscript.

List of Symbols

Roman

$a(i)$, $b(i)$	Parameters of AR, MA, and ARMA models
A_m	Polynomial coefficient
B	Bandwidth
B_e	Equivalent bandwidth
$c_x(k)$	Sample covariance function
$c_{yx}(k)$	Sample cross covariance function
C_n	Coefficients of Trignometric Fourier series
$C_x(k)$	Autocovariance function
$C_{xy}(k)$	Cross covariance function
Cov[ ]	Covariance operator
$d(n)$	Data window
$D(f)$	Data spectral window
e_i	Error in polynomial curve fitting
$E[\ ]$	Expectation operator
E_m	Sum of squared errors
E_{tot}	Total signal energy
f	Cyclic frequency
f_d	Frequency spacing
f_N	Folding frequency, highest frequency component
f_s	Sampling frequency
$f(t)$	Scalar function of variable t
$f_x(\alpha)$, $f(x)$	Probability density function

$f_{xy}(\alpha, \beta)$, $f(x, y)$	Bivariate probability density function
$F_x(\alpha)$, $F(x)$	Probability distribution function
$F_{xy}(\alpha, \beta)$, $F(x, y)$	Bivariate probability distribution function
g	Loss coefficient
$h(t)$, $h(n)$	Impulse response
$H(f)$, $H(\omega)$	Transfer function
$I(f)$, $I(m)$	Periodogram
Im( )	Imaginary part of a complex function
$\Im[\ ]$	Imaginary operator
$K^2(f)$, $K^2(m)$	Magnitude-squared coherence function
$L_i(x)$	Lagrange coefficient function
m	Mean
N	Number of points in a discrete-time signal
p, q	Order of AR, MA, and ARMA processes
P	Signal power, or signal duration
$P[\ ]$	Probability of [ ]
$P_m(x)$	Polynomial function
Re( )	Real part of a complex function
$R_x(k)$	Autocorrelation function
$R_{yx}(k)$	Cross correlation function
$\Re[\ ]$	Real operator
s_p^2	Variance of linear prediction error
$S(f)$, $S(m)$	Power spectral density function
$S_{yx}(f)$, $S_{yx}(m)$	Cross spectral density function
T	Sampling interval
$U(t)$	Unit step function
Var[ ]	Variance operator
$w(k)$	Lag window
$W(f)$	Lag spectral window
$x(t)$, $x(n)$	Time function
$X(f)$, $X(m)$, $X(\omega)$	Fourier transform
Z_n	Coefficients of complex Fourier series

Greek

α	Significance level
$\gamma_x(t_0, t_1)$, $\gamma_x(k)$	Ensemble autocovariance function
$\epsilon(n)$	Linear prediction error

$\delta(t)$	Impulse function, dirac delta function
$\delta(n)$	Unit impulse, Kronecker delta function
λ_i	Energy in a function
$\Lambda_{yx}(f)$	Cospectrum
$\eta(n)$	White noise process
$\xi(\tau)$	Ensemble normalized autocovariance function
Ξ	Gaussian probability distribution function
ρ	Correlation coefficient
$\rho_x(k)$	Normalized autocovariance function
$\rho_{yx}(k)$	Normalized cross covariance function
σ^2	Variance
σ_e	Standard error of estimate
σ_{xy}^2	Covariance
$\phi(f)$	Phase response
$\phi_{yx}(f)$, $\phi_{yx}(m)$	Cross phase spectrum
$\Phi_n(t)$	Orthogonal function set
$\varphi_x(t_0, t_1)$, $\varphi_x(k)$	Ensemble autocorrelation function
$\Psi_{yx}(f)$	Quadrature spectrum
ω	Radian frequency
ω_d	Radian frequency spacing

Operators

$X(f)^*$	Conjugation
$x(n) * y(n)$	Convolution
$\hat{S}(m)$	Sample estimate
$\tilde{S}(m)$	Smoothing
$\bar{x}(n)$	Periodic repetition

Acronyms

ACF	Autocorrelation function
ACVF	Autocovariance function
AIC	Akaike's information criterion
AR	Autoregressive
ARMA	Autoregressive moving average
BT	Blackman–Tukey
CCF	Cross correlation function

CCVF	Cross covariance function
CF	Correlation function
CSD	Cross spectral density
CTFT	Continuous-time Fourier transform
DFT	Discrete Fourier transform
DTFT	Discrete-time Fourier transform
erf	Error function
FPE	Final prediction error
FS	Fourier series
IDFT	Inverse discrete Fourier transform
IDTFT	Inverse discrete-time Fourier transform
LPC	Linear prediction coefficient
MA	Moving average
MEM	Maximum entropy method
MSC	Magnitude-squared coherence
MSE	Mean squared error
NACF	Normalized autocovariance function
NCCF	Normalized cross covariance function
pdf	Probability density function
PDF	Probability distribution function
PL	Process loss
PSD	Power spectral density
TSE	Total squared error
WN	White noise
YW	Yule–Walker

Chapter 1 Introduction and Terminology

1.1 Introduction

Historically, a *signal* meant any set of signs, symbols, or physical gesticulations that transmitted information or messages. The first electronic transmission of information was in the form of Morse code. In the most general sense a signal can be embodied in two forms. In one form, it is some measured or observed behavior or physical property of a phenomenon that contains information about that phenomenon. In the other form, the signal can be generated by a man-made system and have the information encoded. Signals can vary over time or space. Our daily existence is replete with signals and they occur not only in man-made systems but also in human and natural systems. A simple natural signal is the measurement of air temperature over time, as shown in Figure 1.1. Study of the fluctuations in temperature informs us about some characteristics of our environment. A much more complex phenomenon is speech. Speech is intelligence transmitted through a variation over time in the intensity of sound waves. Figure 1.2 shows an example of the intensity of a waveform associated with a typical sentence. Each sound has a different characteristic wave-shape that conveys different information to the listener. In television systems, the signal is the variation in electromagnetic wave intensity that encodes the picture information. In human systems, measurements of heart and skeletal muscle activity in the form of

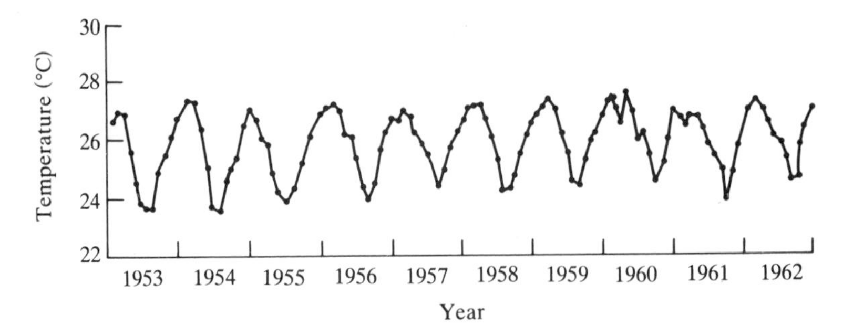

Figure 1.1 The average monthly air temperature at Recife, Brazil. [Adapted from Chatfield (1975), Figure 1.2, with permission.]

electrocardiographic and electromyographic voltages are signals. With respect to these last three examples, the objective of signal analysis is to process these signals in order to extract information concerning the characteristics of the picture, cardiac function, and muscular function. Signal processing has been implemented for a wide variety of applications. Many of them will be mentioned throughout this book. Good sources for other applications are Chen (1988) and Cohen (1986).

A time-dependent signal measured at particular points in time is called a *time series*. This term arose within the field of applied mathematics and initially pertained to the application of probability and statistical analysis to data varying over time. Some of the analyses were performed on economic or astronomic data, such as the Beveridge wheat price index or Wolfer's sunspot numbers [Anderson (1971)]. Many of the techniques that are used currently were devised by mathematicians decades ago. The invention of the computer and now the development of powerful and inexpensive computers has made the application of these techniques very feasible. The availability of inexpensive software of good quality has made their implementation widespread. All of the examples in this book were implemented using inexpensive subroutine libraries or programming environments, such as MATLAB and ILS, that are good for a broad variety of engineering and scientific applications. These libraries are also good from a pedagogical perspective because the algorithms are explained in the accompanying books. Refer to the Preface for a list of some of these libraries and

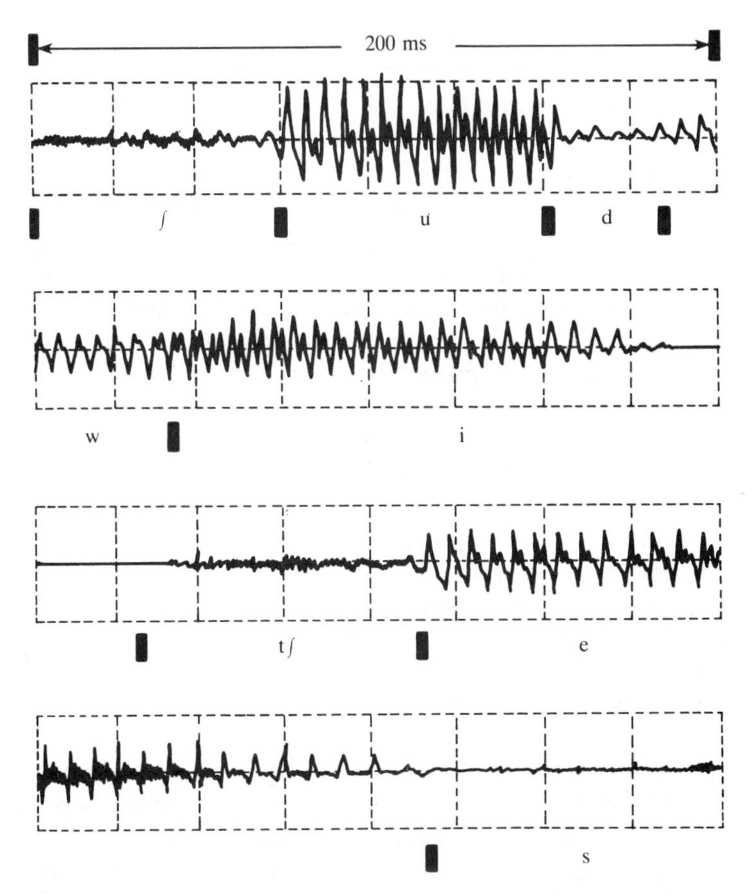

Figure 1.2 An example of a speech waveform illustrating different sounds. The utterance is "should we chase ··· ." [From Oppenheim (1978), Figure 3.3, with permission.]

programming environments. Before beginning a detailed study of the techniques and capabilities of signal or time-series analysis, an overview of terminology and basic properties of signal waveforms is necessary. As with any field, symbols and acronyms are a major component of the terminology, and standard definitions have been utilized as much as possible [Granger (1982) and Jenkins and Watts (1968)]. Other textbooks that will provide complementary information of either an introductory or an advanced level are listed in the reference section.

1.2 Signal Terminology

1.2.1 Domain Types

The *domain* of a signal is the independent variable over which the phenomenon is considered. The domain that is encountered most often is the time domain. Figure 1.3 shows the electrocardiogram (ECG) measured from the abdomen of a pregnant woman. The ECG exists at every instant of time; thus, the ECG evolves in the *continuous time domain*. Signals that have values at a finite set of instants exist in the *discrete time domain*. The temperature plot in Figure 1.1 shows a discrete time signal with average temperatures given each month. There are two types of discrete time signal. If the dependent variable is processed in some way, it is an *aggregate* signal. Processing can be averaging, such as the temperature plot, or summing, such as a plot of daily rainfall. If the dependent variable is not processed but represents only an instantaneous measurement, it is simply called *instantaneous*. The time interval between points, called the *sampling interval*, is very important. The time intervals between successive points in a time series are usually equal. However, there are several applications that require this interval to change. The importance of the sampling interval will be discussed in Chapter 3.

Another domain is the *spatial* domain and usually this has two or three dimensions, in the sense of having two or three independent variables. Images and moving objects have spatial components with these dimensions. Image analysis has become extremely important within the last decade. Applications are quite diverse and include medical imaging of body organs, robotic vision, remote sensing, and inspection of products on an assembly line.

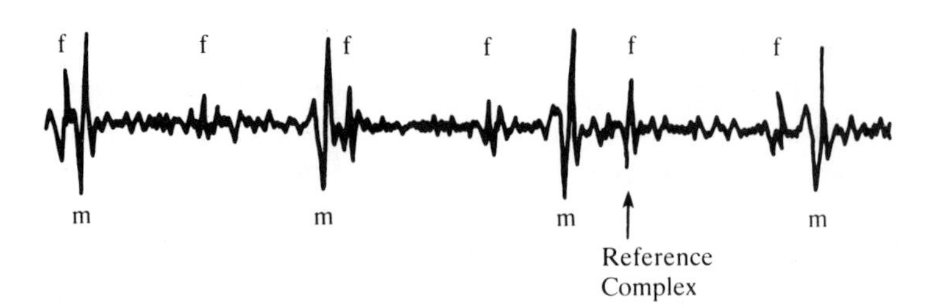

Figure 1.3 The abdominal ECG from a pregnant woman showing the maternal ECG waveform (m) and the fetal ECG waveform (f). [From Inbar (1975), page 73, Figure 8, with permission.]

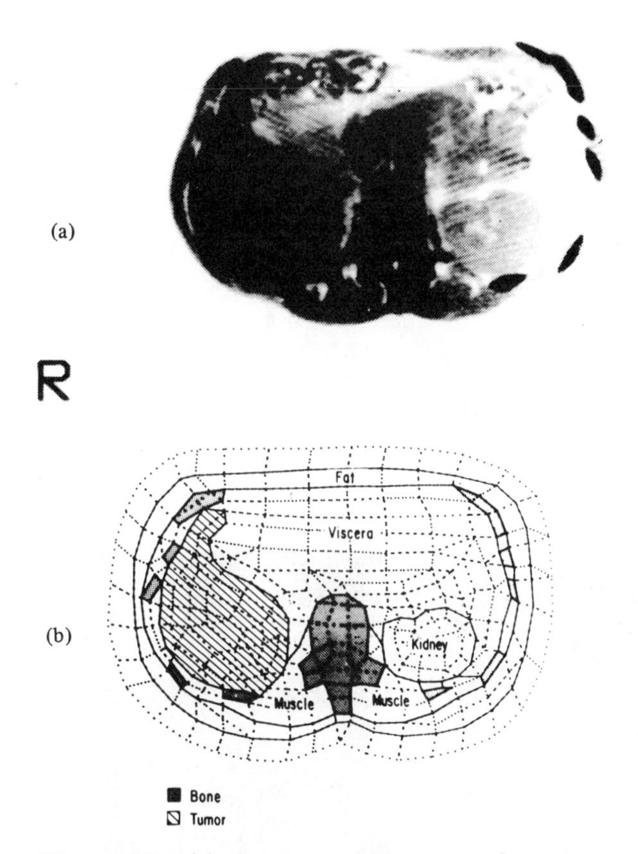

Figure 1.4 (a) A tomographic scan of a cross section of the abdomen. (b) A digitized model of the scan showing the calculated outlines of the major anatomical regions. [From Strohbehn and Douple (1984), Figure 4, with permission.]

The signal is the amount of whiteness (called gray level) or color in the image. Figure 1.4 illustrates the task of locating a tumor in a tomographic scan. In an assembly line the task may be to inspect objects for defects, as in Figure 1.5. The spatial domain can also be discretized for computerized analyses. Image analysis is an extensive topic of study and will not be treated in this book.

1.2.2 Amplitude Types

The amplitude variable, like the time variable, also can have different forms. Most amplitude variables, such as temperature, are

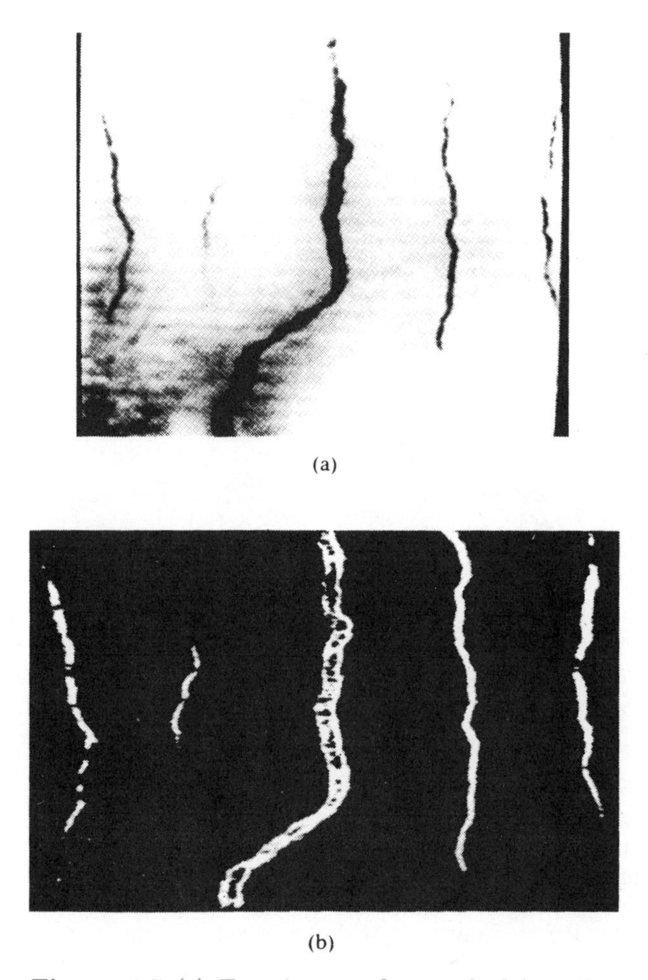

(a)

(b)

Figure 1.5 (a) Test image of a steel slab with imperfections. (b) Processed image with defects located. [From Suresh et al. (1983), Figures 12 and 13, with permission.]

continuous in magnitude. The most pertinent *discrete-amplitude* variations involve counting. An example is the presentation of the number of monthly sales in Figure 1.6. Other phenomena that involve counting are radioactive decay and routing processes (such as in telephone exchanges or other queueing processes). Another type of process exists that has no amplitude value. These are called *point processes* and occur when one is only interested in the time or place of occurrence. The study of neural coding of information

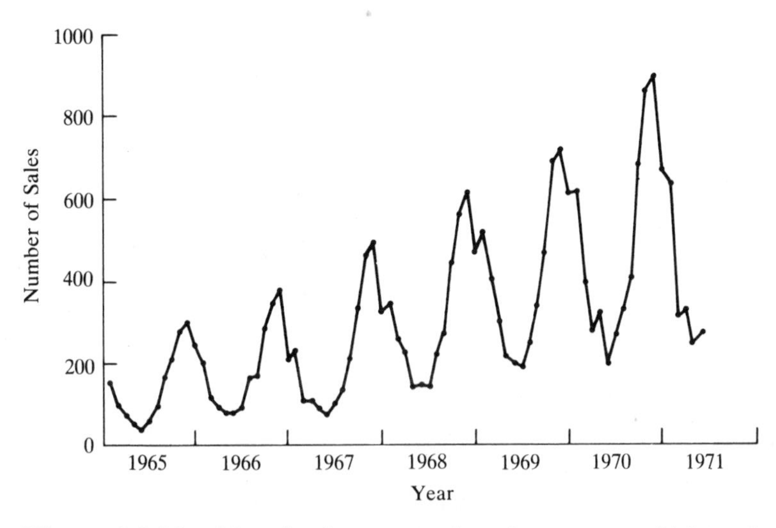

Figure 1.6 Monthly sales from an engineering company. [Adapted from Chatfield (1975), Figure 1.3, with permission.]

involves the mathematics of point processes. Figure 1.7 illustrates the reduction of a measured signal into a point process. The information is encoded in the time interval between occurrences or in the interaction between different channels (neurons).

1.2.3 Basic Signal Forms

There are different general types of signal forms. One concerns periodicity. A signal $x(t)$ is *periodic* if it exists for all time t

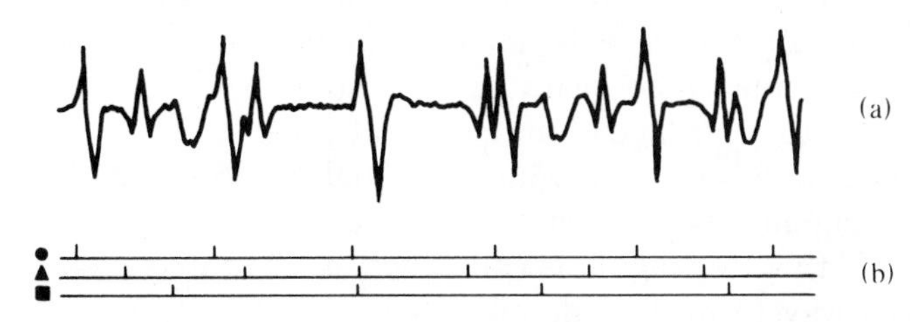

Figure 1.7 (a) An electromyographic signal containing three different waveform complexes. (b) Three impulse trains showing the times of occurrence of these complexes. [From Guiheneuc et al. (1983), Figure 1, with permission.]

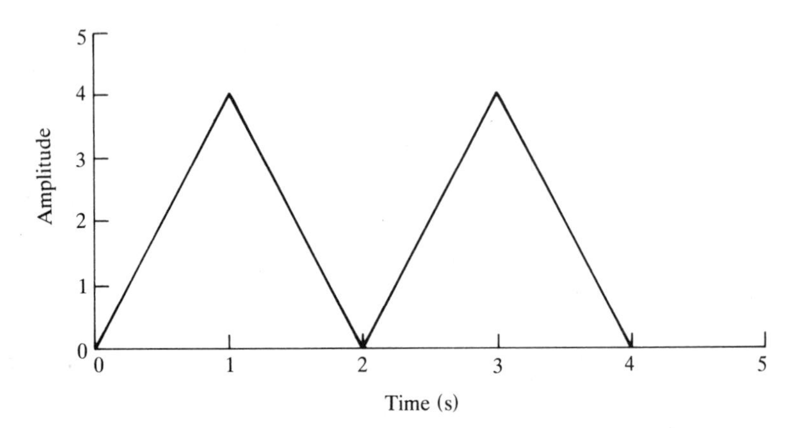

Figure 1.8 Two periods of a periodic triangular waveform.

and

$$x(t) = x(t + T), \tag{1.1}$$

where T is the duration of the period. These signals can be constituted as a summation of periodic waveforms that are harmonically related. The triangular waveform in Figure 1.8 is periodic. Some signals can be constituted as a summation of periodic waveforms that are not harmonically related. The signal itself is not periodic and is called *quasi-periodic*. Most signals are neither periodic nor quasi-periodic and are called *aperiodic*. Aperiodic signals can have very different waveforms as shown in Figures 1.9 and 1.10. Figure 1.9 shows a biomedical signal, an electroencephalogram, with some spike features indicated by dots. Figure 1.10 shows the output voltage of an electrical generator.

The time span over which a signal is defined is also important. If a signal has zero value or is nonexistent during negative time, $t < 0$, then the signal is called *causal*. The unit step function is a causal waveform. It is defined as

$$U(t) = \begin{cases} 1, & t \geq 0, \\ 0, & t < 0. \end{cases} \tag{1.2}$$

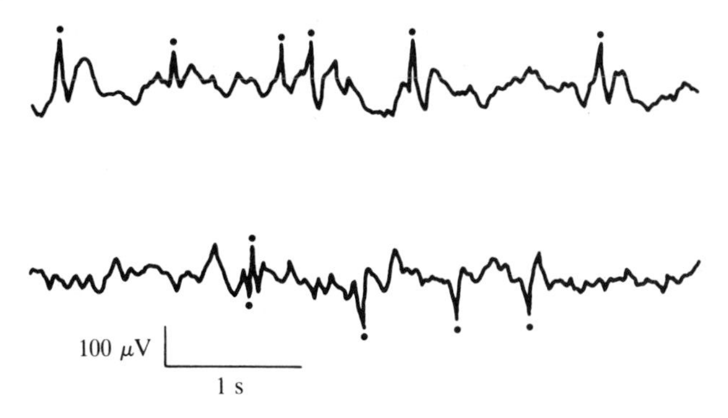

Figure 1.9 Electroencephalographic signal with sharp transients marked with dots. [From Glover et al. (1986), Figure 1, with permission.]

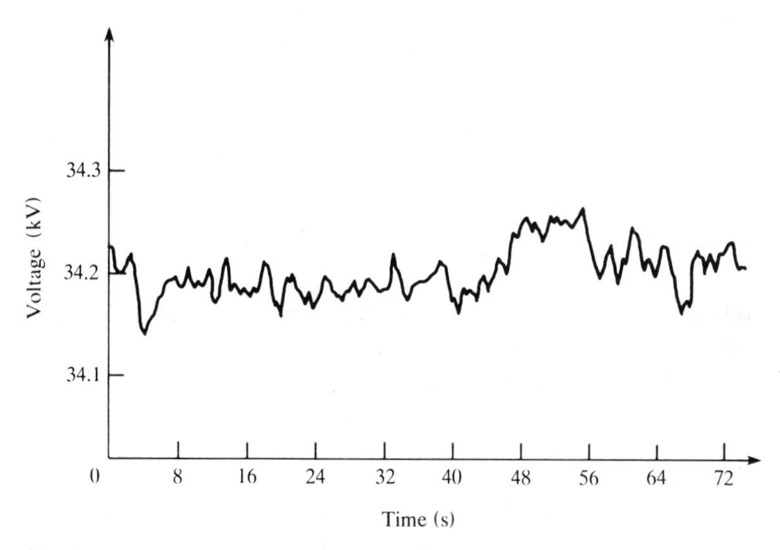

Figure 1.10 Output voltage signal from an electrical generator used for process control. [Adapted from Jenkins and Watts (1968), Figure 1.1, with permission.]

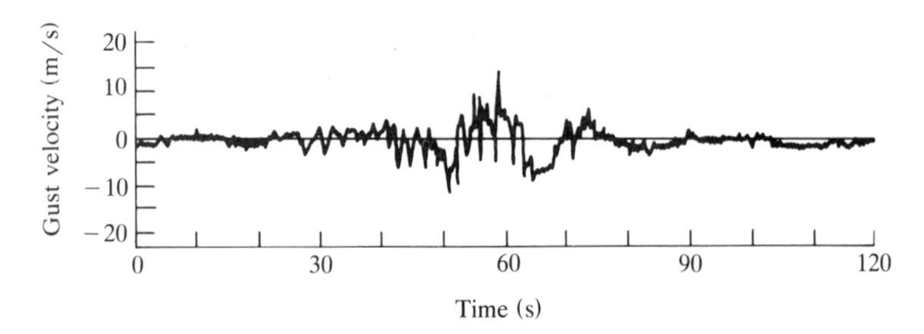

Figure 1.11 Record of a wind gust velocity measurement. [From Bendat and Piersol (1980), Figure 1.3, with permission.]

Any signal can be made causal by multiplying it by $U(t)$. If a signal's magnitude approaches zero after a relatively short time, it is *transient*. An example of a transient waveform is a decaying exponential function that is defined during positive time; that is,

$$x(t) = e^{-at}U(t), \qquad a > 0. \tag{1.3}$$

The wind gust velocity measurement shown in Figure 1.11 is a transient signal.

1.2.4 The Transformed Domain, The Frequency Domain

Other domains for studying signals involve mathematical transformations of the signal. A very important domain over which the information in signals is considered is the *frequency domain*. Knowledge of the distribution of signal strength or power over different frequency components is an essential part of many engineering endeavors. At this time it is best understood in the form of the Fourier series. Recall that any periodic function, with a period of T units, as plotted in Figure 1.8, can be mathematically modeled as an infinite sum of trigonometric functions. The frequency terms in these functions are *harmonics*, integer multiples, of the *fundamental* frequency ω_0. The form is

$$x(t) = C_0 + \sum_{n=1}^{\infty} C_n \cos(n\omega_0 t + \theta_n), \tag{1.4}$$

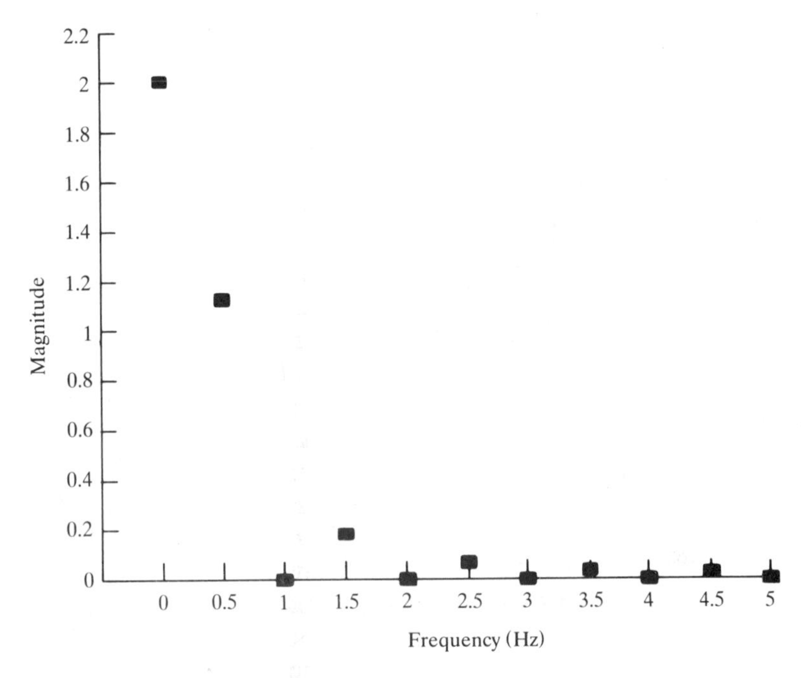

Figure 1.12 Magnitude spectrum of the periodic triangular waveform in Figure 1.8.

where $x(t)$ is the function, $\omega_0 = 2\pi/T$, C_n are the harmonic magnitudes, $n\omega_0$ are the harmonic frequencies, and θ_n are the *phase angles*. The signal can now be studied with the harmonic frequencies assuming the role of the independent variable. Information can be gleaned from the plots of C_n versus ω_n, called the *magnitude spectrum*, and θ_n versus ω_n, called the *phase spectrum*. The magnitude spectrum for the periodic waveform in Figure 1.8 is shown in Figure 1.12. Different signals have different magnitude and phase spectra. Signals that are aperiodic also have a frequency domain representation and are much more prevalent than periodic signals. This entire topic will be studied in great detail under the titles of frequency and spectral analysis.

1.2.5 General Amplitude Properties

There are two important general classes of signal that can be distinguished by waveform structure: deterministic and random. Many signals exist whose future values can be determined with

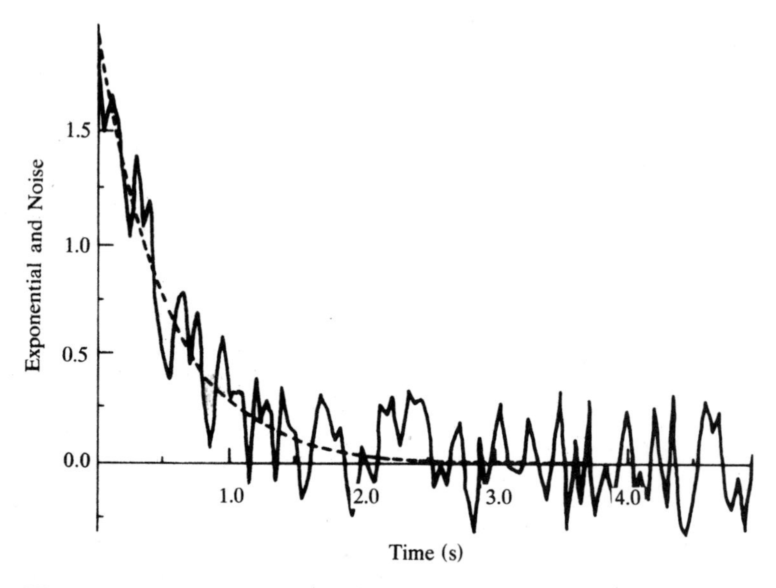

Figure 1.13 Random signal (—) with exponential trend (– –).

certainty if their present value and some parameters are known. These are called *deterministic* and include all waveforms that can be represented with mathematical formulae, such as cosine functions, exponential functions, square waves, and so forth. *Random*, or stochastic, signals are those whose future values cannot be determined with absolute certainty based upon knowing present and past values. Figure 1.13 is an example. Notice there is a decaying exponential trend that is the same as in the deterministic waveform; however, there are no methods to predict the exact amplitude values. Figures 1.9–1.11 are examples of actual random signals. The concepts of probability must be used to describe the properties of these signals.

A property associated with random signals is whether their characteristics change with time. Consider again the temperature signal in Figure 1.1. If one calculates an average and maximum and minimum values over short periods of time, they do not change; this is a *stationary* signal. Contrast this with the trends in the sales quantities in Figure 1.6. Similar calculations show that these parameters change over time; this signal is *nonstationary*. In general, stationary signals have average properties and characteristics that do not change with time, whereas the average properties and characteristics of nonstationary signals do change with time. These concepts will be considered in detail in subsequent chapters.

1.3 Analog-to-Digital Conversion

As mentioned previously, most signals that are encountered in man-made or naturally occurring systems are continuous in time and amplitude. However, because it is desired to perform computerized signal analysis, the signal values must be acquired by the computer. The input process is called *analog-to-digital* (A/D) *conversion*. This process is schematically diagrammed in Figure 1.14. The signal $g(t)$ is measured continuously by a sensor with a transducer. The transducer converts $g(t)$ into an electrical signal $f(t)$. Usually, the transduction process produces a linear relationship between these two signals. The sampler measures $f(t)$ every T time units and converts it into a *discrete time sequence*, $f(nT)$. Typical A/D converters are capable of taking from 0 to 100,000 samples per second. The technology of telecommunications and radar utilizes A/D converters with sampling rates up to 100 MHz. Also inherent in the process is the *quantization* of the magnitude of the signal. Computer memory is composed of words with a finite bit length. Within the hardware of the A/D converter, each sampling (measurement) of the analog signal's magnitude is converted into a digital word of a finite bit length. These integer words are then stored in memory. The word length varies from 4 to 32 bits. For many applications, a 12-bit quantization has sufficient accuracy to ignore the quantization error. For applications requiring extreme precision or analyzing signals with a large *dynamic range*, defined as a range of magnitudes, converters with the longer word lengths are utilized.

For mathematical operations that are implemented in software, the set of numbers is transformed into a floating-point

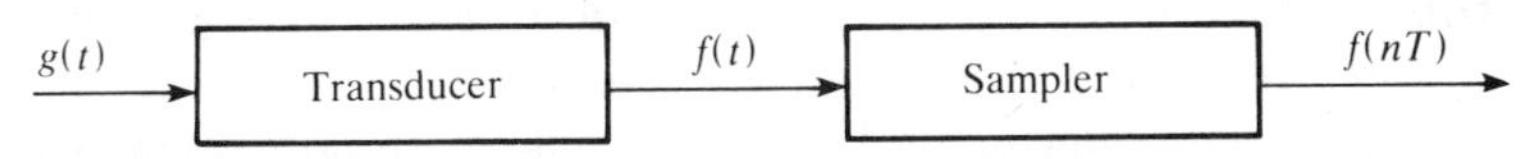

Figure 1.14 Process to convert an analog signal to a discrete-time signal.

representation that have proper units such as voltage, force, degrees, and so forth. For mathematical operations that are implemented in hardware, such as in mathematical coprocessors or in special-purpose digital signal processing chips, the set of numbers remains in integer form but word lengths are increased up to 80 bits.

1.4 Measures of Signal Properties

There are many measures of signal properties that are used to extract information from or to study the characteristics of signals. A few of the useful simple ones will be defined in this section. Many others will be defined as the analytic techniques are studied throughout the text. Initially, both the continuous-time and discrete-time versions will be defined.

1.4.1 Time Domain

The predominant number of measures quantize some property of signal magnitude as it varies over time or space. The simplest measures are the maximum and minimum values. Another measure that everyone uses intuitively is the average value. The magnitude of the *time average* is defined as

$$x_{av} = \frac{1}{P}\int_{O}^{P} x(t)\, dt \tag{1.5}$$

in continuous time and

$$x_{av} = \frac{1}{N}\sum_{n=1}^{N} x(nT) \tag{1.6}$$

in discrete time, where P is the time duration, N the number of data points, and T the sampling interval.

Signal energy and power are also important parameters. They provide a major classification of signals and sometimes determine the types of analyses that can be applied [Cooper and McGillem

(1967)]. *Energy* is defined as

$$E = \int_{-\infty}^{\infty} x^2(t)\, dt \tag{1.7}$$

or, in discrete time,

$$E = T \sum_{n=-\infty}^{\infty} x^2(nT). \tag{1.8}$$

An *energy signal* is one in which the energy is finite. Examples are pulse signals and transient signals, such as the wind gust velocity measurement in Figure 1.11. Sometimes signal energy is infinite, as in periodic waveforms such as the triangular waveform in Figure 1.8. However, for many of these signals the power can be finite. *Power* is energy averaged over time and is defined as

$$P = \lim_{T \to \infty} \frac{1}{2T} \int_{-T}^{T} x^2(t)\, dt \tag{1.9}$$

or, in discrete time,

$$P = \frac{1}{2N+1} \sum_{n=-N}^{N} x^2(nT). \tag{1.10}$$

Signals with nonzero and finite power are called *power signals*. The class of periodic functions always have finite power.

1.4.2 Frequency Domain

Power and energy as they are distributed over frequency are also important measures. Again, periodic signals will be used to exemplify these measures. From elementary calculus, the power in constant and sinusoidal waveforms is known. The power in the average component with magnitude C_0 is C_0^2/T. For the sinusoidal components with amplitude C_1, the power is $C_1^2/2$. Thus, for a periodic signal, the power P_M within the first M harmonics is

$$P_M = \frac{C_0^2}{T} + 0.5 \sum_{m=1}^{M} C_m^2. \tag{1.11}$$

This is called the *integrated power*. A plot of P_M versus harmonic frequency is called the *integrated power spectrum*. More will be studied about frequency-domain measures in subsequent chapters.

References

Anderson, T. (1971). *The Statistical Analysis of Time Series*. John Wiley and Sons, New York.

Bendat, J. and A. Piersol (1980). *Engineering Applications of Correlation and Spectral Analysis*. John Wiley and Sons, New York.

Cadzow, J. (1987). *Foundations of Digital Signal Processing and Data Analysis*. Macmillan Publishing Co., New York.

Chatfield, C. (1975). *The Analysis of Time Series: Theory and Practice*. John Wiley and Sons, New York.

Chen, C. (1988). *Signal Processing Handbook*. Marcel Dekker, New York.

Cohen, A. (1986) *Biomedical Signal Processing*, Vol. 2: *Compression and Automatic Recognition*. CRC Press, Boca Raton, FL.

Cooper, G. and C. McGillem (1967). *Methods of Signal and Systems Analysis*. Holt, Rinehart and Winston, New York.

Glover, J., P. Ktonas, N. Raghavan, J. Urunuela, S. Velamuri, and E. Reilly (1986 IEEE). A multichannel signal processor for the detection of epileptogenic sharp transients in the EEG. *IEEE Trans. Biomed. Engrg.* 33:1121–1128.

Granger, C. (1982). Acronyms in time series analysis (ATSA). *J. Time Series Analysis* 3:103–107.

Guiheneuc, P., J. Calamel, C. Doncarli, D. Gitton, and C. Michel (1983). Automatic detection and pattern recognition of single motor unit potentials in needle EMG. In *Computer-Aided Electromyography* (J. Desmedt, ed.). S. Karger, Basel.

Inbar, G. (1975). *Signal Analysis and Pattern Recognition in Biomedical Engineering*. John Wiley and Sons, New York.

Jenkins, G. and D. Watts (1968). *Spectral Analysis and Its Applications*. Holden-Day, San Francisco.

Oppenheim A. (1978) *Applications of Digital Signal Processing*. Prentice-Hall, Englewood Cliffs, NJ.

Schwartz, M. and L. Shaw (1975). *Signal Processing: Discrete Spectral Analysis, Detection, and Estimation*. McGraw-Hill Book Co., New York.

Strohbehn, J. and E. Douple (1984 IEEE). Hyperthermia and cancer therapy: A review of biomedical engineering contributions and challenges. *IEEE Trans. Biomed. Engrg*. 31:779–787.

Suresh, B., R. Fundakowski, T. Levitt, and J. Overland (1983 IEEE). A real-time automated visual inspection system for hot steel slabs. *IEEE Trans. Pattern Anal. Machine Intell*. 5:563–572.

Chapter 2 Empirical Modeling and Approximation

2.1 Introduction

Many situations exist when it is necessary to discover or develop a relationship between two measured variables. This occurs in the study of physics, biology, economics, engineering, and so forth. Often, however, neither a priori nor theoretical knowledge is available regarding these variables, or their relationship is very complicated. Examples include the relationship between the heights of parents and children, cigarette smoking and cancer, operating temperature and rotational speed in an electric motor, and resistivity and deformation in a strain gauge transducer. Thus, one must resort to *empirical modeling* of the relationship. Techniques for empirical modeling have been available for quite some time and are alternatively called *curve fitting* in engineering, *regression analysis* in statistics, or time series *forecasting* in economics. The fundamental principles for modeling are the same in all of these areas.

We examine now a few applications that are interesting to physiologists, engineers, or educators. In Figure 2.1 are plotted two sets of measurements from walking studies relating step length to body height in men and women. The plot of the measured data points is called a *scatter diagram*. There is clearly an increasing trend in the data even though there may be several values of step length for a certain height. Straight lines should be good approximations for these sets of data and they are also drawn in the figure. Notice that for a certain height, no step-length values for either men or women are predicted exactly. There is some error. How does this occur? There are two possibilities; there is some error in the measurements and more independent variables are necessary for a more accurate approximation. Perhaps if step length is modeled as a function of both height and age, the modeling could be more accurate. Thus the model becomes more complex. Let us restrict ourselves to bivariate relationships. Figure 2.2 shows the data relating the electrical activity in a muscle and the muscular force generated when the muscle's nerve is stimulated. Notice that this data has a curvilinear trend, so a polynomial curve would be more suitable than a linear one. A suitable quadratic model is also shown in the figure. Figure 2.3 shows another second-order relationship from protein data.

A capability provided by models is the ability to estimate values of the dependent variable for desired values of the independent variable when the actual measurements are not available. For instance, in Figure 2.3, one can estimate that a protein concentra-

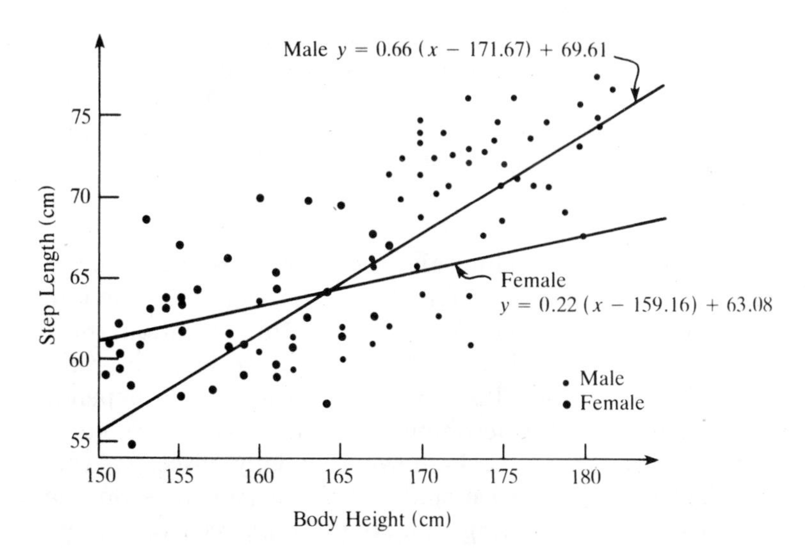

Figure 2.1 Scatter diagrams and linear models relating walking step length and body height in men and women. [Adapted from Hirokawa and Matsumara (1987), Figure 11, with permission.]

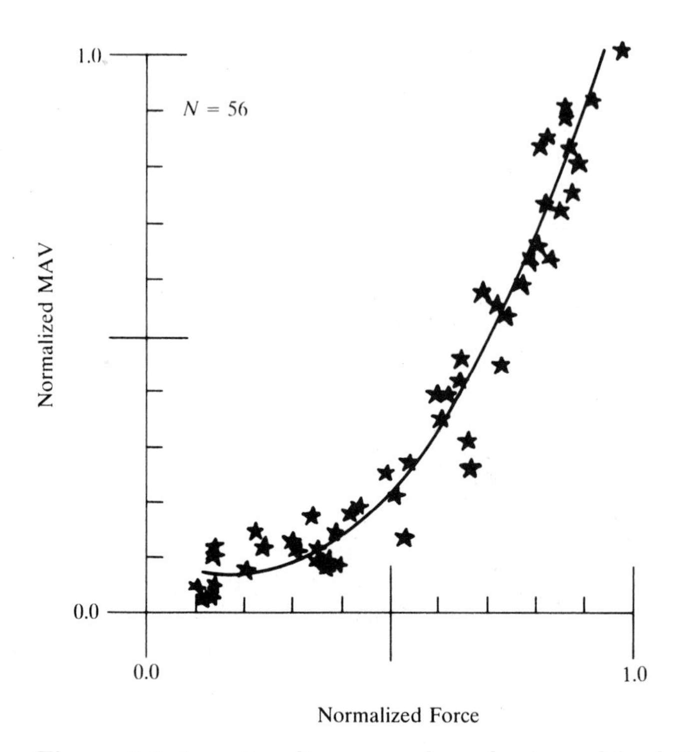

Figure 2.2 A scatter diagram and quadratic model relating normalized electrical activity in a muscle, MAV, and the normalized force, F, generated by the muscle when its nerve is stimulated. The model is $\text{MAV} = 0.12 - 0.60F + 1.57F^2$. [Adapted from Solomonow et al. (1986), Figure 7, with permission.]

tion of 8 g/dl will produce an oncotic pressure of 30 Torr. This is a common situation when using tables, for example, when looking up values of trigonometric functions. In this situation the available information or data is correct and accurate but one still needs to determine accurately the unavailable values of a dependent variable. Techniques for accomplishing this task are called *interpolation*. Finally, Figure 2.4 shows the daily viscosity values of a chemical product. Notice that the viscosity values fluctuate cyclically over time. Handling this type of data requires different techniques that will be considered in subsequent chapters. In this chapter we will focus on developing linear and curvilinear relationships between two measured variables for the purposes of modeling or interpolation.

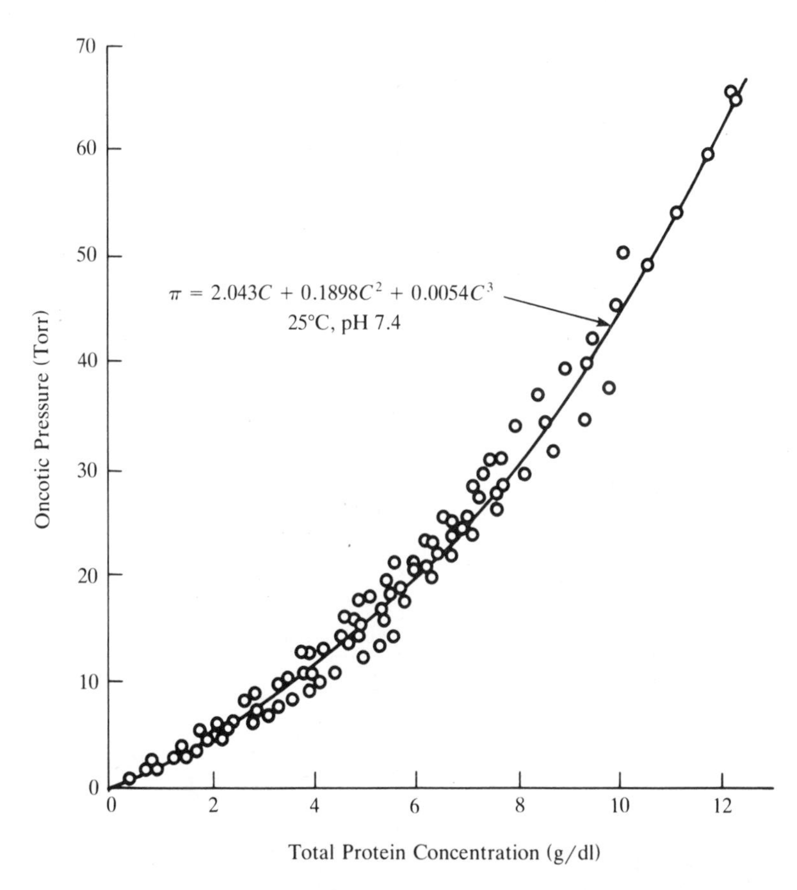

Figure 2.3 The scatter diagram and quadratic model relating protein osmotic (oncotic) pressure and protein concentration. [Adapted from Roselli et al. (1980), with permission.]

2.2 Model Development

Whatever variables are being examined, they exist in pairs of values of the independent and dependent variables and a qualitative relationship can be appreciated by examining the Cartesian plot of the point pairs. Figure 2.5 shows a scatter diagram and two models for some stress–strain data [Dorn and McCracken (1972)]. The points seem to have a straight-line trend, and a line that seems to be an appropriate model is drawn. Notice that there is an error between the ordinate values and the model. The error for one of the

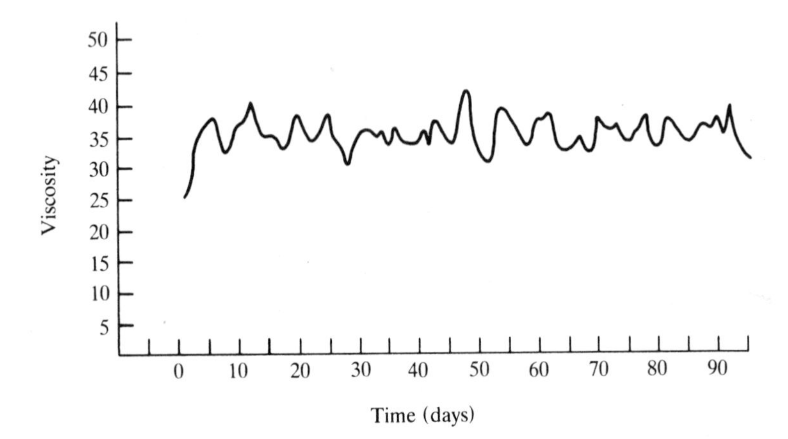

Figure 2.4 Daily reading of viscosity of chemical product. [From Bowerman and O'Connell (1987), Figure 2.11, with permission.]

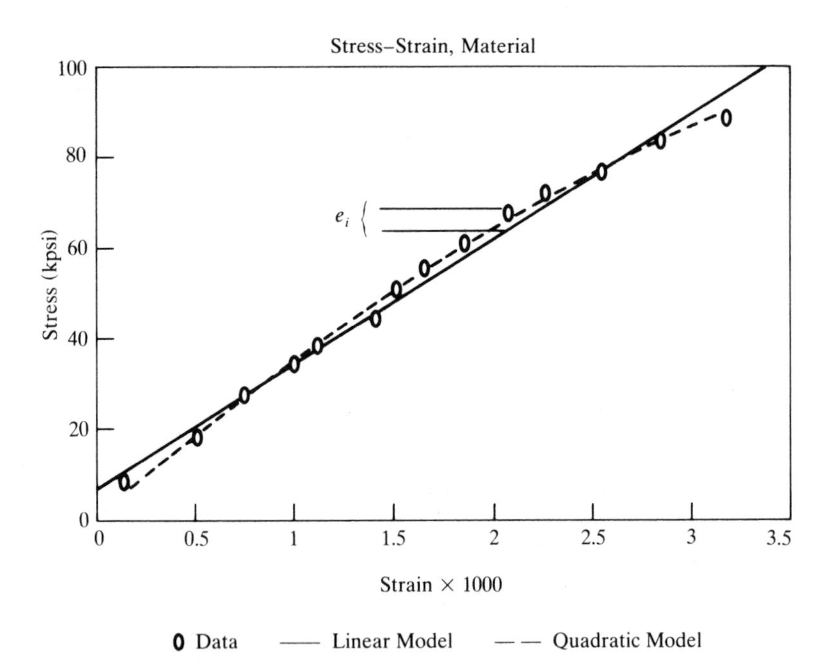

Figure 2.5 The scatter plot and two models of stress–strain data for a certain material.

point pairs is indicated as e_i. Some of the errors are smaller than others. If the line's slope is changed slightly, the errors change, but the model still appears appropriate. The obvious desire is of course for the line to pass *as close as possible* to all the data points. Thus, some definition of model accuracy must be made. An empirical relationship can be determined quantitatively through the use of curve-fitting techniques. An essential notion with these techniques is that the measure of accuracy or error is part of the model development. There are many error measures, but the most general and easily applicable are those that involve the minimization of the total or average squared error between the observed data and the proposed model. The resultant model then satisfies the *least squared error* criterion.

Consider now the modeling of a data set with polynomial functions, $y = f(x) = a + bx + cx^2 + \cdots$. The straight line is the simplest function and will be used to demonstrate the general concept of model development. Higher-order functions will then be used to improve the modeling but the general concept and procedure are the same. Begin by considering the data in Figure 2.5. The straight line with parameters a and b is

$$y = a + bx. \tag{2.1}$$

The value $\hat{y}_i$ estimated by equation (2.1) for point pair (y_i, x_i) is

$$\hat{y}_i = a + bx_i. \tag{2.2}$$

The error of estimation, e_i, is then

$$e_i = y_i - \hat{y}_i = y_i - a - bx_i. \tag{2.3}$$

The total squared error, E_m, for this model is

$$E_m = \sum_{i=1}^{N} e_i^2 = \sum_{i=1}^{N} (y_i - a - bx_i)^2, \tag{2.4}$$

where N is the number of data points. The quantity E_m is a measure of how well the line fits the entire set of points and is sometimes called the *sum of squared residuals*. E_m will be zero if and only if each of the points is on the line, and the farther the points are, on average, from the line, the larger the errors will become. The least-square error criterion *selects the parameters a and b so that the total squared error is as small as possible*. (Note that it is assumed that the surface produced by E_m has a well-defined minimum.)

To accomplish this, the usual procedure for minimizing a function of several variables is performed. The two first partial derivatives of equation (2.4) with respect to the unknown parameters (a and b) are equated to zero. The first equation is derived in detail. The derivation and verification of the second one is left as an exercise for the reader. Remember that the operations of summation and differentiation are commutable (interchangeable). Taking the first partial derivative yields

$$\frac{\partial E_m}{\partial a} = \frac{\partial}{\partial a} \sum_{i=1}^{N} (y_i - a - bx_i)^2 = 0$$

$$= -2 \sum_{i=1}^{N} (y_i - a - bx_i). \tag{2.5}$$

After rearranging by collecting terms, with the unknown terms on the right-hand side of the equation, the result is

$$\sum_{i=1}^{N} y_i = aN + b \sum_{i=1}^{N} x_i. \tag{2.6}$$

Similarly, the solution of the second minimization yields the equation

$$\sum_{i=1}^{N} y_i x_i = a \sum_{i=1}^{N} x_i + b \sum_{i=1}^{N} x_i^2. \tag{2.7}$$

The solution of this set of simultaneous equations is a standard procedure.

Example 2.1

Fit the stress versus strain data of Figure 2.5 with a straight-line model by the method of least squares. The data and

Table 2.1 Stress [y_i (psi × 10^3)] versus Strain [x_i (strain × 10^{-3})]

i	y_i	x_i
1	8.37	0.15
2	17.9	0.52
3	27.8	0.76
4	34.2	1.01
5	38.8	1.12
6	44.8	1.42
7	51.3	1.52
8	55.5	1.66
9	61.3	1.86
10	67.5	2.08
11	72.1	2.27
12	76.9	2.56
13	83.5	2.86
14	88.9	3.19

are listed in Table 2.1 and the necessary sums are

$$\sum_{i=1}^{14} y_i = 728.87, \qquad \sum_{i=1}^{14} x_i = 22.98,$$

$$\sum_{i=1}^{14} x_i^2 = 48.08, \qquad \sum_{i=1}^{14} y_i x_i = 1480.76.$$

As you study this example, verify the results with a calculator. Solution of equations (2.6) and (2.7) produces $a = 6.99$ and $b = 27.46$. The resulting equation is

$$y = 6.99 + 27.46x \tag{2.8}$$

and is also plotted in Figure 2.5 with the measured data. Notice that this is a good approximation but that the data plot has some curvature. Perhaps a quadratic or higher-order model is more appropriate. The squared error is 103.

2.3 Generalized Least Squares

Viewed in somewhat more general terms, the method of least squares is simply a process for finding the best possible values for a set of $m + 1$ unknown coefficients, $a, b, c, \ldots,$ for an mth-order model. The general nonlinear model is

$$y = a + bx + cx^2 + dx^3 + \cdots . \tag{2.9}$$

The number of data points still exceeds the number of unknowns, that is $N > m + 1$. The same general principle is utilized; the total squared error of the model is minimized to find the unknown coefficients. The individual error, e_i, is then

$$e_i = y_i - \hat{y}_i = y_i - a - bx_i - cx_i^2 - \cdots . \tag{2.10}$$

The total squared error, E_m, for this general model becomes

$$E_m = \sum_{i=1}^{N} e_i^2 = \sum_{i=1}^{N} (y_i - a - bx_i - cx_i^2 - \cdots)^2. \tag{2.11}$$

To minimize E_m, obtain the first partial derivatives of equation (2.11) with respect to the $m + 1$ unknown parameters $(a, b, \ldots)$ and equate them to zero. For the coefficient b the procedure is

$$\frac{\partial E_m}{\partial b} = \frac{\partial}{\partial b} \sum_{i=1}^{N} (y_i - a - bx_i - cx_i^2 - \cdots)^2 = 0$$

$$= -2 \sum_{i=1}^{N} (y_i - a - bx_i - cx_i^2 - \cdots)x_i. \tag{2.12}$$

The results of the partial derivatives will yield $m + 1$ equations similar in form to equation (2.12). The only difference will be in the power of x_i on the extreme right end. Then one must solve the simultaneous equations for the parameters. For a quadratic model, $m = 2$, after rearranging terms and collecting unknowns on the right-hand side of the equation, the equations are

$$\sum_{i=1}^{N} y_i = aN + b\sum_{i=1}^{N} x_i + c\sum_{i=1}^{N} x_i^2, \tag{2.13}$$

$$\sum_{i=1}^{N} y_i x_i = a\sum_{i=1}^{N} x_i + b\sum_{i=1}^{N} x_i^2 + c\sum_{i=1}^{N} x_i^3, \tag{2.14}$$

$$\sum_{i=1}^{N} y_i x_i^2 = a\sum_{i=1}^{N} x_i^2 + b\sum_{i=1}^{N} x_i^3 + c\sum_{i=1}^{N} x_i^4. \tag{2.15}$$

We have, thus, obtained a system of $m + 1$ simultaneous linear equations, called *normal* equations, in the $m + 1$ unknowns $(a, b, c, \dots)$, whose solution is now a routine matter. These equations can easily be structured into the matrix form

$$\begin{bmatrix} N & \sum_{i=1}^{N} x_i & \sum_{i=1}^{N} x_i^2 \\ \sum_{i=1}^{N} x_i & \sum_{i=1}^{N} x_i^2 & \sum_{i=1}^{N} x_i^3 \\ \sum_{i=1}^{N} x_i^2 & \sum_{i=1}^{N} x_i^3 & \sum_{i=1}^{N} x_i^4 \end{bmatrix} \begin{bmatrix} a \\ b \\ c \end{bmatrix} = \begin{bmatrix} \sum_{i=1}^{N} y_i \\ \sum_{i=1}^{N} y_i x_i \\ \sum_{i=1}^{N} y_i x_i^2 \end{bmatrix}. \tag{2.16}$$

Several numerical techniques and computer subroutine algorithms exist to solve this system of equations.

Example 2.2

Because the stress–strain data seemed to have some nonlinear trend, it will be modeled with the quadratic model from equations (2.13)–(2.16). The additional sums required are

$$\sum_{i=1}^{14} y_i x_i^2 = 3{,}433.7, \qquad \sum_{i=1}^{14} x_i^2 = 48.1,$$

$$\sum_{i=1}^{14} x_i^3 = 113.7, \qquad \sum_{i=1}^{14} x_i^4 = 290.7.$$

The solution of these equations yields $a = 0.77$, $b = 37.5$, and $c = -2.98$ for the model

$$y = 0.77 + 37.5x - 2.98x^2.$$

This is plotted also in Figure 2.5. The squared error E_2 is 27.4. It is obvious that the quadratic model is a better model. The next question is, how can this be quantitated. Initially, it was stated that the squared error is also the criterion of "goodness of fit." Comparing the errors of the two equations shows that the quadratic model is much superior.

2.4 Generalities

A problem in empirical curve approximation is the establishment of a criterion to decide the limit in model complexity. Reconsider the previous two examples. The stress–strain relationship is definitely nonlinear; however, is the quadratic model sufficient or is a cubic model needed? In general, the total squared error approaches zero as m increases. A perfect fit for the data results when $m + 1$ equals N and the resulting curve satisfies all the data points. However, the polynomial may fluctuate considerably between data points and may represent more artifact and inaccuracies than the general properties desired. In many situations a lower-order model is more appropriate. This will certainly be the case when measuring experimental data that theoretically should have a straight-line relationship but fails to show it because of errors in observation or measurement. To minimize the fluctuations, the degree of the polynomial should be much less than the number of data points. Judgment and knowledge of the phenomena being modeled are important in the decision making. Important also is the trend in E_m versus model order characteristics. Large decreases indicate that significant additional terms are being created, whereas small decreases reflect insignificant and unnecessary improvements. The error characteristic for modeling the stress–strain relationship is shown in Figure 2.6. Notice that the error decreases appreciably as m increases from 1 to 2; thereafter the decrease is slow. This indicates that a model with $m = 2$ is

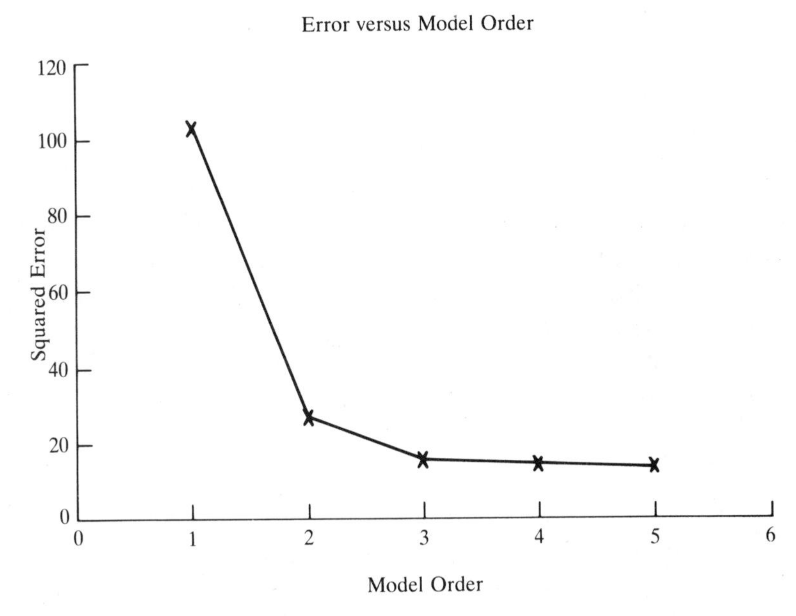

Figure 2.6 The total squared error–model order characteristic for the stress–strain data.

probably sufficient. An approximate indicator is the magnitude of the additional coefficients. If they are small enough to make the effect of their respective term negligible, then the additional complexity is unnecessary. For instance, consider the fifth-order model in the following equation:

$$y = 3.43 + 32.51x - 8.83x^2 + 10.97x^3 - 4.75x^4 + 0.63x^5. \tag{2.17}$$

Notice how the coefficients of the fourth- and fifth-order terms tend to become smaller. Another error measure is the square root of the average squared error, called the *standard error of the estimate* and defined by

$$\sigma_e = \sqrt{\frac{E_m}{N - (m + 1)}}. \tag{2.18}$$

It is also used to compare the accuracy of different models [Lapin (1983)].

Solution of these simultaneous linear equations is rather routine and many methods are available. Matrix methods using Cramér's rule or numerical methods such as Gauss elimination are solution approaches. Although low-order models usually pose no difficulty, numerical difficulties can be encountered when inverting the matrices formed for a high-order model or when the matrices are ill-conditioned. There are many engineering-oriented numerical methods texts that explain these numerical techniques. For instance, refer to Hamming (1962), Pearson (1986), and Wylie (1975). Almost every scientific subroutine library contains these subroutines.

2.5 Models from Linearization

Many types of nonlinear relationships cannot be modeled adequately with polynomial functions. These include natural phenomena like bacterial or radioactive decay, which have exponentially behaving characteristics. Regression analysis can be applied to data of these types if the relationship is linearizable through some mathematical transformation [Bowerman and O'Connell (1987) and Chatterjee and Price (1977)]. The exponential characteristic is one of the classical types. Consider the exponential model

$$y = \alpha e^{\beta x}. \tag{2.19}$$

Applying the logarithm to both sides of equation (2.19) yields

$$\ln y = \ln \alpha + \beta x \tag{2.20}$$

Making the substitutions

$$w = \ln y, \qquad a = \ln \alpha, \quad \text{and} \quad b = \beta,$$

produces the linearized model

$$w = a + bx, \tag{2.21}$$

where the data point pairs are $w_i = \ln y_i$ and x_i. Thus, the logarithmic transformation linearized the exponential model. It will be seen in general that logarithmic transformations can linearize all multiplicative models.

Example 2.3

An interesting example is the data set concerning the survival of marine bacteria being subjected to X-ray radiation. The number of surviving bacteria and the duration of exposure time are listed in Table 2.2 and plotted in Figure 2.7 [Chatterjee and Price (1977)]. The values of the linearized variable w are also listed in the table. Applying the regression equations to the data points w_i and x_i yields $a = 5.973$ and $b = -0.218$. Thus $\alpha = e^a = 392.68$ and $\beta = b = -0.218$. This model is also plotted in Figure 2.7 and shows good correspondence.

Table 2.2 Surviving Bacteria [y_i (number $\times 10^{-2}$)] versus Time [x_i (min)]

i	y_i	x_i	w_i
1	355	1	5.87
2	211	2	5.35
3	197	3	5.28
4	166	4	5.11
5	142	5	4.96
6	106	6	4.66
7	104	7	4.64
8	60	8	4.09
9	56	9	4.02
10	38	10	3.64
11	36	11	3.58
12	32	12	3.47
13	21	13	3.04
14	19	14	2.94
15	15	15	2.71

Another multiplicative model is the power-law relationship

$$y = \alpha x^{\beta}. \qquad (2.22)$$

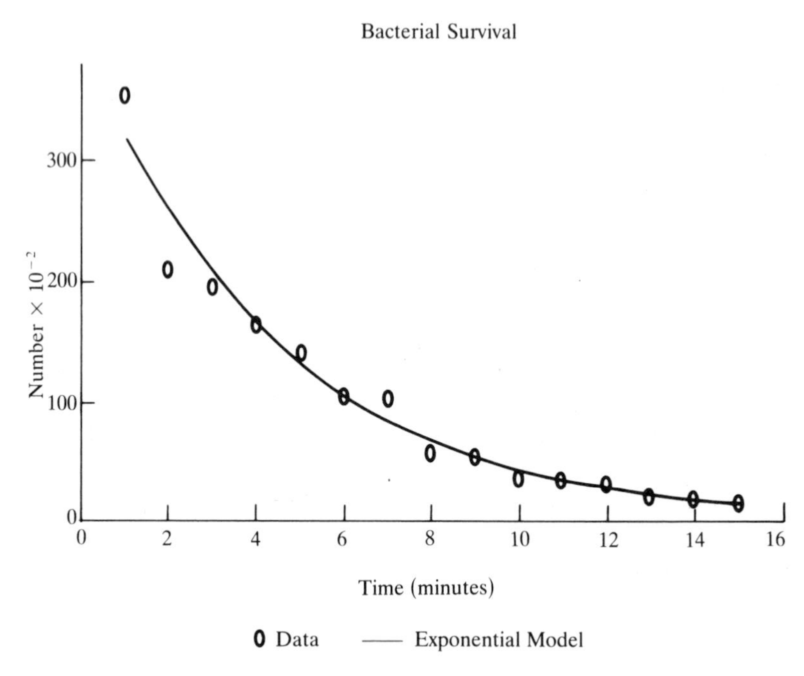

Figure 2.7 The number of surviving bacteria (in units of 100) is plotted against x-ray exposure time (in minutes).

The application of this model is very similar to Example 2.3. Other types of transformation exist if they are single-valued. Sometimes it can be suggested from the model itself. For instance, the model

$$y = a + b \ln x \tag{2.23}$$

is easily amenable to the transformation $z = \ln x$. Other situations that present themselves are not so obvious. Consider the model

$$y = \frac{x}{ax - b}. \tag{2.24}$$

The hyperbolic transformations $w = y^{-1}$ and $z = x^{-1}$ nicely produce the linear model

$$w = a + bz. \tag{2.25}$$

Additional applications of these types of transformation can be found in the exercises for this chapter.

Example 2.4

Another very useful multiplicative model is the product exponential form,

$$y = \alpha x e^{\beta x}.$$

The transformation necessary is simply to use the ratio of the two variables as the independent variable. The model now becomes

$$w = \frac{y}{x} = \alpha e^{\beta x}.$$

An appropriate application is the results of a compression test of a concrete cylinder. The stress–strain data are plotted in Figure 2.8

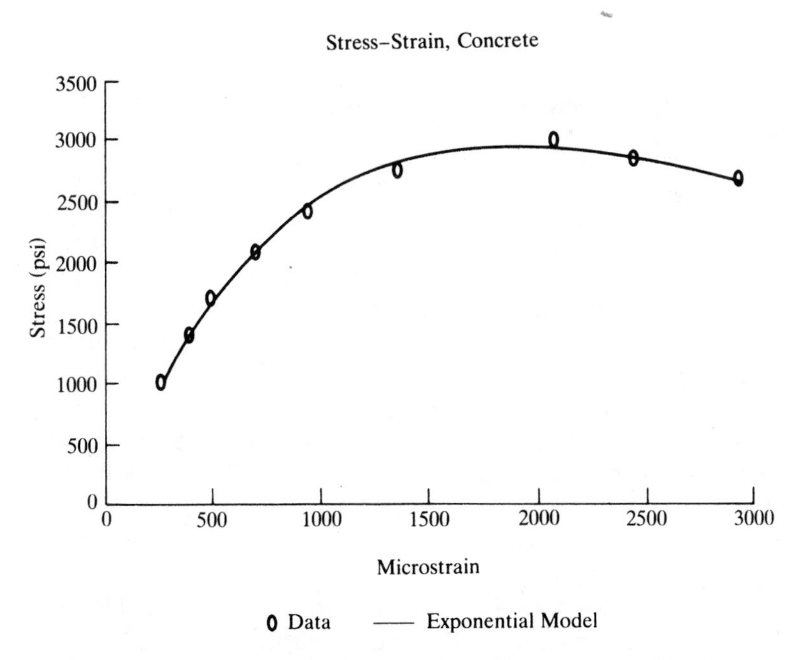

Figure 2.8 Stress–strain characteristic for a concrete block.

Table 2.3 Stress [y_i (psi)] versus Microstrain (x_i)

i	y_i	x_i	w_i
1	1025	265	3.86
2	1400	400	3.50
3	1710	500	3.42
4	2080	700	2.97
5	2425	950	2.55
6	2760	1360	2.03
7	3005	2080	1.44
8	2850	2450	1.16
9	2675	2940	0.91

and the data are listed in Table 2.3 [James, Smith and Wolford (1985)]. The resulting equation is

$$y = 4.41 \times e^{-540.62x}.$$

Verify this solution as an exercise.

2.6 Orthogonal Polynomials

As has been shown in Section 2.4, when it is necessary to increase the order of a model, a new matrix solution of increased dimension must be undertaken. There is a faster and more efficient method for determining the coefficients of the model when the interval or difference between values of the independent variable are equally spaced. This method requires the use of orthogonal polynomials. The results will be the same but the complexity and labor involved in the least-square procedure that has just been described can be significantly reduced. In addition, the concept of orthogonality is important to learn in itself because it is an essential component in other areas. The mathematics of orthogonal functions defined when the independent variable is continuous is reviewed in Appendix 2.1.

First we define a set of polynomial functions $\{P_m(x_i)\}$, where x_i is the set of discrete independent variables, m is the order of the polynomial, N is the number of data points, and $0 \leq m \leq N - 1$.

The model or approximation for the data becomes

$$y = f(x) = A_0 P_0(x) + A_1 P_1(x) + \cdots + A_m P_m(x). \quad (2.26)$$

The squared error is defined as

$$E_m = \sum_{i=1}^{N} (f(x_i) - y_i)^2,$$

and substituting the model's values yields

$$E_m = \sum_{i=1}^{N} (A_0 P_0(x_i) + A_1 P_1(x_i) + \cdots + A_m P_m(x_i) - y_i)^2. \quad (2.27)$$

Differentiating equation (2.27) in order to minimize the error produces the set of equations

$$\frac{\partial E_m}{\partial A_j} = \sum_{i=1}^{N} (A_0 P_0(x_i) + A_1 P_1(x_i) + \cdots + A_m P_m(x_i) - y_i) P_j(x_i) = 0, \quad (2.28)$$

where $0 \leq j \leq m$. Distributing the summations and collecting terms produces equations of the form

$$\sum_{i=1}^{N} y_i P_j(x_i) = A_0 \sum_{i=1}^{N} P_0(x_i) P_j(x_i) + A_1 \sum_{i=1}^{N} P_1(x_i) P_j(x_i) + \cdots + A_m \sum_{i=1}^{N} P_m(x_i) P_j(x_i). \quad (2.29)$$

There are, in general, $m + 1$ simultaneous equations and the solution is quite formidable. By making the function set $\{P_m(x)\}$ satisfy orthogonality conditions, the solution for the coefficients

A_m becomes straightforward. The orthogonality conditions are

$$\sum_{i=1}^{N} P_m(x_i) P_j(x_i) = \begin{cases} 0 & \text{for } m \neq j, \\ \lambda_j & \text{for } m = j. \end{cases} \tag{2.30}$$

The λ_j equal the energy in the function $P_j(x_i)$. This reduces equation (2.29) to the form

$$\sum_{i=1}^{N} y_i P_j(x_i) = A_j \sum_{i=1}^{N} P_j^2(x_i). \tag{2.31}$$

Thus, the coefficients can now be directly solved by equations of the form

$$A_j = \frac{\sum_{i=1}^{N} y_i P_j(x_i)}{\sum_{i=1}^{N} P_j^2(x_i)}, \qquad 0 \leq j \leq m. \tag{2.32}$$

The polynomials for discrete data that satisfy this orthogonality condition are the *gram polynomials* and have the general form

$$P_j(z) = \sum_{i=0}^{j} (-1)^i \frac{(i+j)!(N-i-1)!z!}{i!i!(j-i)!(N-1)!(z-i)!}, \tag{2.33}$$

when z is an integer whose lower bound is zero. Some particular polynomials are

$$P_0(z) = 1,$$

$$P_1(z) = 1 - 2\frac{z}{N-1},$$

$$P_2(z) = 1 - 6\frac{z}{N-1} + 6\frac{z(z-1)}{(N-1)(N-2)}.$$

The energies for this set are, in general,

$$\lambda_j = \frac{(N+j)!(N-j-1)!}{(2j+1)(N-1)!(N-1)!}. \tag{2.34}$$

For some particular values they are

$$\lambda_0 = N,$$

$$\lambda_1 = \frac{N(N+1)}{3(N-1)},$$

$$\lambda_2 = \frac{N(N+1)(N+2)}{5(N-1)(N-2)}.$$

Because of the orthogonality conditions, the total energy E_{tot} and the error E_m have simplified forms:

$$E_{\text{tot}} = \sum_{j=0}^{\infty} A_j^2 \lambda_j \tag{2.35}$$

and

$$E_m = \sum_{i=1}^{N} y(x_i)^2 - \sum_{j=0}^{m} \lambda_j A_j^2. \tag{2.36}$$

Equation (2.35) is one form of Parseval's theorem, which will be encountered several times in this book. This set of polynomials can be applied to model any equally spaced data set by the simple linear transformation

$$z = \frac{(x - x_0)}{h}, \tag{2.37}$$

where x_0 is the smallest value of the independent variable and h is the spacing. An example will illustrate the methodology.

Example 2.5

When a muscle is active, it generates electrical activity that can be easily measured. This measurement is called the electromyogram (EMG). It is now desired to examine the relationship between a quantification of this electrical activity and the load the muscle

Table 2.4 Muscle Activity [y_i (mV)] versus Load [x_i (lb)] and Parameter Values

i	y_i	x_i	$P_1(x_i)$	$P_2(x_i)$	y_iP_1	y_iP_2
1	10.27	0.00	1.00	1.00	10.27	10.27
2	14.07	1.25	0.67	0.00	9.38	0.00
3	20.94	2.50	0.33	−0.60	6.98	−12.56
4	24.14	3.75	0.00	−0.80	0.00	−19.31
5	22.5	5.00	−0.33	−0.60	−7.50	−13.50
6	24.91	6.25	−0.67	0.00	−16.61	0.00
7	27.06	7.50	−1.00	1.00	−27.06	27.06
Σ	143.89				−24.54	−8.08

supports. The quantification is the average absolute value (AAV). To study this, a person was seated in a chair with the right foot parallel to but not touching the floor. The person kept the ankle joint at 90° while weights of different magnitudes were placed upon it. Table 2.4 lists the AAV (in millivolts) and the corresponding load (in pounds) [Shiavi (1969)]. The data are plotted in Figure 2.9.

Because orthogonal polynomials were to be used, there are equal increments of 1.25 lb between successive loads. Polynomials of up to fourth order are to be used. For seven points the polynomial equations are

$$P_0(z) = 1,$$

$$P_1(z) = 1 - \tfrac{1}{3}z,$$

$$P_2(z) = 1 - \tfrac{6}{5}z + \tfrac{1}{5}z^2,$$

$$P_3(z) = 1 - \tfrac{10}{3}z + \tfrac{3}{2}z^2 - \tfrac{7}{6}z^3,$$

$$P_4(z) = 1 - \tfrac{59}{6}z + \tfrac{311}{36}z^2 - \tfrac{7}{3}z^3 + \tfrac{7}{36}z^4.$$

Let us now derive, in particular, the second-order model that is given in general by the equation

$$y(z) = A_0P_0(z) + A_1P_1(z) + A_2P_2(z).$$

The important sums are given in Table 2.4 and the energies in

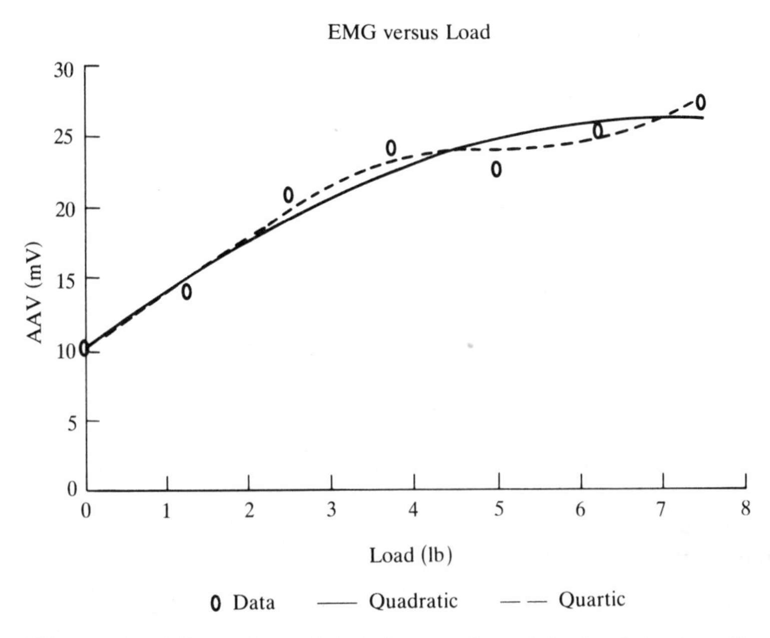

Figure 2.9 The scatter plot and several models for data describing the relationship between AAV (in millivolts) and muscle load (in pounds) in a human muscle.

Table 2.5. For the second-order term, equation (2.31) becomes

$$A_2 = \frac{\sum_{i=1}^{7} y_i P_2(x_i)}{\sum_{i=1}^{7} P_2^2(x_i)} = \frac{\sum_{i=1}^{7} y_i P_2(x_i)}{\lambda_2} = \frac{-8.08}{3.36} = -2.40,$$

where

$$\lambda_2 = \frac{N(N+1)(N+2)}{5(N-1)(N-2)} = \frac{7 \cdot 8 \cdot 9}{5 \cdot 6 \cdot 5} = 3.36.$$

Table 2.5 Model Parameters

j	A_j	λ_j	E_j
0	20.56	7.00	
1	−7.89	3.11	32.89
2	−2.40	3.36	13.12
3	−0.73	6.00	9.91
4	0.53	17.11	5.04

The parameters for the zeroth- and first-order polynomials are calculated similarly. The squared error is

$$E_2 = \sum_{i=1}^{7} y(x_i)^2 - \sum_{j=0}^{2} \lambda_j A_j^2$$

$$= 3{,}184 - 7 \cdot 423 - 3.11 \cdot 62 - 3.36 \cdot 5.76$$

$$= 13.12.$$

The associated energies, coefficients and squared errors for models up to the fourth order are listed in Table 2.5.

The second-order model is given by the equation

$$y(z) = A_0 P_0(z) + A_1 P_1(z) + A_2 P_2(z)$$

$$= 20.56 P_0(z) - 7.89 P_1(z) - 2.39 P_2(z)$$

$$= 10.28 + 5.5z - 0.48z^2$$

after substituting for A_i and $P_i(z)$. The next step is to account for the change in scale of the independent variable with parameters $x_0 = 0$ and $h = 1.25$. The previous equation now becomes

$$y(x) = 10.28 + 4.4x - 0.27x^2.$$

2.7 Interpolation and Extrapolation

After a suitable model of a data set has been created, an additional benefit has been gained. Now one can estimate the value of the dependent variable for any desired value of the independent variable. When the value of the independent variable lies within the range of magnitude of the dependent variable, the estimation is called *interpolation*. Again, we consider Figure 2.9. Although the EMG intensity for a weight of 5.5 lb was not measured, it can be estimated to be 24 mV. This is a very common procedure for instrument calibration, filling in for missing data points, and esti-

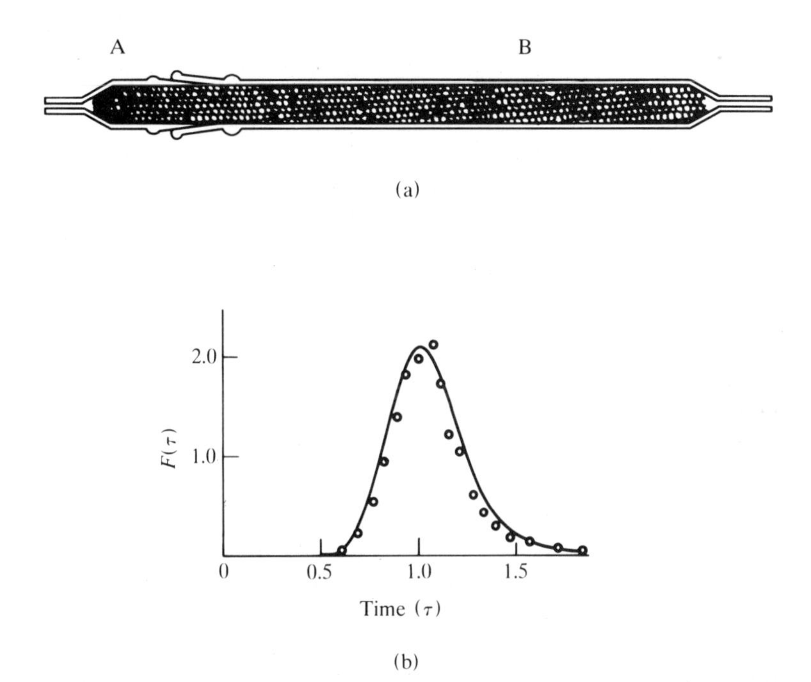

Figure 2.10 (a) Schematic dispersion of a dye in a glass-bead column at the beginning (A) and at an intermediary position of its transversal (B). (b) Dye concentration $F(\tau)$ at the exit point versus time τ: measured data (○); fitted model (—). [Adapted from Sheppard (1962), Figures 63 and 65, with permission.]

mation in general when using discrete data measurements. An application is the study of the dispersion of tracer dye in a nonhomogeneous fluid medium. Figure 2.10(a) shows the schematic of a column filled with glass beads in which a bolus of dye is injected at point A. The dye is dispersed in its travel (B) and its concentration is measured at the exit point of the column. A model is fitted to the measured data and does not represent the data well [Figure 2.10(b)] therefore, using interpolation to estimate nonmeasured values of concentration would be more suitable.

The estimation or prediction of a value of the dependent variable when the magnitude of the independent variable is outside the range of the measured data is called *extrapolation*. This is often used in engineering and very often in economics and business management. For instance, in the management of factory production it is necessary to predict future production demands in order to maintain an inventory of sufficient raw materials and basic components and to have sufficient trained personnel. In the business and economic fields this prediction is also called *forecasting*.

Several polynomials-based methods exist for performing interpolation. There are several major differences between these methods of interpolation, which will be described in this section, and the curve-fitting methods. First, it is assumed that the data are accurate and that the curve needs to have zero error at the acquired data points. This necessarily means that for N points a polynomial of order $N - 1$ will be produced. For even a moderate value of N, a curve with large fluctuations can be produced as illustrated in Figure 2.11. These fluctuations can be inordinately large and thus it is necessary to use a low-order polynomial to estimate the magnitude of the unknown function. The obvious implication here is that there will not be a characteristic curve produced that will represent the entire range of magnitudes, but just a lesser range as defined by the series of N points. The second difference arises in the need to calculate the coefficients without matrix operations for data in which the independent variable is not equally spaced. (This set of polynomials is called the *Lagrange polynomials*.) Another difference is that certain conditions may be imposed on the polynomials. (A set in which this happens is called *spline* functions.) Both of these basic methods will be studied in this section, and additional material can be found in the numerical

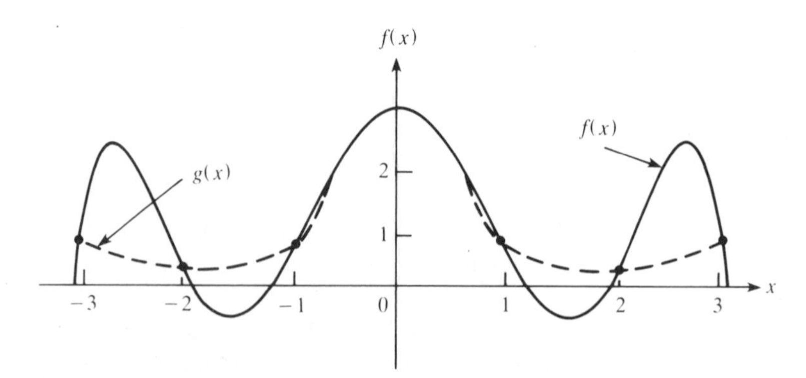

Figure 2.11 Graphical illustration of the possible inaccuracies inherent in interpolation; $g(x)$ represents the ideal function and $f(x)$ represents the interpolator polynomial.

methods literature [Chapra and Canale (1988) and Al-Khafaji and Tooley (1986)].

2.7.1 Lagrange Polynomials

Given $N = m + 1$ data points (x_i, y_i) and an unknown functional relationship $y = g(x)$, the Lagrange interpolation function is defined as

$$\hat{y} = \hat{g}(x) = f(x) = L_0(x)f(x_0) + L_1(x)f(x_1)$$

$$+ \cdots + L_m(x)f(x_m),$$

or

$$f(x) = \sum_{i=0}^{m} L_i(x)f(x_i), \qquad x_0 \le x \le x_m. \tag{2.38}$$

Refer again to Figure 2.11. If $m = 3$ and $x_0 = -1$ and $x_2 = 2$, then $f(x)$ is defined over the interval $-1 \le x \le 2$. The coefficient functions $L_i(x)$ have special properties, which are

$$L_i(x_j) = \begin{cases} 0 & \text{for } i \ne j, \\ 1 & \text{for } i = j, \end{cases} \tag{2.39}$$

and

$$\sum_{i=0}^{m} L_i(x) = 1. \tag{2.40}$$

They have a special factored form that directly produces these properties. The first condition of equation (2.39) is produced if

$$L_i(x) = C_i(x - x_0)(x - x_1)\cdots(x - x_m), \quad \text{excluding } (x - x_i), \tag{2.41}$$

where C_i is a proportionality coefficient. Then for the second condition to be met,

$$C_i = \frac{L_i(x_i)}{(x_i - x_0)(x_i - x_1) \cdots (x_i - x_m)}$$

$$= \frac{1}{(x_i - x_0)(x_i - x_1) \cdots (x_i - x_m)}. \tag{2.42}$$

Thus, the coefficient functions have the form

$$L_i(x) = \frac{(x - x_0)(x - x_1) \cdots (x - x_m)}{(x_i - x_0)(x_i - x_1) \cdots (x_i - x_m)} = \prod_{\substack{j=0 \\ j \neq i}}^{m} \frac{x - x_j}{x_i - x_j},$$

$$i = 0, 1 \ldots m. \tag{2.43}$$

Note that the number of factors in the numerator and denominator, and hence the order of each polynomial, is m. The explicit interpolation function is obtained by combining equation (2.38) with equation (2.43) and is

$$\hat{y} = f(x) = \sum_{i=0}^{m} \prod_{\substack{j=0 \\ j \neq i}}^{m} \frac{x - x_j}{x_i - x_j} y_i. \tag{2.44}$$

This is a powerful function because it allows interpolation with unevenly spaced measurements of the independent variable without using matrix operations. Refer to the least squares method in Section 2.3.

Example 2.6

For the stress versus strain measurements in Example 2.2, develop a second-order interpolation function for the data points [(1.01, 34.2,), (1.12, 38.8), (1.42, 44.8)]. Compare this function with the regression equation. How does the estimate of the value of stress differ for $x = 1.30$?

Because $N = 3$, $m = 2$, and the coefficient functions are found using equation (2.43):

$$L_0(x) = \prod_{\substack{j=0 \\ j \neq i}}^{2} \frac{x - x_j}{x_i - x_j}$$

$$= \frac{(x - x_1)(x - x_2)}{(x_0 - x_1)(x_0 - x_2)}$$

$$= \frac{(x - 1.12)(x - 1.42)}{(1.01 - 1.12)(1.01 - 1.42)}$$

$$= \frac{x^2 - 2.54x + 1.59}{0.0451}$$

$$= 22.173(x^2 - 2.54x + 1.59).$$

Similarly,

$$L_1(x) = \frac{(x - x_0)(x - x_2)}{(x_1 - x_0)(x_1 - x_2)}$$

$$= \frac{(x - 1.01)(x - 1.42)}{(1.12 - 1.01)(1.12 - 1.42)}$$

$$= \frac{x^2 - 2.43x + 1.434}{-0.033},$$

$$L_2(x) = \frac{(x - x_0)(x - x_1)}{(x_2 - x_0)(x_2 - x_1)}$$

$$= \frac{(x - 1.01)(x - 1.12)}{(1.42 - 1.01)(1.42 - 1.12)}$$

$$= \frac{x^2 - 2.13x + 1.131}{0.123}.$$

The polynomial is

$$\hat{y} = f(x) = L_0(x)y_0 + L_1(x)y_1 + L_2(x)y_2.$$

Combining all the terms produces

$$\hat{y} = -53.215x^2 + 115.166x - 68.374, \qquad 1.01 \le x \le 1.42.$$

Compare this equation with the one developed in Example 2.2. Notice that the coefficients are different in magnitude. As expected for the three points used for the coefficient calculation, the ordinate values are exact values in the data. Compare now for $x = 1.3$, the Lagrange method estimates a value of 43.41, whereas the regression approach estimates a value of 44.48, a 2.5% difference.

By now the reader has probably realized that for interpolating between $m + 1$ points using the Lagrange method, an mth-

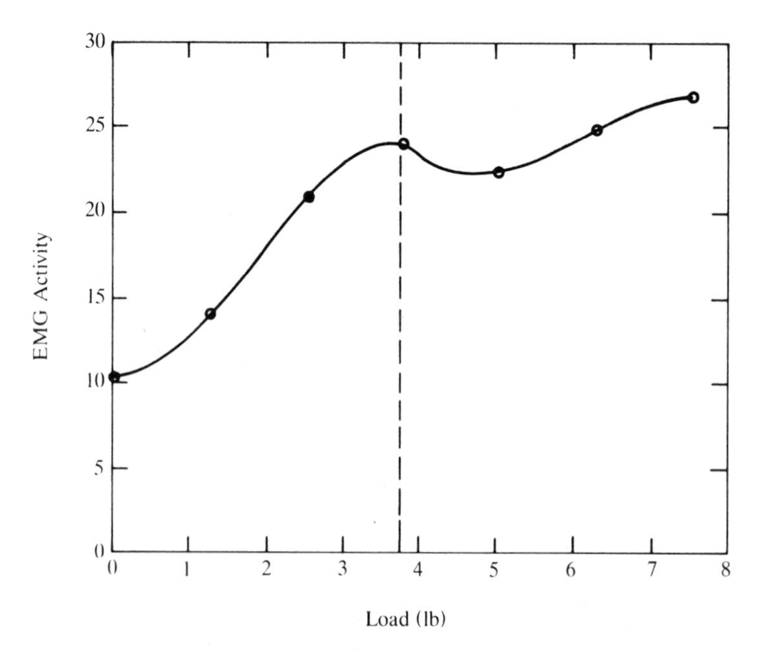

Figure 2.12 The EMG versus load data are plotted with the load range divided into two regions. Each region has a separate Lagrange polynomial curve.

order polynomial is produced. For a large data set, this would produce a high-order polynomial with the likelihood of containing large fluctuation errors. How is this situation handled? The approach is to segment the data set into regions and determine polynomials for each region. Consider again the muscle activity versus load data in Example 2.4, replotted in Figure 2.12. Assume that the measurements are accurate and divide the range of the load variable into two regions. Each region contains four points and two third-order Lagrange polynomials can be utilized to estimate the muscle activity. This has been done and the resulting curve is plotted also in Figure 2.12. The points are interpolated with a load increment of 0.25 lb. Many applications with time series involve

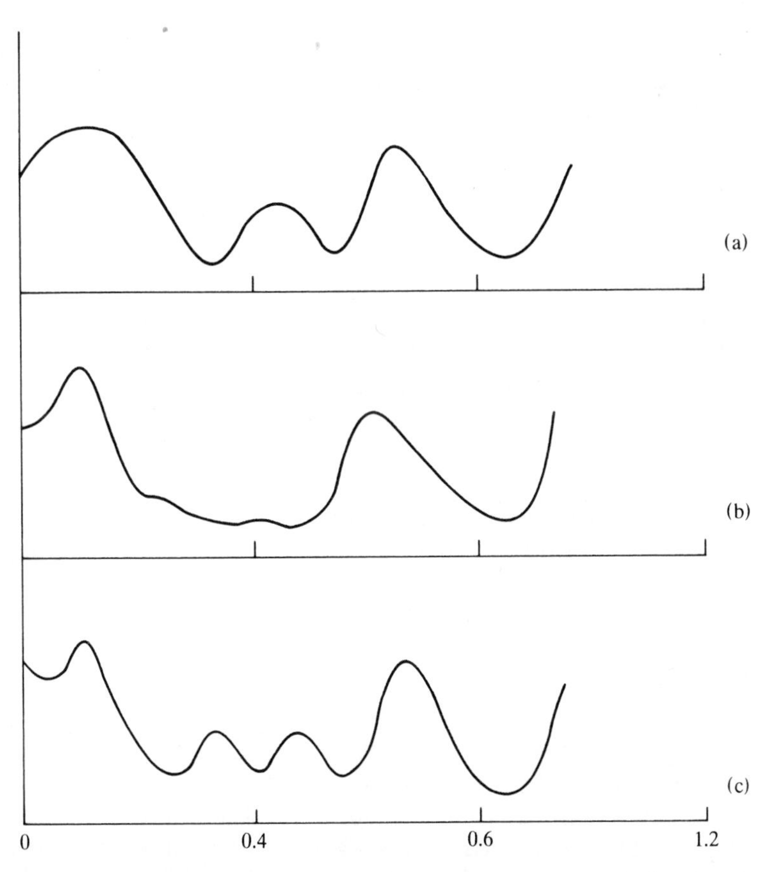

Figure 2.13 Three segments of a random quasi-periodic signal, each with a different duration, are shown in (a), (b), and (c).

thousands of signal points and hence hundreds of segments. Consider the need in human locomotion for producing an average of quasi-periodic signals when successive periods are not exactly equal. Figures 2.13(a)–2.13(c) show the muscle signal from successive walking strides (the duration of each is different) [Shiavi and Green (1983)]. The sampling rate is 250 samples per second and the duration of periods ranges between 0.95 s and 1.0 s. Before averaging, each signal must be converted into an equivalent signal with 256 points. This was done using cubic Lagrange interpolation polynomials.

The errors associated with Lagrange interpolation depend on many factors. These include inaccurate data and truncation. A truncation error is incurred when the order of the interpolation polynomial is lower than that of the true data. In general, the errors are difficult to quantitate and there are not any good approximation schemes available. One can appreciate the effect of inaccurate data because the polynomials fit the acquired data exactly. For instance, if a particular y_i contains a positive measurement error, the polynomial fitting the section of data containing it will be shifted upward. Because the Lagrange method is similar to the Newton divided difference method, a truncation error can be approximated using the error term from the Newton method [Al-Khafaji and Tooley (1986)]. However, the error depends upon the mth-order derivative of the unknown function and thus the error term is not helpful. One must use good judgment and ensure that the data are accurate and noise-free.

2.7.2 Spline Interpolation

The *spline functions* are another set of mth-order polynomials that are used as an interpolation method. These are also a succession of curves $f_i(x)$ of the same order. However, a major difference between spline functions and the other interpolation methods is that an $f_i(x)$ exists only between data points x_i and x_{i+1} as schematically shown in Figure 2.14. Successive curves have common endpoints, the data points, called *knots*. Functional conditions are imposed between successive polynomial curves in order to develop enough equations to solve for the unknown coefficients. The conditions are that successive curves must have the first $m - 1$ derivatives equal at the common knot. The result is that the range of magnitudes in the resulting functional approximation is

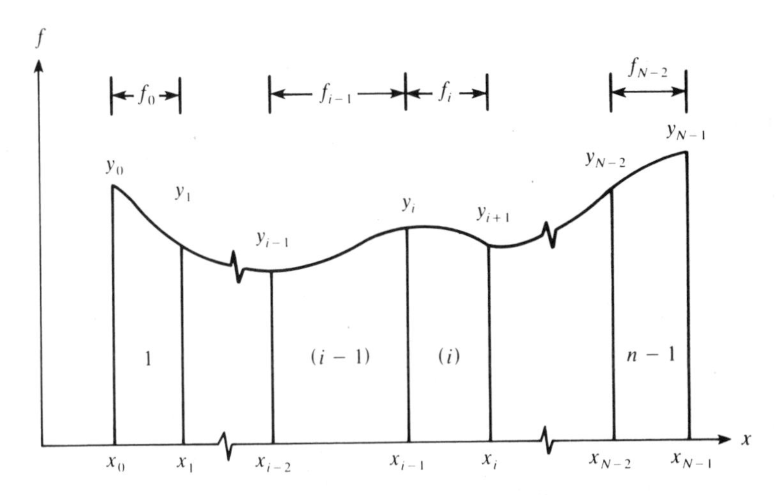

Figure 2.14 Schematic for spline interpolation for a set of N data points.

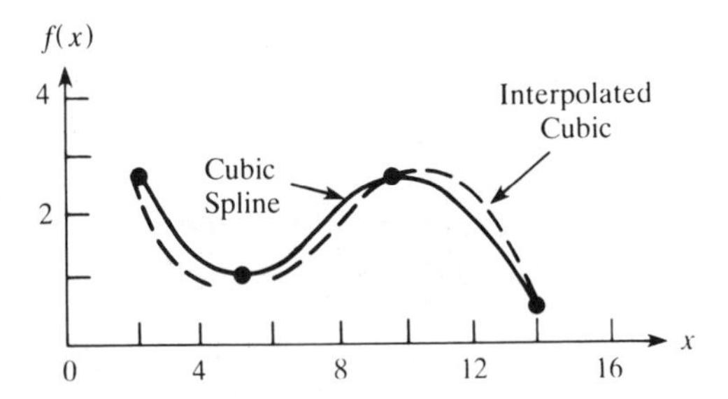

Figure 2.15 Schematic of a fit of four points with a cubic spline and an interpolating cubic polynomial.

less than that produced by the other interpolation methods. Figure 2.15 illustrates this.

2.7.2.1 Conceptual Development

The most-used spline functions are the cubic splines. The development of their equations is quite laborious but not conceptually different from the other-order splines. In order to understand the procedure more easily, the equations will be developed in detail for the quadratic spline and an example will be given.

The general equation for a quadratic spline in interval i is

$$f_i(x) = a_i + b_i x + c_i x^2, \qquad 0 \le i \le N-2,\ x_i \le x \le x_{i+1}. \quad (2.45)$$

For N data points there are $N-1$ equations and therefore $3(N-1)$ unknown coefficients. The conditions for solving these coefficients are produced as follows:

1. *The functional values must be equal at the interior knots.* That is,

$$\begin{aligned} f_i(x_i) &= a_i + b_i x_i + c_i x_i^2, \\ &= f_{i-1}(x_i) = a_{i-1} + b_{i-1}x_i + c_{i-1}x_i^2, \qquad 1 \le i \le N-2, \end{aligned} \quad (2.46)$$

   and produces $2(N-2)$ conditions.
2. *The first and last functions must pass through the endpoints.*

$$\begin{aligned} f_0(x_0) &= a_0 + b_0 x_0 + c_0 x_0^2, \\ f_{N-2}(x_{N-1}) &= a_{N-2} + b_{N-2}x_{N-1} + c_{N-2}x_{N-1}^2. \end{aligned} \quad (2.47)$$

   Two additional conditions are produced.
3. *The first derivatives at the interior knots must be equal.* That is, the first derivatives of equations (2.46) are equal,

$$b_i + 2c_i x_i = b_{i-1} + c_{i-1}x_i, \qquad 1 \le i \le N-2. \quad (2.48)$$

   This produces $N-2$ additional conditions. Now $3N-4$ conditions exist and one more must be sought.
4. The additional condition is that *the second derivative of $f_0(x_0)$ is zero. This means that c_0 is zero.*

Example 2.7

For the concrete block data in Example 2.4, the quadratic splines will be developed to interpolate a value of stress for a microstrain of 2200. Use the last four points: (1360, 2760), (2080, 3005), (2450, 2850), and (2940, 2675).

For four points, three intervals are formed and the necessary equations are

$$f_0(x) = a_0 + b_0 x + c_0 x^2,$$

$$f_1(x) = a_1 + b_1 x + c_1 x^2,$$

$$f_2(x) = a_2 + b_2 x + c_2 x^2.$$

There are nine unknown parameters that must be determined by the four sets of conditions. Condition 1 states that the equations must be equal at the interior knots and produces four equations:

$$3005 = a_0 + b_0 \quad 2080 + c_0 \quad 2080^2;$$

$$3005 = a_1 + b_1 \quad 2080 + c_1 \quad 2080^2;$$

$$2850 = a_1 + b_1 \quad 2450 + c_1 \quad 2450^2;$$

$$2850 = a_2 + b_2 \quad 2450 + c_2 \quad 2450^2.$$

Condition 2 states that the first and last functions must pass through the endpoints and

$$2760 = a_0 + b_0 \quad 1360 + c_0 \quad 1360^2,$$

$$2675 = a_2 + b_2 \quad 2940 + c_2 \quad 2940^2.$$

Two more equations are produced because the first derivatives must be equal at the interior knots and

$$b_0 + c_0 \quad 4160 = b_1 + c_1 \quad 4160,$$

$$b_1 + c_1 \quad 4900 = b_2 + c_2 \quad 4900.$$

The final condition stipulates that the second derivative of $f_0(x)$ is zero or $c_0 = 0$. Thus, now there are actually eight equations and eight unknowns. Solving them yields

$$(a_0, b_0, c_0) = (2.2972 \times 10^3, 3.4028 \times 10^{-1}, 0),$$

$$(a_1, b_1, c_1) = (-6.5800 \times 10^3, 8.8761, -2.0519 \times 10^{-3}),$$

$$(a_2, b_2, c_2) = (-5.7714 \times 10^3, 6.7489, -1.3184 \times 10^{-3}).$$

The strain value of 2200 is within the first interval, so that using the $f_1(x)$ spline the interpolated stress value is 3016.29 psi.

2.7.2.2 Cubic Splines

For the cubic splines the objective is to develop a set of third-order polynomials for the same intervals as illustrated in Figure 2.14. The equations now have the form

$$f_i(x) = a_i + b_i x + c_i x^2 + d_i x^3, \qquad x_i \le x \le x_{i+1}, \quad (2.49)$$

and there are $4(N-1)$ unknown coefficients. The equations for solving these coefficients are developed by expanding the list of conditions stated in the previous section. The additional equations are produced by imposing equality constraints on the second derivatives. The totality of conditions is the following:

1. Functions of successive intervals must be equal at their common knot, $2(N-2)$ conditions.
2. The first and last functions must pass through the endpoints, 2 conditions.
3. The first and second derivatives of functions must be equal at their common knots, $2(N-2)$ conditions.
4. The second derivatives at the endpoints are zero, 2 conditions.

The specification of condition 4 leads to what are defined as *natural splines*. The $4(N-1)$ conditions stated are sufficient for the production of the natural cubic splines for any data set. It is straightforward to solve this set of equations that produces a $4(N-1) \times 4(N-1)$ matrix equation. However there is an alternative approach that is more complex to derive but results in requiring the solution of only $N-2$ equations.

The alternative method involves incorporating Lagrange polynomials into the formulation of the spline functions. This incorporation is contained in the first step, which is based upon the fact that a cubic function has a linear second derivative. For any interval the second derivative, $f_i''(x)$, can be written as a linear interpolation of the second derivatives at the knots or

$$f_i''(x) = \frac{x - x_i}{x_{i-1} - x_i} f''(x_{i-1}) + \frac{x - x_{i-1}}{x_i - x_{i-1}} f''(x_i). \tag{2.50}$$

This expression is integrated twice to produce an expression for $f_i(x)$. The integration will produce two constants of integration. Expressions for these two constants can be found by invoking the equality conditions at the two knots. If these operations are performed, the resulting cubic equation is

$$\begin{aligned} f_i(x) &= \frac{f''(x_{i-1})}{6(x_i - x_{i-1})}(x_i - x)^3 + \frac{f''(x_i)}{6(x_i - x_{i-1})}(x - x_{i-1})^3 \\ &+ \left[\frac{f(x_{i-1})}{x_i - x_{i-1}} - \frac{f''(x_{i-1})(x_i - x_{i-1})}{6}\right](x_i - x) \\ &+ \left[\frac{f(x_i)}{x_i - x_{i-1}} - \frac{f''(x_i)(x_i - x_{i-1})}{6}\right](x - x_{i-1}). \end{aligned} \tag{2.51}$$

This equation is much more complex than equation (2.49) but only contains two unknown quantities, $f''(x_{i-1})$ and $f''(x_i)$. These two quantities are resolved by invoking the continuity of derivatives at the knots. The derivative of equation (2.51) is derived for the intervals i and $i - 1$ and the derivatives are equated at $x = x_i$ because

$$f_{i-1}'(x_i) = f_i'(x_i). \tag{2.52}$$

This results in the equation

$$\begin{aligned} &(x_i - x_{i-1}) f''(x_{i-1}) + 2(x_{i+1} - x_{i-1}) f''(x_i) + (x_{i+1} - x_i) f''(x_{i+1}) \\ &= \frac{6}{x_{i+1} - x_i}[f(x_{i+1}) - f(x_i)] + \frac{6}{x_i - x_{i-1}}[f(x_{i-1}) - f(x_i)]. \end{aligned} \tag{2.53}$$

Equation (2.53) is written for all interior knots and results in $N - 2$ equations. Because natural splines are used, $f''(x_0) = f''(x_{N-1}) = 0$, only $N - 2$ second derivatives are unknown and the system of equations can be solved. The results are inserted into the polynomial equations of equation (2.51) and the cubic spline functions are formed.

Example 2.8

The nine data points of the concrete block data listed in Example 2.4 are used to find a set of eight natural cubic spline functions. The desired values of the independent variable, microstrain, are chosen in order to produce a set of values equally spaced from 600 to 2800 at increments of 200. The results are plotted in Figure 2.16. Compare this plot with the model presented in Figure

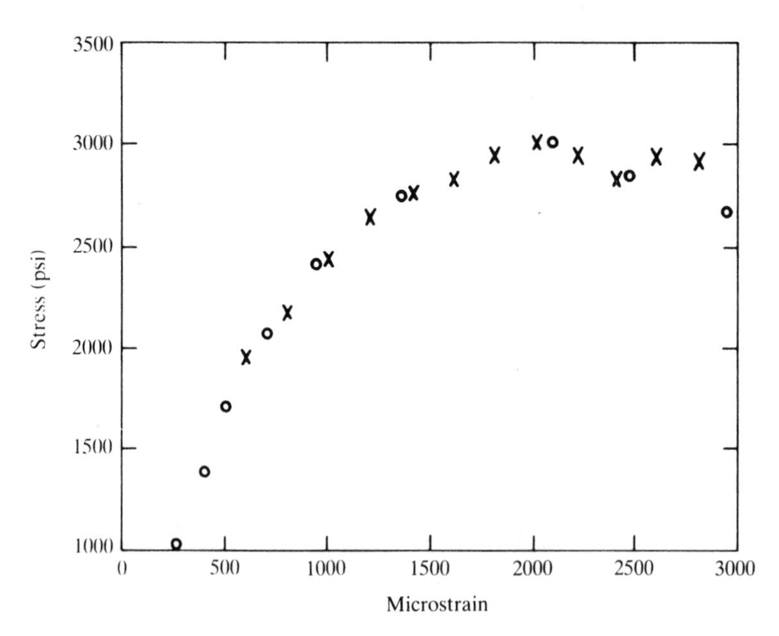

Figure 2.16 Plotted are the stress versus microstrain data points (○) of a concrete block and the points interpolated (×) using cubic splines.

2.8. Notice how there is much more curvature incorporated in the set of points produced by the spline functions.

2.8 Overview

There are many techniques for empirical modeling and approximation of values of unknown functions. All of them have their origin in the field of numerical methods and have been utilized extensively in engineering. One must be aware of the underlying assumptions and basic principles of development in order to implement them properly. One main distinction between curve fitting and interpolation is that in the former set of techniques, inaccurate or noisy measurements were used. Another important distinction is that interpolation techniques produce a set of estimated points, whereas curve fitting actually produces a functional model of the measurements. Some software sources for techniques presented in this chapter are listed in Table 2.6. Refer to the Preface for citations.

Recently, the use of interpolation techniques has been extended to changing the number of points in a measured time series. When the number of resulting points has been increased, the procedure is called *interpolation*; when the number of resulting points is decreased, the procedure is called *decimation*. Because this approach actually changes the sampling rate of the signal, and because different techniques produce different values, the entire approach can be interpreted as a *filtering procedure*. This is a more advanced coverage, reserved for advanced courses in digital

Table 2.6 Curve Fitting and Interpolation Software

	MATLAB	NR	Peerless	SPA
Normal Equations	×	×	×	×
Orthogonal polynomials				×
Lagrange polynomials		×	×	
Spline functions	×	×		

NR: *Numerical Recipes* [Press et al. (1986)].
SPA: Sterns and David (1988).

signal processing. Consult textbooks such as DeFatta, Lucas, and Hodgkiss (1988) for detailed explanations.

References

Ahmed, N. and K. Rao (1975). *Orthogonal Transforms for Digital Signal Processing*. Springer-Verlag, New York.

Al-Khafaji, A. and J. Tooley (1986). *Numerical Methods in Engineering Practice*. Holt, Rinehart and Winston, New York.

Bowerman, B. and R. O'Connell (1987). *Time Series Forecasting*. Dunbury Press, Boston.

Chapra, S. and R. Canale (1988). *Numerical Methods for Engineers*. McGraw-Hill Co., New York.

Chatterjee, S. and B. Price (1977). *Regression Analysis by Example*. John Wiley and Sons, New York.

de Coulon, F. (1986). *Signal Theory and Processing*. Artech House, Dedham, MA.

DeFatta, D., J. Lucas, and W. Hodgkiss (1988). *Digital Signal Processing: A System Design Approach*. John Wiley and Sons, New York.

Dorn, W. and D. McCracken (1972). *Numerical Methods with FORTRAN IV Case Studies*. John Wiley and Sons, New York.

Hamming, R. (1962). *Numerical Methods for Scientists and Engineers*. McGraw-Hill Book Co., New York.

Hirokawa, S. and K. Matsumara (1987). Gait analysis using a measuring walkway for temporal and distance factors. *Med. Biol. Engrg. Comput.* 25:577–582.

James, M., G. Smith, and J. Wolford (1985). *Applied Numerical Methods for Digital Computation*. Harper and Row Publishers, New York.

Lapin, L. (1983). *Probability and Statistics for Modern Engineering*. Brooks/Cole Engineering Division, Monterey, CA.

Pearson, C. (1986). *Numerical Methods in Engineering and Science*. Van Nostrand Reinhold Co., New York.

Roselli, R., R. Parker, K. Brigham, and T. Harris (1980). Relations between oncotic pressure and protein concentration for sheep plasma and lung lymph. *The Physiologist* 23:75.

Sheppard, C. (1962). *Basic Principles of the Tracer Method*. John Wiley and Sons, New York.

Shiavi, R. (1969). A proposed electromyographic technique for testing the involvement of the tibialis anterior muscle in a neuromuscular disorder. Master's thesis, Drexel Institute of Technology.

Shiavi, R. and N. Green (1983). Ensemble averaging of locomotor electromyographic patterns using interpolation. *Med. Biol. Engrg. Comput.* 21:573–578.

Solomonow, M., A. Guzzi, R. Baratta, H. Shoji, and R. D'Ambrosia (1986). EMG-Force model of the elbow's antagonistic muscle pair. *Am. J. Phys. Med.* 65:223–244.

Wylie, C. (1975). *Advanced Engineering Mathematics*. McGraw-Hill Book Co., New York.

Exercises

2.1 Derive the equation (summation formula) to estimate the average value of a set of data points using the least squares criteria.

2.2 Derive equation (2.7), the second of the normal equations for solving the linear model.

2.3 Data relating the volume rate of water discharging from a cooling system in a factory and the resultant height of a stationary water column gauge are listed in Table E2.3. What is the linear model for this relationship?

Table E2.3

i	y_i (m^3/s)	x_i (cm)
1	15.55	−23
2	15.46	−22
3	20.07	−16
4	21.99	−16
5	36.11	14
6	59.82	33
7	86.58	46
8	110.96	69
9	136.52	88
10	204.40	120
11	232.87	136
12	492.50	220
13	1412.48	400

2.4 An industrial engineer wishes to establish a relationship between the cost of producing of a batch of laminated wafers and the size of the run. The data listed in Table E2.4 have been gathered from previous runs. [Adapted from Lapin (1983), page 327, with permission.]

Table E2.4

Size	Cost ($)	Size	Cost ($)
1550	17,224	786	10,536
2175	24,095	1234	14,444
852	11,314	1505	15,888
1213	13,474	1616	18,949
2120	22,186	1264	13,055
3050	29,349	3089	31,237
1128	15,982	1963	22,215
1215	14,459	2033	21,384
1518	16,497	1414	17,510
2207	23,483	1467	18,012

a. What is a suitable regression equation for this relationship?
b. What is the estimated cost of a run of 2000 units?
c. How are the data and model to be changed if one were interested in the relationship between cost per unit and size of the run?

2.5 The data concerning the number of service calls to repair computer units and the time (duration) to repair them are tabulated (Table E2.5).
a. Make the scatter plot and calculate the parameters of a second-order model.
b. Plot the model overlaying the data. Now develop a third-order model. Is it much better than the quadratic model? Why?

Table E2.5

Units (#)	Time (min)	Units (#)	Time (min)
1	23	10	154
2	29	10	166
3	49	11	162
4	64	11	174
4	74	12	180
5	87	12	176
6	96	14	179
6	97	16	193
7	109	17	193
8	119	18	195
9	149	18	198
9	145	20	205

2.6 Verify the fitting of the product exponential model in Example 2.4.

2.7 Derive the normal equations analogous to those in Section 2.3 for the power-law model $y = \alpha x^{\beta}$.

2.8 The nonlinear equation

$$k = a\frac{x}{b + x}, \qquad 0 \leq x \leq \infty,$$

is often called the saturation–growth-rate model. It adequately describes the relationship between the population growth rate k of many species and the food supply x. Sketch the curve and explain a reason for this name. Linearize the model.

2.9 Derive the normal equations for the function $y = a(\tan x)^b$.

2.10 It is desired to determine the relationship between mass density ρ (slug/ft^3), and altitude above sea level, h (ft), using the exponential model

$$\rho = \alpha e^{\beta h}.$$

The data are listed in Table E2.10. [Adapted from James, Smith, and Wolford (1985), page 357, with permission.]

Table E2.10

h	$\rho \times 10^6$	h	$\rho \times 10^6$
0	2,377*	15,000	1,497
1,000	2,308	20,000	1,267*
2,000	2,241	30,000	891
4,000	2,117	40,000	587*
6,000	1,987	50,000	364
10,000	1,755*	60,000	224*

a. Plot the data and find the model parameters, using a calculator and the data points marked with an asterisk. Plot the model.
b. Find the model parameters using all the data and a computer program. Is this set of parameters better than those found in part a? Why?
c. What is the estimated (interpolated) air density at 35,000 ft? What is the predicted (extrapolated) air density at 80,000 ft?

2.11 For equation (2.24), which requires that both variables have a hyperbolic transformation, plot the function with $a = 0.5$ and $b = -1.0$.

2.12 For the function

$$y = \frac{e^{a+bx}}{1 + e^{a+bx}},$$

with $a = 0.5$ and $b = 2.0$, do the following:
a. Plot the function for $0 \le x \le 3$.
b. Find the linearizing transformation.

2.13 In lung research the relationship between the lymph to plasma protein ratio, LP, and pore size R is approximated by the equation

$$\mathrm{LP} = a + \frac{b}{R} + \frac{c}{R^2}.$$

a. What transformation will permit a polynomial regression?
b. Experimental data measured are listed in Table E2.13. (1) Find the model coefficients. (2) Plot the data and the model.

Table E2.13

LP	R	LP	R
0.824	36	0.599	76
0.731	46	0.576	86
0.671	56	0.557	96
0.630	66		

2.14 Verify the solution found in Example 2.4.

2.15 Prove Parseval's theorem, equation (2.34), starting with equation (2.26). Using equations (2.29) and (2.31) could be helpful.

2.16 Verify the final equations for $y(z)$ and $y(x)$ in Example 2.5.

2.17 Fit a second-order polynomial to the data set in Table E2.17 using orthogonal polynomials.
a. What are the coefficients A_i?
b. What is the $f(z)$ equation?
c. What is the $f(x)$ equation?

Table E2.17

x_i	0	0.5	1.0	1.5	2.0	2.5	3.0
$f(x_i)$	1	3	6	10	18	24	35

2.18 A metal pole for supporting a street light was installed erect with a height of 574 in. It has been struck by an automobile in an accident and is badly misshapen. It is necessary to derive an equation for this shape in order to analyze the impact energies. In order to keep the model single-valued, the pole has been laid on its side. The x coordinate represents the vertical coordinate of a spot on the pole and the y direction the lateral coordinate of a spot. The base is at position (0, 0). The data are in Table E2.18.

Table E2.18

y_i	x_i	y_i	x_i
0	0	−137	240
−24	24	−151	264
−43	48	−166	288
−59	72	−180	312
−71	96	−191	336
−82	120	−196	360
−91	144	−193	384
−101	168	−176	408
−111	192	−141	432
−124	216	−81	456

a. Plot a scatter diagram of the points.
b. What order polynomial seems sufficient from visual examination? Why?
c. Derive coefficients and squared errors necessary to create a fourth-order model using orthogonal polynomials.
d. What is the sufficient order for the model? Why?
e. Plot the model equation overlying the data?

2.19 Perform Exercise 2.5a with the data for even numbers of units repaired, using orthogonal polynomials. Are the resulting models similar?

2.20 For the water discharge data in Exercise 2.3, do the following:
a. Find the Lagrange polynomial factors, $L_i(x)$, for the ranges $14 \leq x \leq 46$ and $-22 \leq x \leq 14$.
b. Find $f(x)$ for the range $14 \leq x \leq 46$.

2.21 Derive a linear interpolation formula for interpolating between the point pairs (x_i, y_i) and (x_{i+1}, y_{i+1}) with

$$y = Ay_i + By_{i+1},$$

where A and B are functions of x, x_i, and x_{i+1}. Show that this form is the same as that given by a first-order Lagrange formula.

2.22 Use the EMG versus load data in Example 2.5.

a. Divide the load range into three regions.
b. In the first region derive the Lagrange interpolating polynomial.
c. What EMG magnitudes does it produce for load values of 0.5, 1.0, and 1.5 lb?

2.23 Using the results in Exercise 2.22, predict the EMG magnitude for a load of 3.5 lb. Notice that this load value is outside the design range of the polynomial. This is extrapolation. How does the extrapolated value compare to the measured value?

2.24 In Table E2.24 are listed measurements on the force–displacement characteristics of an automobile spring.

Table E2.24

Force (N $\times$ 10^4)	Displacement (m)
10	0.1
20	0.17
30	0.24
40	0.34
50	0.39
60	0.42
70	0.43

a. Plot the data.
b. Divide the displacement range into two or three regions.
c. Use a computer algorithm to interpolate for one value of displacement within each of the measured intervals.
d. Plot the measured and interpolated values.
e. Use a computer algorithm to interpolate over the range $0.1 \leq x \leq 0.45$ and produce a set of interpolated values that have displacement evenly spaced at intervals of 0.05 m.

2.25 For the natural cubic splines, write the general equations that satisfy the four conditions in Section 2.7.2.2.

2.26 For the alternative cubic spline method, derive equation (2.51) from equation (2.49) and the general first derivative equality condition, equation (2.50).

2.27 For the concrete block example, fit a set of quadratic splines to the last three data points and interpolate the stress for a microstrain of 2800. Show that the coefficients are

$$(a_0, b_0, c_0) = (3.8764 \times 10^3, -0.41891, 0)$$

$$(a_1, b_1, c_1) = (4.6331 \times 10^3, -1.0367, 1{,}2607 \times 10^{-4})$$

and that $f_1(2800) = 2718.8$.

2.28 Fit quadratic spline functions to the first three points of the EMG versus load data. Specifically, what are the equations satisfying conditions 1–4 in Section 2.7.2.1; solve these equations.

2.29 For the impact data listed in Exercise 2.18, you are asked to fit natural cubic splines.

a. For $0 \le i \le 3$, write the equations for conditions 1, 2, and 3.

b. For $i = 0$ and $N - 1$, write the equations for condition 4.

2.30 A nonlinear resistor has the experimental current–voltage data listed in Table E2.30.

Table E2.30

Current, i	1.00	0.50	0.25	−0.25	−0.50	−1.00
Voltage, v	200.0	40.0	12.5	12.5	−40.0	−200.0

a. Determine and plot the characteristic curve using a fifth-order polynomial.

b. Determine and plot the characteristic curve using a spline function; let $\Delta i = 0.2$.

c. Are there any differences?

2.31 The kinematic viscosity of water is tabulated in Table E2.31.

a. Plot this data and find ν for $T = 53$ using a linear interpolation.

b. Repeat part a using a second-order Lagrange interpolator.

Table E2.31

T, (°F)	40	50	60	70	80
ν (10^{-5}, ft^2/s)	1.66	1.41	1.22	1.06	0.93

c. Repeat part a using a second-order spline interpolator.
d. What are the percentage differences between each method?

2.32 Values of complicated functions are often tabulated and interpolation required to find the untabulated values. In order to demonstrate the utility of this procedure use the saturation–growth rate characteristic for microbial kinetics that is given as

$$k = 1.23\frac{x}{22.18 + x}, \qquad x \geq 0.$$

a. Using $N = 5$, $\Delta x = 4$, and the Lagrange polynomials, what is the interpolated value for $f(10)$? How does it compare to the value given by the model?
b. Do part a using spline functions.

Appendix

Appendix 2.1 Properties of Orthogonal Function Sets

There are several function sets that can be used to describe deterministic signals in continuous time over a finite time range with duration P. They can be described with a series of weighted functions such as

$$f(t) = A_0\Phi_0(t) + A_1\Phi_1(t) + \cdots + A_m\Phi_m(t) + \cdots$$

$$= \sum_{m=0}^{\infty} A_m\Phi_m(t). \qquad \text{(A2.1)}$$

The members of these sets have an orthogonality property similar to that of the orthogonal polynomials. The Fourier series is comprised of one of these sets. The function set $\{\Phi_m(t)\}$ is orthogonal if

$$\int_0^P \Phi_m(t)\Phi_n(t)\,dt = \begin{cases} \lambda_m, & \text{for } m = n, \\ 0, & \text{for } m \neq n. \end{cases} \qquad \text{(A2.2)}$$

The usual range for n is $0 \le n \le \infty$; sometimes the range is $-\infty \le n \le \infty$ and equation (A2.1) is defined over this extended range. The coefficients A_n are found by using the least squares error principle. Define an approximation of the function $f(t)$ using a finite number of terms, $M + 1$, as

$$\hat{f}(t) = A_0\Phi_0(t) + A_1\Phi_1(t) + \cdots + A_M\Phi_M(t)$$

$$= \sum_{m=0}^{M} A_m\Phi_m(t). \qquad \text{(A2.3)}$$

The squared error for the approximation is

$$E_M = \int_0^P \left(f(t) - \hat{f}(t)\right)^2 dt = \int_0^P \left(f(t) - \sum_{m=0}^{M} A_m\Phi_m(t)\right)^2 dt. \qquad \text{(A2.4)}$$

Performing the indicated operations gives

$$E_M = \int_0^P \left[f(t)^2 - 2f(t) \sum_{m=0}^{M} A_m\Phi_m(t) + \left(\sum_{m=0}^{M} A_m\Phi_m(t) \right)^2 \right] dt,$$

and taking the partial derivative with respect to a particular A_m yields

$$\frac{\partial E_M}{\partial A_M} = -2\int_0^P f(t)\Phi_m(t)\, dt + 2\int_0^P \left(\sum_{n=0}^{M} A_n\Phi_n(t) \right) \Phi_m(t)\, dt = 0.$$

Rearranging the preceding equation and using the orthogonality conditions produces

$$\int_0^P f(t)\Phi_m(t)\, dt = \sum_{n=0}^{M} \int_0^P A_n\Phi_n(t)\Phi_m(t)\, dt = A_m\lambda_m,$$

or

$$A_m = \frac{1}{\lambda_m} \int_0^P f(t)\Phi_m(t)\, dt, \quad \text{for } 0 \le m \le \infty. \qquad \text{(A2.5)}$$

Thus, as with the orthogonal polynomials, there is a simple method to calculate the coefficients in a series. The range of A_m would be $\infty \leq m \leq \infty$ when the function set is also defined over that range. Parseval's theorem states that the total energy for a waveform is

$$E_{tot} = \sum_{m=-\infty}^{\infty} A_m^2 \lambda_m. \tag{A2.6}$$

There are many orthogonal function sets that can be used to approximate finite energy signals. The continuous sets include the Legendre, Laguerre, and Hermite functions [de Coulon (1986)]. More recently, discontinuous sets composed of square and rectangular waveforms, the Walsh and Radamaker functions, have become utilized [Ahmed and Rao (1975)].

Chapter 3 Fourier Analysis

3.1 Introduction

Knowledge of the cyclic or oscillating activity in various physical and biological phenomena and in engineering systems has been recognized as essential information for many decades. In fact, interest in determining sinusoidal components in measured data through modeling began at least several centuries ago and was known as harmonic decomposition. The formal development of Fourier analysis dates back to the beginning of the eighteenth century and the creative work of Jean Fourier, who first developed the mathematical theory that enabled the determination of the frequency composition of mathematically expressible waveforms. This is the theory of Fourier transforms and has become widely used in engineering and science. During the middle of the nineteenth century, numerical techniques were developed to determine the harmonic content of measured signals. Bloomfield (1976) summarizes a short history of these developments.

One of the older phenomena that has been studied is the changes in the intensity of light emitted from a variable star. A portion of such a signal is plotted in Figure 3.1. It is oscillatory, with a period of approximately 25 days. Astronomers theorized that knowledge of the frequency content of this light variation could yield not only general astronomical knowledge but also information about the star's creation [Whittaker and Robinson, (1967)]. An-

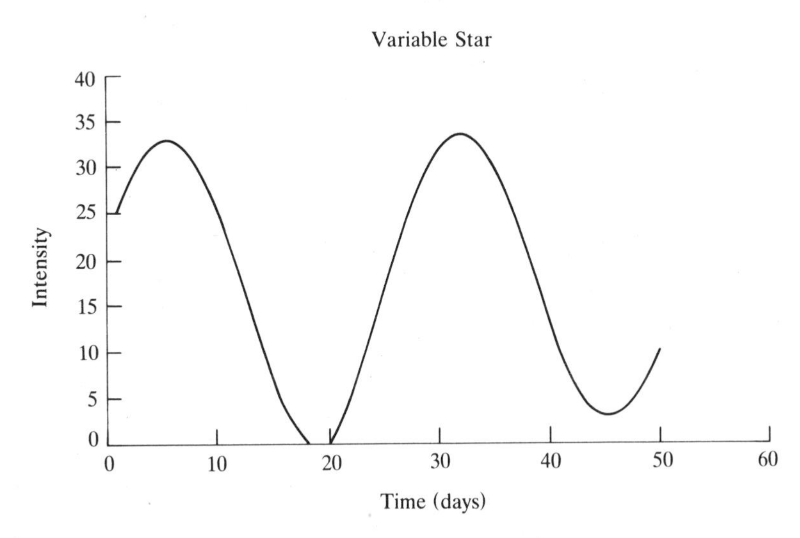

Figure 3.1 Brightness changes of a variable star during 50 days.

other phenomenon that is usually recognized as oscillatory in nature is vibration. Accelerometers are used to measure the intensity and the frequencies of vibration, and testing them is necessary in order to determine their accuracy. Figure 3.2 shows the electrical output of an accelerometer that has been perturbed sinusoidally [Licht, Andersen, and Jensen, (1987)]. Examination of the waveform reveals that the response is not a pure sinusoid. Fourier analysis will indicate what other frequency components, in addition to the perturbation frequency, are present. A biomedical application is the analysis of electroencephalographic (EEG) waveforms to determine the state of mental alertness (i.e., drowsiness, deep sleeping, thinking) of an individual as determined by differences in frequency content [Bodenstein and Praetorius (1977)]. Figure 3.3 shows a variety of EEG waveforms for a variety of conditions.

For different types of signal, there are different versions of Fourier analysis. Applications with deterministic periodic signals, such as in Figure 3.2, will require a Fourier series expansion; those with deterministic aperiodic signals, such as in Figure 3.1, will require a Fourier transform. The most complicated are those applications with random signals, such as in Figure 3.3. They require spectral analysis and will be treated in subsequent chapters. For all these signal types there is the discrete-time version of Fourier analysis, which is utilized throughout this book.

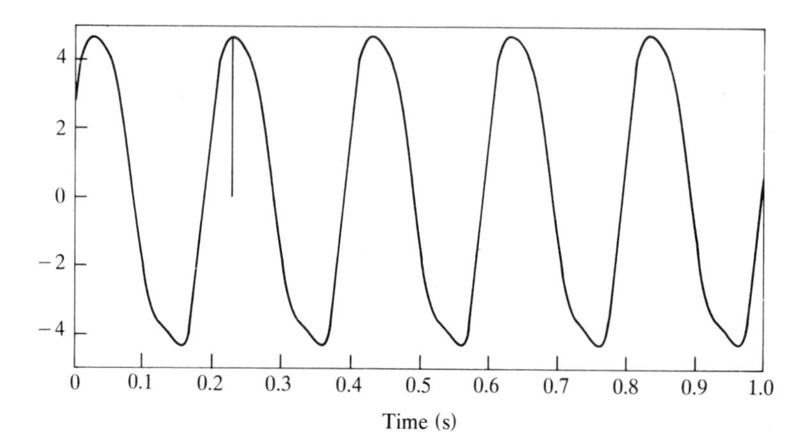

Figure 3.2 Oscillatory response of an accelerometer to a sinusoidal perturbation of 4.954 Hz with amplitude of 250 microstrain. [Adapted from Licht, Andersen, and Jensen (1987), Figure 6, with permission.]

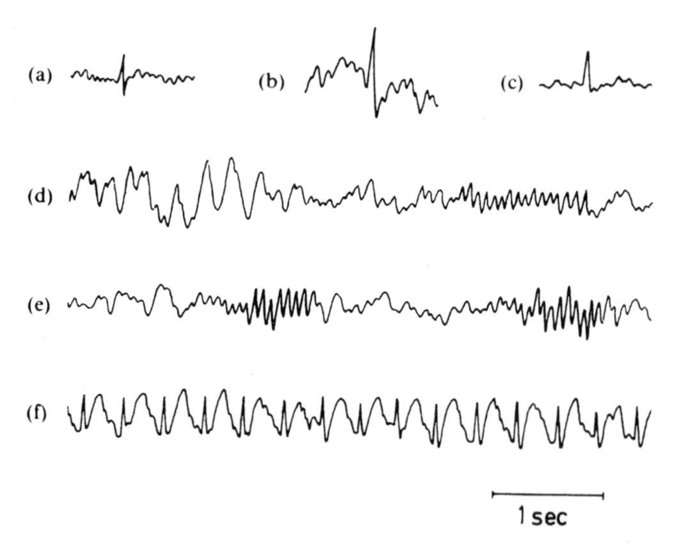

Figure 3.3 Some examples of elementary EEG patterns occurring during various conditions: (a)–(c) transients [(a) biphasic spike, (b) biphasic sharp wave, (c) monophasic sharp wave]; (d) and (e) adult, light sleep; (f) epilepsy. [From Bodenstein and Praetorius (1977 IEEE), Figure 2, with permission.]

The goal of this chapter is to provide a comprehensive review of the techniques for implementing discrete-time Fourier analysis. This is valuable because these techniques are applied to deterministic signals and are fundamental for developing spectral analysis. It is presumed that readers have had exposure to Fourier series and transforms through course work or experience. A review of these topics is provided in the appendices of this chapter, for the reader's convenience. If a comprehensive treatment is desired, please consult the reference listing. There are several books that treat Fourier transforms in continuous time extensively, such as those of Papoulis (1962) and Bracewell (1978). However, there are many books on signals and systems that treat Fourier transforms in discrete time. Some of them are listed in the reference section. Although all of the referenced books treat many aspects of the Fourier transform, an attempt has been made to classify them according to the aspects that they seem to emphasize.

3.2 Overview of Fourier Transform Relationships

There are several forms of Fourier transform where both continuous and discrete aspects of the time and frequency domains are considered. Understanding their relationships is important when learning the procedures for implementing discrete-time Fourier analysis. The relationships will be explained using the *Poisson sum formula* [Papoulis (1977)].

3.2.1 Continuous versus Discrete Frequency

The Poisson sum formula is developed using the impulse function and the *continuous-time Fourier transform* (CTFT) pair

$$X(\omega) = \int_{-\infty}^{\infty} x(t) e^{-j\omega t}\, dt,$$
$$x(t) = \frac{1}{2\pi}\int_{-\infty}^{\infty} X(\omega) e^{j\omega t}\, d\omega. \tag{3.1}$$

The methodology is to use a periodic train of impulses, $\bar{\delta}(t)$, as an

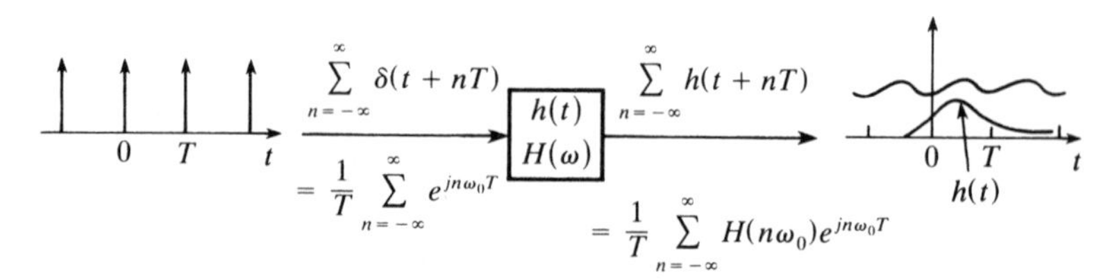

Figure 3.4 Schematic linear system response to a periodic impulse train. [Adapted from Papoulis (1977), Figure 1-23, with permission.]

input to a system with an impulse response $h(t)$, where

$$\bar{\delta}(t) = \sum_{n=-\infty}^{\infty} \delta(t + nP). \tag{3.2}$$

This is shown schematically in Figure 3.4. Because the impulse train is periodic with period P, it can be represented by a Fourier series. The *Fourier series* (FS) for a periodic function $f(t)$ with period P is

$$f(t) = \sum_{n=-\infty}^{\infty} z_n e^{jn\omega_0 t}, \qquad z_n = \frac{1}{P} \int_{t_1}^{t_1+P} f(t) e^{-jn\omega_0 t} \, dt, \tag{3.3}$$

where t_1 is an arbitrary time, z_n are the complex coefficients, and ω_0 is the fundamental frequency $2\pi/P$. Thus, for the impulse train the Fourier series coefficients are

$$z_n = \frac{1}{P} \int_{-P/2}^{P/2} \delta(t) e^{-jn\omega_0 t} \, dt = \frac{1}{P}. \tag{3.4}$$

Remember that the integral of equation (3.4) is evaluated with the sifting property of the delta function. Thus,

$$\bar{\delta}(t) = \frac{1}{P} \sum_{n=-\infty}^{\infty} e^{jn\omega_0 t}. \tag{3.5}$$

The obvious system output is

$$y(t) = \sum_{n=-\infty}^{\infty} h(t + nP) = \bar{h}(t). \tag{3.6}$$

An equivalent form of the output based on the right-hand side of equation (3.5) provides the essential relationship between the CTFT and the FS. For one term of the input, the transform of the output is

$$H(\omega)\frac{2\pi}{P}\delta(\omega - n\omega_0). \tag{3.7}$$

Thus, an alternate form of the output is

$$\begin{aligned} \bar{h}(t) &= \frac{1}{P}\sum_{n=-\infty}^{\infty}\int_{-\infty}^{\infty} H(\nu)\delta(\nu - n\omega_0)e^{j\nu t}\,d\nu \\ &= \frac{1}{P}\sum_{n=-\infty}^{\infty} H(n\omega_0)e^{jn\omega_0 t}. \end{aligned} \tag{3.8}$$

This is the Poisson sum formula for a time function. Consider now that $h(t)$ is restricted to $0 \leq t \leq P$, then $\bar{h}(t)$ is a conventional periodic function. Compare equation (3.8) with the definition of the FS in equation (3.3). The Fourier series coefficients are proportional to the sampled values of the Fourier transform of the aperiodic function $h(t)$.

3.2.2 Continuous versus Discrete Time

The frequency domain version of this summation can also be found. Starting again with the continuous time signal $x(t)$, form the summation of its transforms at a frequency interval $\omega_1 = 2\pi/T_1$.

$$\bar{X}(\omega) = \sum_{n=-\infty}^{\infty} X(\omega + n\omega_1) \tag{3.9}$$

and is plotted schematically in Figure 3.5. Following a procedure similar to that used in the previous paragraph, it can be shown that

$$\bar{X}(\omega) = \frac{2\pi}{\omega_1}\sum_{n=-\infty}^{\infty} x(nT_1)e^{-j\omega nT_1}. \tag{3.10}$$

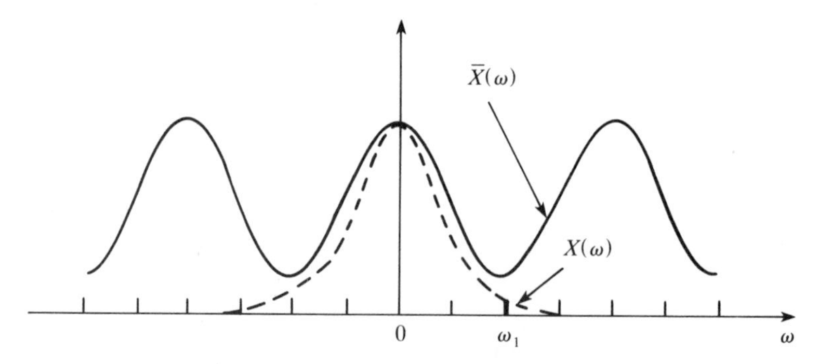

Figure 3.5 Schematic of summation of periodically shifted Fourier transforms of a signal. [Adapted from Papoulis (1977), Figure 3-14, with permission.]

The right-hand side of this equation contains the weighted summation of a set of sampled values of $x(t)$ with a *sampling interval* of T_1 or, equivalently, with a *sampling frequency* of $f_s = 1/T_1$. The *discrete-time Fourier Transform* (DTFT) of a signal sampled every T time units is defined as

$$X_{\mathrm{DT}}(\omega) = T \sum_{n=-\infty}^{\infty} x(nT) e^{-j\omega nT}. \tag{3.11}$$

Thus, a sampled function has a Fourier transform that is periodic in frequency with the repetition occurring at integer multiples of the sampling frequency. The inverse DTFT is defined as

$$x(nT) = \frac{1}{2\pi} \int_{-\omega_N}^{\omega_N} X_{\mathrm{DT}}(\omega) e^{j\omega nT} \, d\omega, \tag{3.12}$$

where $2\omega_N = 2\pi f_s$. Examine Figure 3.5 again. For this integration to be unique, none of the periodic repetitions of $X(\omega)$ can overlap into other regions of the frequency domain. Thus, in order for the DTFT to represent the CTFT accurately, $x(t)$ must be *band-limited*, that is,

$$X(\omega) = 0 \quad \text{for } |\omega| \geq \omega_N. \tag{3.13}$$

This is consistent with the Shannon sampling theorem, which states that the sampling rate of a signal, $f_s = 1/T$, must be twice the highest frequency value, f_N, at which a signal has energy; otherwise an error called *aliasing* occurs [Cadzow (1987)]. This minimum sampling frequency, $2f_N$, is called the *Nyquist rate*.

3.2.3 Discrete Time and Frequency

For actual computer calculation, both the time and frequency domains must be discretized. In addition, the number of sample points N must be a finite number. Define the duration of the signal as $P = NT$. The frequency domain is discretized at integer multiples of the inverse of the signal duration, or

$$\omega = \frac{2\pi m}{P} = \frac{2\pi m}{NT} = m\omega_d, \tag{3.14}$$

where ω_d is called the frequency spacing. Now the fourth version of the Fourier transform, the *discrete Fourier transform* (DFT) is derived by truncating the DTFT and discretizing the frequency domain and is defined as

$$\begin{aligned} X_{\text{DFT}}(m\omega_d) &= T\sum_{n=0}^{N-1} x(nT)\exp(-j\omega nT) \\ &= T\sum_{n=0}^{N-1} x(nT)\exp\left(\frac{-j2\pi mnT}{NT}\right). \end{aligned} \tag{3.15}$$

Note carefully that because the frequency domain has been discretized, the signal in the time domain has become periodic. This is consistent with the Poisson sum formula in equation (3.10). Understanding these periodic repetitions is important when implementing the DFT and they will be studied in detail in the next section. The *inverse DFT* (IDFT) is defined as

$$x(nT) = \frac{1}{NT}\sum_{m=0}^{N-1} X(m\omega_d)e^{j2\pi mn/N}. \tag{3.16}$$

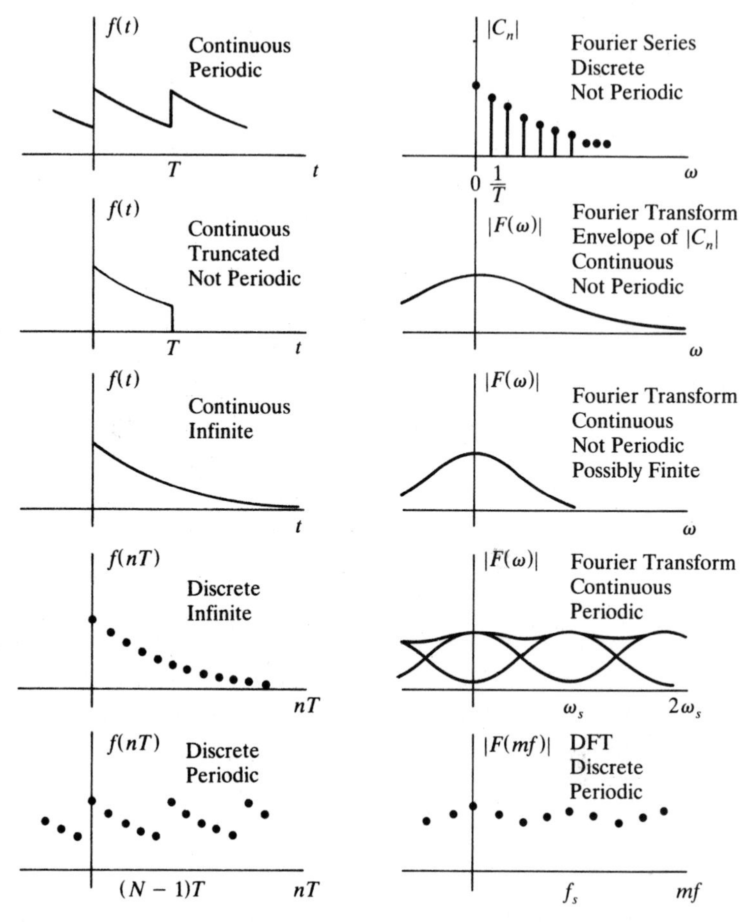

Figure 3.6 Summary of the major characteristics of the various versions of the Fourier transform. [Adapted from Childers and Durling (1975) Figure 2.14, with permission.]

Remembering the various versions of the Fourier transform and the effects of discretizing the time and/or frequency domains can be confusing initially. Figure 3.6 is provided as a schematical summary to help with this. Because the computation of the Fourier transform is the focus of this chapter, the DFT will be dealt with in the remaining sections. Other textbooks that treat these relationships in detail are Brigham (1988) and Marple (1987).

3.3 Discrete Fourier Transform

3.3.1 Definition Continued

Because both domains of the DFT are discretized with sampling interval T and frequency spacing $\omega_d = 2\pi f_d$, these two parameters are dropped from the arguments in equation (3.16). A formal definition of the DFT–IDFT pair that is often used is

$$X_{\mathrm{DFT}}(m) = T \sum_{n=0}^{N-1} x(n) \exp\left(-\frac{j2\pi mn}{N}\right),$$
$$x(n) = \frac{1}{NT} \sum_{m=0}^{N-1} X(m) \exp\left(\frac{j2\pi mn}{N}\right). \tag{3.17}$$

A matter of preference is whether to use the radian or cyclic frequency scale. From here onwards the cyclic frequency scale will be used.

In the previous section it was shown mathematically that when the frequency of the DTFT is discretized, $x(n)$ becomes periodic with period P. This is graphically summarized in Figure 3.7. Essentially, the signal has become periodic inadvertently. In fact, many texts derive a version of equation (3.17) from the definition of the Fourier series with $X_{\mathrm{FS}}(m) = z_m$ and using the rectangular rule approximation for integration. Then

$$X_{\mathrm{FS}}(m) = \frac{1}{P} \int_0^P x(t) e^{-jn\omega_0 t}\, dt$$
$$\approx \frac{1}{NT} \sum_{n=0}^{N-1} x(nT) \exp\left(-\frac{j2\pi mnT}{NT}\right) T,$$

or

$$X_{\mathrm{FS}}(m) = \frac{1}{N} \sum_{n=0}^{N-1} x(n) \exp\left(-\frac{j2\pi mn}{N}\right). \tag{3.18}$$

Notice that the only difference between the two definitions of a discrete time Fourier transform is the scaling factor. The former definition, equation (3.17), is used most often in the signal pro-

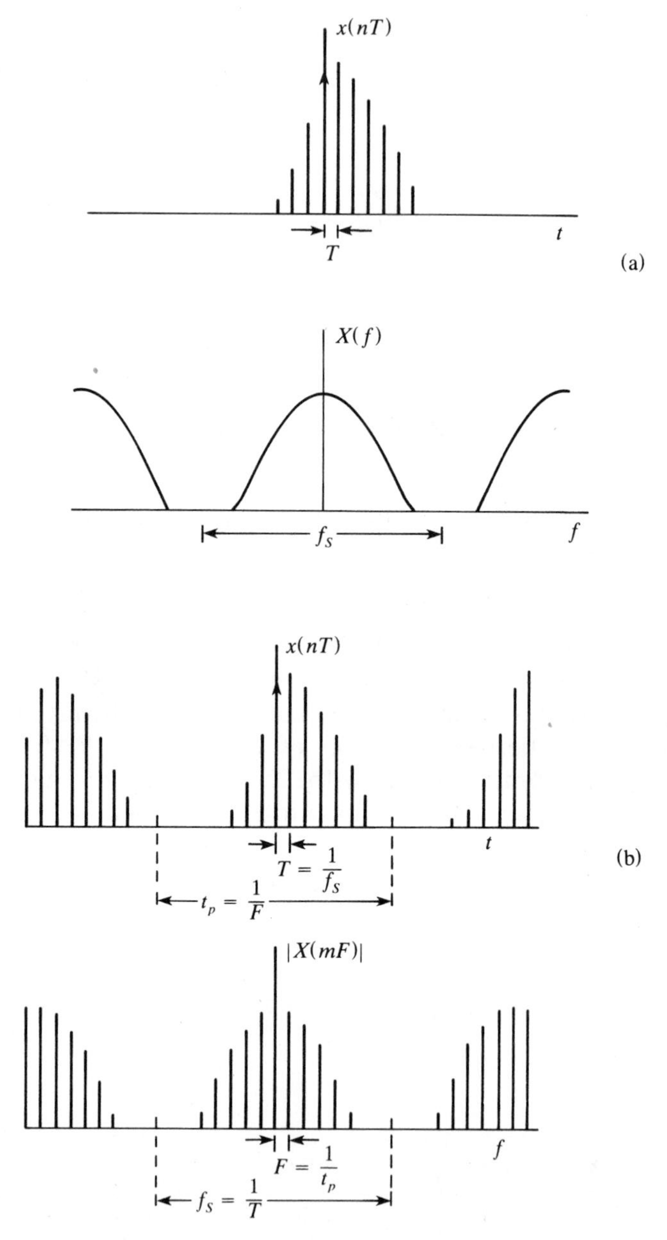

Figure 3.7 Periodic repetitions: (a) spectral repetition from sampling in the time domain; (b) signal repetition from discretizing the frequency domain.

cessing literature. One must be aware of the definition being implemented in an algorithm when interpreting the results of a computation. For instance, in the definition of equation (3.18), $X_{FS}(0)$ is the average of the samples, whereas the other definition will produce a weighted sum of the sampled values. The IDFT corresponding to equation (3.18) is

$$x(n) = \sum_{m=-N/2}^{N/2-1} X_{FS}(m)\exp\left(\frac{j2\pi mn}{N}\right)$$

$$= \sum_{m=0}^{N-1} X_{FS}(m)\exp\left(\frac{j2\pi mn}{N}\right) \tag{3.19}$$

when N is even. In the first definition, equation (3.17), the IDFT contains the factor $1/NT$. Detailed derivations of the DFT–IDFT pair, their properties, important theorems, and transform pairs can be found in many texts. Refer to the appropriate bibliographic section in this chapter for some references. The definition in equation (3.17) will be utilized in this text.

3.3.2 Partial Summary of DFT Properties and Theorems

Several theorems and properties of the DFT that pertain to real sequences will be summarized briefly because they are utilized in various sections of this book. The convolution relationships are often called cyclic convolutions because both the signal and its DFT have periodic repetitions.

a. *Linearity.* The DFT of a sum of signals equals the sum of the individuals DFTs. For two time sequences $x(n)$ and $y(n)$,

$$X_{x+y}(m) = T\sum_{n=0}^{N-1} (ax(n) + by(n))\exp\left(\frac{-j2\pi mn}{N}\right)$$

$$= T\sum_{n=0}^{N-1} ax(n)\exp\left(\frac{-j2\pi mn}{N}\right)$$

$$+ T\sum_{n=0}^{N-1} by(n)\exp\left(\frac{-j2\pi mn}{N}\right)$$

$$= aX(m) + bY(m). \tag{3.20}$$

b. *Periodicity.* $X(m) = X(m + kN)$, k is an integer.

$$X(m + kN) = T \sum_{n=0}^{N-1} x(n) \exp\left(\frac{-j2\pi n(m + kN)}{N} \right)$$

$$= T \sum_{n=0}^{N-1} x(n) \exp\left(\frac{-j2\pi nm}{N} \right) \exp\left(\frac{-j2\pi nkN}{N} \right)$$

$$= T \sum_{n=0}^{N-1} x(n) \exp\left(\frac{-j2\pi nm}{N} \right) \exp(-j2\pi nk).$$

Because $e^{-j2\pi nk} = 1$ for k and n being integers,

$$X(m + kN) = T \sum_{n=0}^{N-1} x(n) \exp\left(\frac{-j2\pi nm}{N} \right) 1 = X(m). \quad (3.21)$$

c. *Conjugate symmetry.* $X(N - m) = X^*(m)$.

$$X(N - m) = T \sum_{n=0}^{N-1} x(n) \exp\left(\frac{-j2\pi n(N - m)}{N} \right)$$

$$= T \sum_{n=0}^{N-1} x(n) \exp\left(\frac{j2\pi nm}{N} \right) \exp\left(\frac{-j2\pi nN}{N} \right)$$

$$= T \sum_{n=0}^{N-1} x(n) \exp\left(\frac{j2\pi nm}{N} \right) \exp(-j2\pi n)$$

$$= T \sum_{n=0}^{N-1} x(n) \exp\left(\frac{j2\pi nm}{N} \right) 1 = X^*(m). \quad (3.22)$$

d. *Time-shift theorem.*

$$y(n) = x(n - k) \Leftrightarrow Y(m) = X(m) \exp(-j2\pi mk/N).$$

$$Y(m) = T \sum_{n=0}^{N-1} x(n - k) \exp\left(\frac{-j2\pi mn}{N} \right)$$

$$= T \sum_{l=-k}^{N-1-k} x(l) \exp\left(\frac{-j2\pi lm}{N} \right) \exp\left(\frac{-j2\pi mk}{N} \right),$$

with the substitution $l = n - k$. Because of the periodicity occurring in the discretized domains, the complex exponentials are periodic and $x(l) = x(l + N)$.

$$Y(m) = \exp\left(\frac{-j2\pi mk}{N}\right) T \sum_{l=0}^{N-1} x(l) \exp\left(\frac{-j2\pi lm}{N}\right)$$

$$= \exp\left(\frac{-j2\pi mk}{N}\right) X(m). \tag{3.23}$$

e. *Convolution in frequency theorem.* $x(n)y(n) \Rightarrow X(m) * Y(m)$.

$$\text{DFT}[x(n)y(n)] = T \sum_{n=0}^{N-1} x(n)y(n) \exp\left(\frac{-j2\pi nm}{N}\right).$$

Substituting for $y(n)$ its DFT with summing index k and changing the order of summation gives

$$\text{DFT}[x(n)y(n)]$$

$$= T \sum_{n=0}^{N-1} x(n) \cdot \frac{1}{NT} \sum_{k=-N/2}^{N/2-1} Y(k) \exp\left(\frac{j2\pi kn}{N}\right) \exp\left(\frac{-j2\pi nm}{N}\right)$$

$$= \frac{1}{NT} \sum_{k=-N/2}^{N/2-1} Y(k) \cdot T \sum_{n=0}^{N-1} x(n) \exp\left(-\frac{j2\pi n(m-k)}{N}\right)$$

$$= \frac{1}{NT} \sum_{k=-N/2}^{N/2-1} Y(k) \cdot X(m-k)$$

$$= X(m) * Y(m). \tag{3.24}$$

f. *Convolution in time theorem.* $X(m)Y(m) \Rightarrow x(n) * y(n)$.

$$\text{DFT}[x(n) * y(n)]$$

$$= T \sum_{n=0}^{N-1} \left(T \sum_{k=0}^{N-1} x(k)y(n-k) \right) \exp\left(\frac{-j2\pi nm}{N}\right).$$

Rearranging the summations and using the time-shift theorem yields

$$\begin{aligned}\mathrm{DFT}[x(n)*y(n)] &= T\sum_{k=0}^{N-1} x(k)\left(T\sum_{n=0}^{N-1} y(n-k)\exp\left[\frac{-j2\pi nm}{N}\right]\right)\\ &= T\sum_{k=0}^{N-1} x(k)Y(m)\exp\left(\frac{-j2\pi km}{N}\right)\\ &= Y(m)T\sum_{k=0}^{N-1} x(k)\exp\left(\frac{-j2\pi km}{N}\right) = Y(m)X(m). \quad (3.25)\end{aligned}$$

3.4 Fourier Analysis

The DFT is used computationally to perform Fourier analysis on signals. There are, in general, two types of algorithm. One is direct implementation of either equation (3.17) or equation (3.18), using either complex arithmetic or cosine–sine equivalent. These are rather slow computationally and clever algorithms using lookup tables were used to increase the speed of calculation of trigonometric functions. Fortunately, much faster algorithms were developed and have been routinely implemented during the last 20 years. The fast Fourier transform (FFT) algorithm as introduced by Cooley and Tukey has revolutionized the utility of the DFT. Interestingly, the earliest development and usage of fast DFT algorithms occurred around 1805 and is attributable to Gauss. Heideman, Johnson, and Burrus (1984) have written a history of the development of these algorithms. The radix 2 FFT is the most often used algorithm and is described in almost every signals and systems textbook; its software implementation is available in most signal processing libraries. There are also mixed radix and other specialized fast algorithms for calculating the DFT [Blahut (1985) and Elliot and Rao, (1982)]. Usage of the DFT is so prevalent that special hardware signal processors that implement some of these algorithms are also available. The theory and description of these algorithms and their software is widespread and can easily be found in the IEEE special publications [IEEE (1979)] and textbooks on algorithms listed in the reference section. Whatever algorithms are used, they provide the same information and the same care must be taken to implement them. The next several sections describe the details necessary for implementation. They are necessary because of the properties of the DFT and the possible errors that may occur because signals of finite duration are being analyzed. The book by Brigham (1988) explains comprehensively all the details of the properties and errors involved with the DFT. Several of the many current applications are described in Section 3.6.

3.4.1 Frequency Range and Scaling

The DFT is uniquely defined for signals band-limited by the folding frequency range, that is, $-f_N \le f \le f_N$, where $f_N = \frac{1}{2}f_s = 1/2T$. Refer again to Figure 3.7. In addition, because of conjugate symmetry (refer to Section 3.3.2, item c), the spectrum in the negative frequency region is just the complex conjugate of that in the positive region; therefore, the frequency range with unique spectral values is only $0 \le f \le f_N$. Because the frequency spacing is $f_d = 1/NT$, the range of frequency numbers is $0 \le m \le N/2$ with the specific frequency values being $f = mf_d$. These details will be exemplified on a small data set.

Example 3.1

The signal intensity measurements, starting at midnight, of ionospheric reflections from the E-layer are listed in Table 3.1. The signal is plotted in Figure 3.8. The DFT was calculated directly using equation (3.17). The real and imaginary parts of $X(m)$ are plotted in Figure 3.9 for frequency numbers in the range $0 \le m < N - 1$, $N = 12$, and are tabulated in Appendix 3.1. The symmetry line for the conjugate symmetry is also shown in the figure. The magnitude and phase spectra were calculated with the polar coordinate transformation

$$|X(f)| = \sqrt{\mathrm{Re}(f)^2 + \mathrm{Im}(f)^2} \quad \text{and} \quad \theta = -a\tan\frac{\mathrm{Im}(f)}{\mathrm{Re}(f)}$$

and are plotted in Figure 3.10 for the positive frequency range. The sampling interval used is $T = \frac{1}{24}$ day. Thus, the frequencies on the frequency axis are $f = m(1/NT) = m2$ cycle/day.

Table 3.1 Ionospheric Reflections

Time (hr)	Intensity
0	6
1	20
2	28
3	8
4	1
5	−7
6	20
7	6
8	7
9	−14
10	−19
11	−12

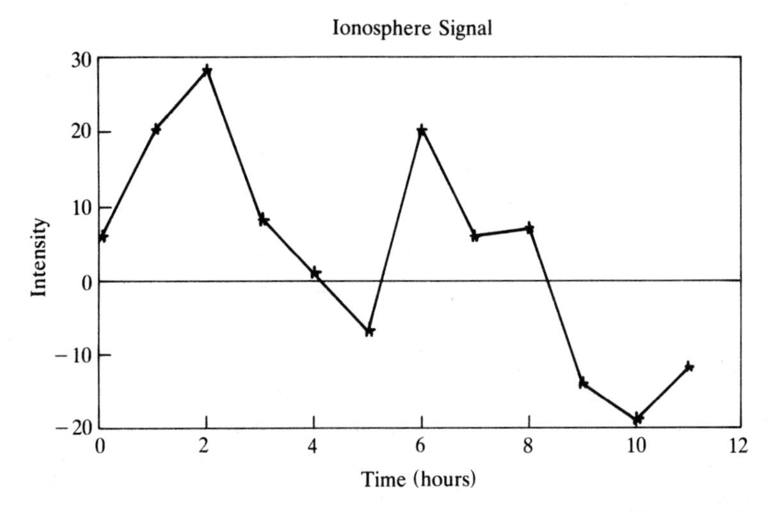

Figure 3.8 The intensity of ionospheric reflections for a 12-hr period.

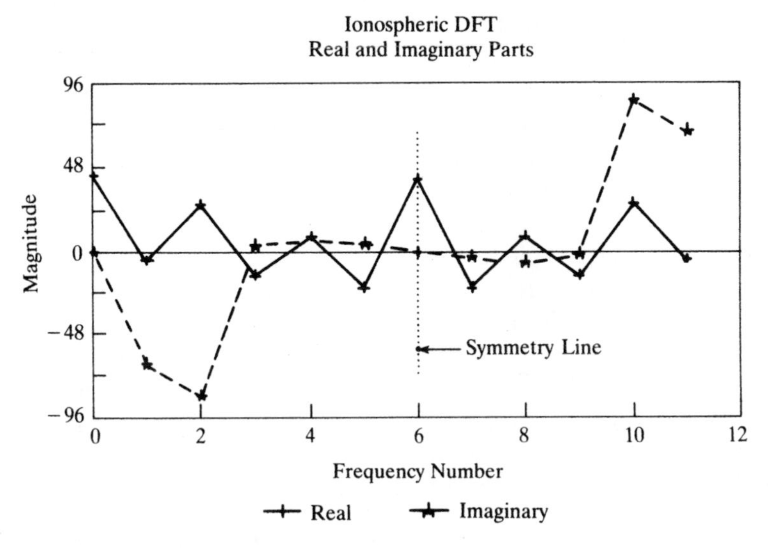

Figure 3.9 The real and imaginary parts of the DFT of the ionospheric signal are plotted versus frequency number. The dotted vertical line indicates the point of symmetry about the Nyquist frequency number.

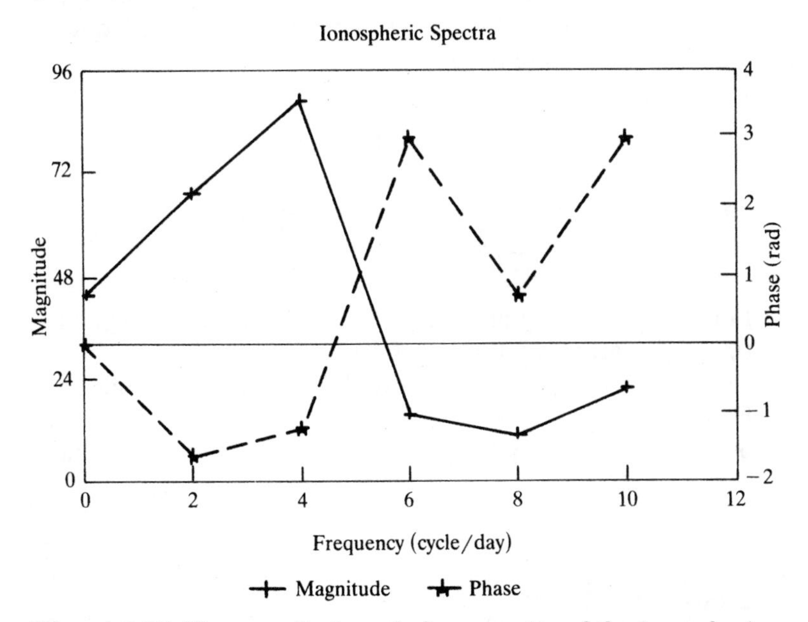

Figure 3.10 The magnitude and phase spectra of the ionospheric signal.

3.4.2 The Effect of Discretizing Frequency

One of the main concerns with the DFT is that it provides values of $X(f)$ only at a specific set of frequencies. What if an important value of the continuous spectrum existed at a frequency $f \neq mf_d$. This value would not be appreciated. This error is called the *picket fence effect* in the older literature. The analogy is that when one looks at a scene through a picket fence, one only sees equally spaced parts of the total scene. Naturally, an obvious solution is to increase the number of signal points. This would not only increase N, and hence decrease the frequency spacing, but would also increase the accuracy of the DFT. What if this were not possible? Fortunately, the frequency spacing can be reduced by *zero padding*. This is adding a string of 0-valued data points to the end of a signal. If M zeros were added to increase the signal length to LT, the new frequency spacing is

$$f_d' = \frac{1}{LT} = \frac{1}{(N+M)T}$$

and the DFT is calculated for frequency values $f = kf_d' = k(1/LT)$. The effect of the zero padding on the magnitude of the DFT can be appreciated directly by using the defining equation [equation (3.17)]. The DFT for L points become

$$\begin{aligned} X_L(k) &= T \sum_{n=0}^{L-1} x(n) \exp\left(\frac{-j2\pi kn}{L}\right) \\ &= T \sum_{n=0}^{N-1} x(n) \exp\left(\frac{-j2\pi kn}{L}\right), \qquad 0 \le k \le \frac{L}{2}. \qquad (3.26) \end{aligned}$$

For the same frequency in the unpadded and zero-padded calculations, $f = m(1/NT) = k(1/LT)$, and the exponents have the same value. Thus, $X_L(k) = X(m)$ and there is no change in accuracy. Essentially, this is an *interpolation* procedure. The effect of zero padding on the calculations is best shown by an example.

Example 3.2

Calculate again the spectra of the ionospheric signal, with the number of signal points increased to 24 by padding with 12 zeros. The resulting spectra are shown in Figure 3.11. The fre-

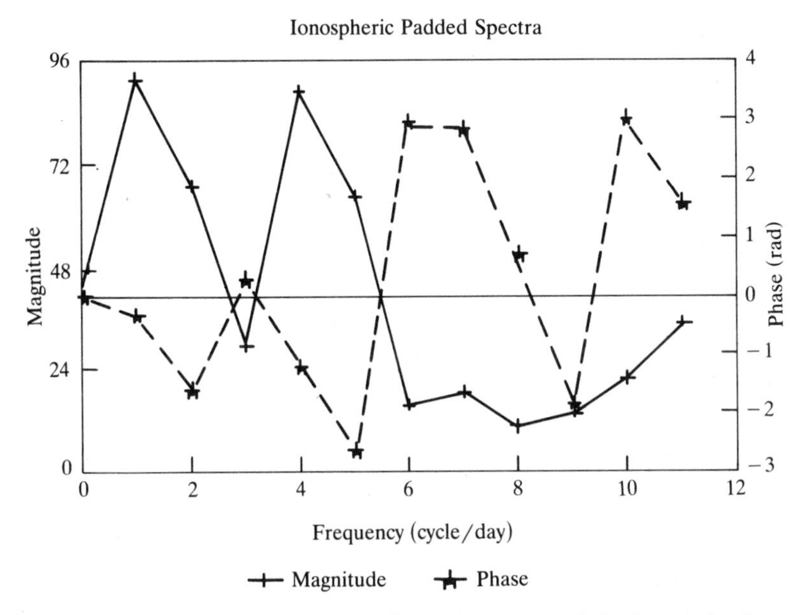

Figure 3.11 The magnitude and phase spectra of the ionospheric signal after zero padding.

quency range has not changed because the sampling interval has not changed. Because the number of signal points has doubled, the number of spectral points has doubled. Notice that now the spectrum has a large magnitude at 1 cycle per day. This was not evident with $f_d = 2$. However, the values of the magnitude and phase spectra at the frequencies used in previous example have not changed.

An important additional use for zero padding has developed because of the implementation of fast algorithms. They often restrict the number of signal points to specific sets of values. For instance, the radix 2 FFT algorithm must have $N = 2^l$, where l is an integer. If the required number of signal points is not available, simply padding with zeros will suffice.

3.4.3 The Effect of Truncation

Another major source of error in calculating the DFT is a consequence of truncation. This is implicit because a finite sample is being made from a theoretically infinite duration signal. The truncation is represented mathematically by having the measured signal $y(n)$ be equal to the actual signal $x(n)$ multiplied by function called a *rectangular data window*, $d_R(n)$. That is,

$$y(n) = x(n)d_R(n), \tag{3.27}$$

where

$$d_R(n) = \begin{cases} 1 & \text{for } 0 \leq n \leq N-1, \\ 0 & \text{for other } n. \end{cases}$$

There are several types of *data windows*, $d(n)$, that will be discussed and, in general,

$$y(n) = x(n)d(n). \tag{3.28}$$

Some data windows are plotted in Figure 3.12. It is known from the convolution theorem that the Fourier transforms of the three time sequences in equation (3.28) have a special relationship. However, it must be remembered that this still occurs while the frequency domain is continuous. The frequency domain convolution is rewritten for this situation and is

$$Y(f) = X(f) * D(f), \tag{3.29}$$

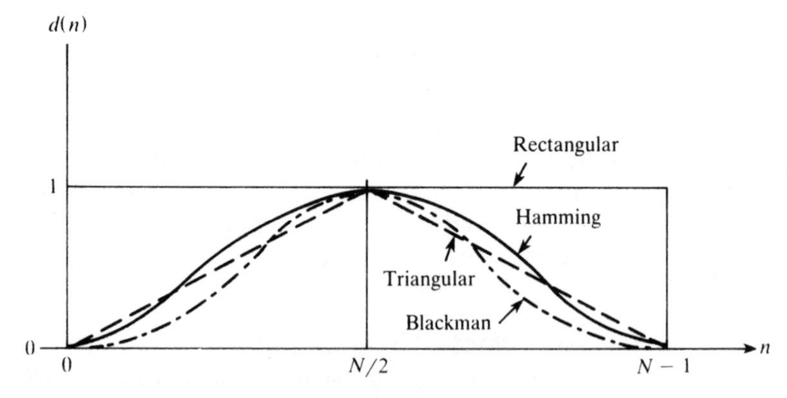

Figure 3.12 Plots of several data windows.

where $D(f)$ is called a *spectral window*. The DTFT of the rectangular data window is well documented and is

$$D_R(f) \doteq \frac{\sin(\pi f N T)}{\sin(\pi f T)} e^{-j\pi f N T}. \qquad (3.30)$$

Thus, in general, the DFT that is calculated, $Y(m)$, is not equal to $X(m)$ because of the convolution in equation (3.29). The error, or discrepancy, caused by the truncation effect is called *leakage error*. The convolution causes signal energy that exists at a given frequency in $X(m)$ to cause nonzero values of $Y(m)$ at frequencies in which no energy of $x(n)$ exists. The characteristics of the leakage error depend upon the spectral window. An illustration of this is presented in the next example.

Example 3.3

Let $x(n)$ be a cosine sequence. Its transform is a sum of two delta functions and the transform pair is

$$\cos(2\pi f_0 n T) \Leftrightarrow 0.5\delta(f - f_0) + 0.5\delta(f + f_0). \qquad (3.31)$$

Figure 3.13 shows the signal and DFT for $y(n)$ and its transform when P is an integral number of periods of $1/f_0$. As expected for $f_0 = \frac{1}{8}$, $N = 32$, and $T = 1$, the unit impulse exists at frequency number 4, $m = 4$. Notice what happens when $f_0 \neq m f_d$. When $f_0 = 1/9.143$, $Y(m)$ becomes much different and is shown in Figure 3.14. There are two discrepancies between this DFT and the theoretical one. First of all, there are nonzero magnitudes of $Y(m)$ at frequencies at which there is no signal component. This is the effect of the leakage error of the rectangular window. The second discrepancy is that the magnitude of $Y(m)$ surrounding the actual frequency is less than 0.5. This happens because the DFT is not calculated at the exact frequency, that is, the spectral window is not evaluated at its peak. This discrepancy will always occur and is called *scalloping loss*. This is not an important factor in situations

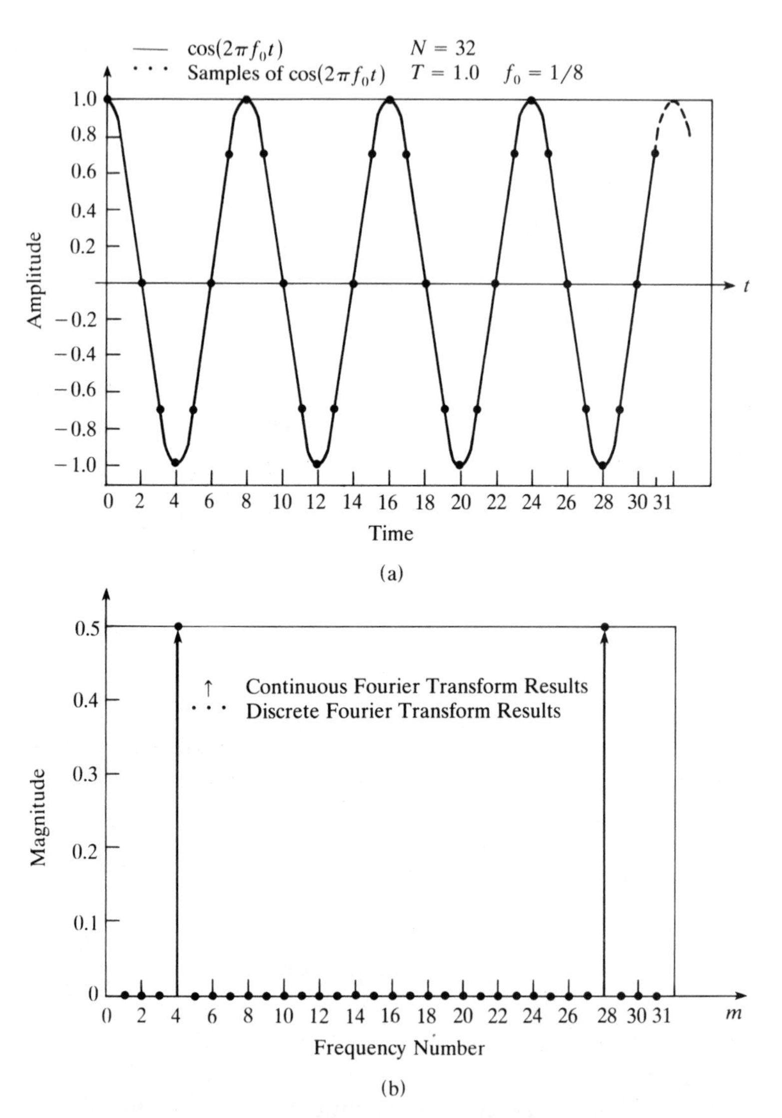

Figure 3.13 (a) Cosine sequence, $N = 32$, with integral multiples of periods and (b) its DFT versus frequency numbers. [Adapted from Brigham (1988), Figure 9-6, with permission.]

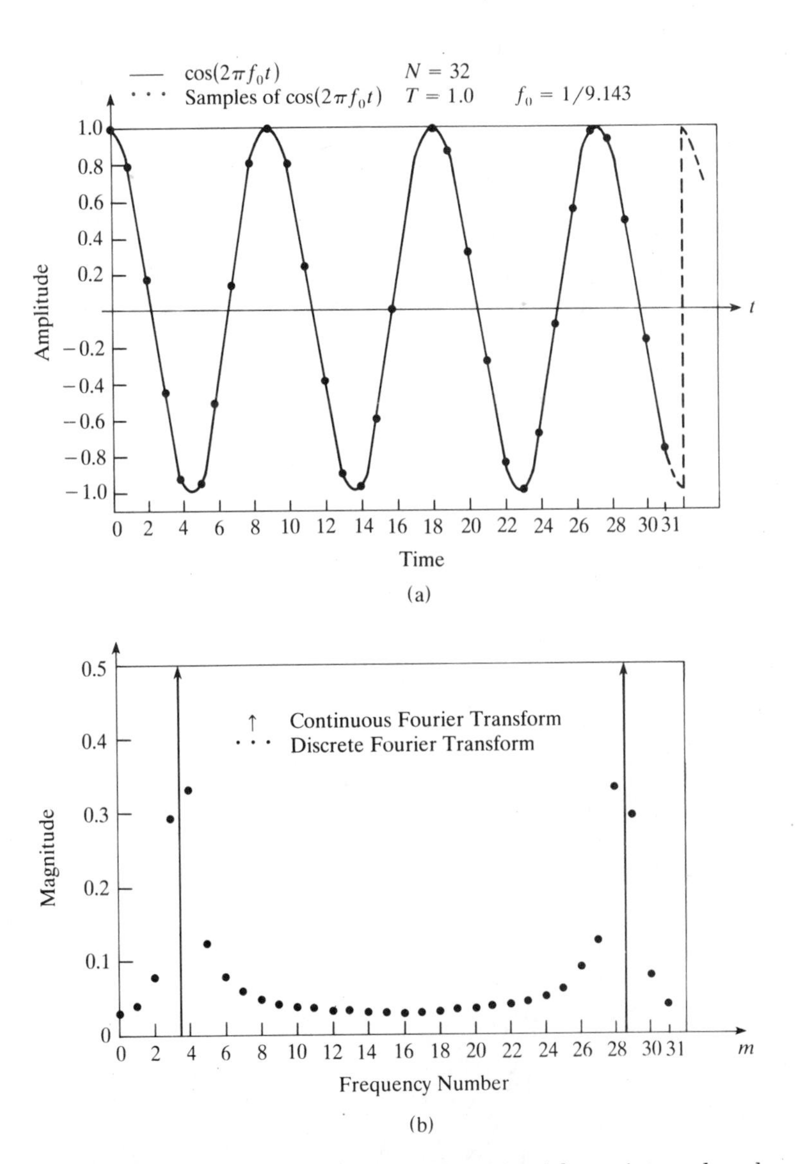

Figure 3.14 (a) Cosine sequence, $N = 32$, with nonintegral multiples of periods and (b) its DFT versus frequency numbers. [Adapted from Brigham (1988), Figure 9-7, with permission.]

when a continuum of frequencies exists; that is, $X(f)$ is continuous. One benefit is that without some leakage, the occurrence of this frequency component would not be evident in the DFT.

Leakage error is unacceptable and must be minimized. The procedure for minimizing leakage can be understood by considering the shape of the magnitude of $D_R(f)$. It is an even function and

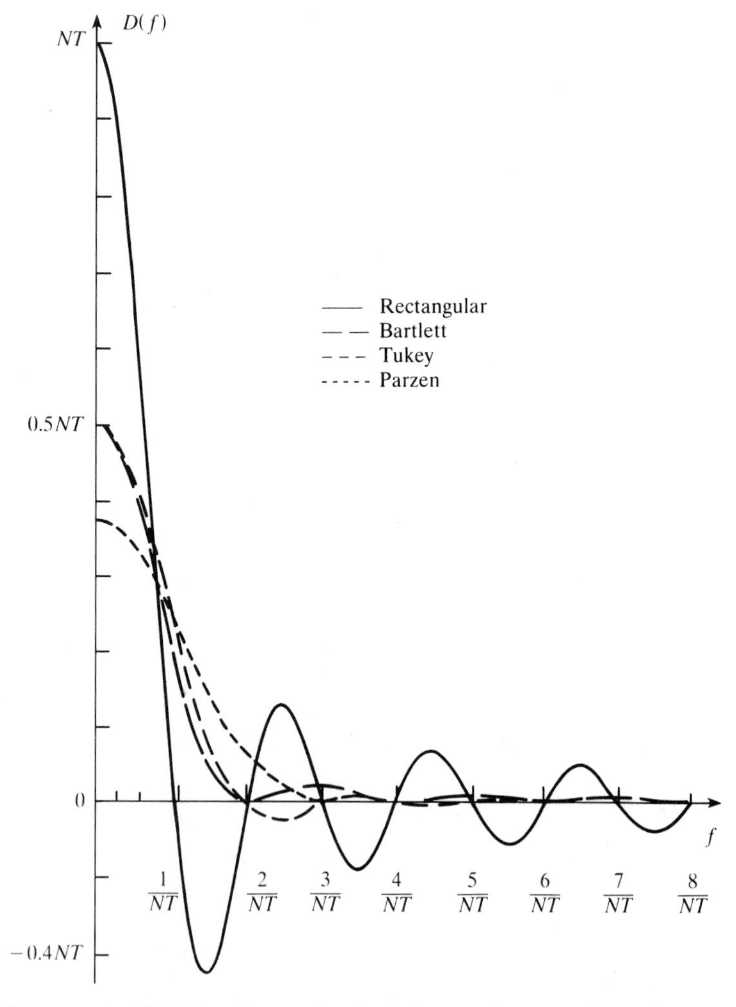

Figure 3.15 Plots of magnitudes of several spectral windows.

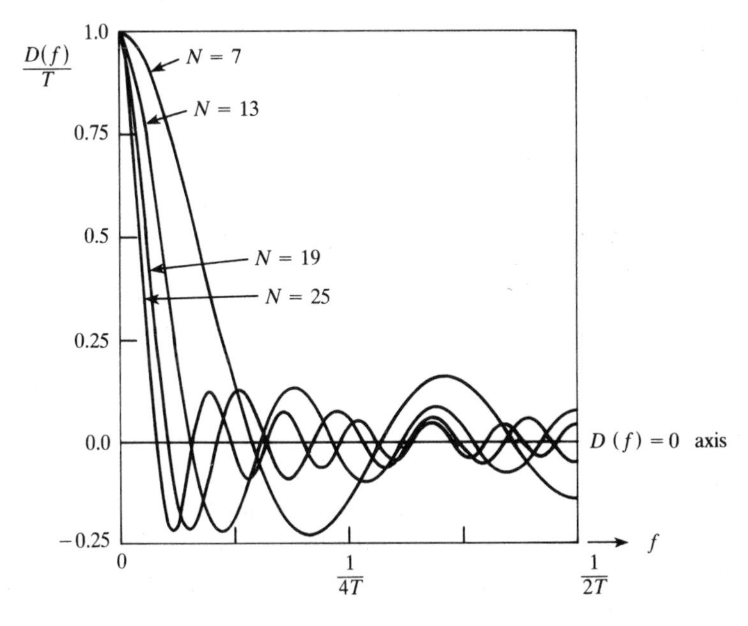

Figure 3.16 The rectangular spectral window for various values of signal points. [Adapted from Roberts and Mullis (1987), Figure 6.3.3, with permission.]

only the positive-frequency portion is plotted in Figure 3.15. There are many crossings of the frequency axis and the portions of $D_R(f)$ between crossings are called *lobes*. The lobe centered around the zero frequency is called the *main lobe*, and the other lobes are called *side lobes*. The convolution operation spreads the weighting effect of side lobes over the entire frequency range. One way to minimize their effect is to increase N. Figure 3.16 shows the rectangular spectral window for various values of N. As N increases, the width of the main lobe decreases and the magnitude of the side lobes decreases at any frequency value.

3.4.4 Windowing

Because reducing the spread of the leakage error by increasing the number of signal points is not usually possible, another method is necessary. An effective and well-used method is to change the shape of the data window. The purpose is to produce spectral windows whose side lobes contain much less energy than the main lobe. Several spectral windows are also plotted in Figure 3.15 and their Fourier transforms are tabulated in Appendix 3.5. It is evident from studying the figure that the magnitudes of the side lobes of these other spectral windows fit this specification. The quantification of this property is the *side-lobe level*, that is, the ratio of the magnitudes of the largest side lobe to the main lobe. These are also listed in the table in Appendix 3.5. The use of the other windows should reduce the leakage error considerably.

Example 3.4

One of the commonly used windows is the Hanning, or Tukey, window; it is sometimes called the cosine bell window for an obvious reason. Its formula is

$$d_T(n) = 0.5\left(1 - \cos\left(\frac{2\pi n}{N-1}\right)\right); \qquad 0 \le n \le N-1. \quad (3.32)$$

Its side-lobe level is -31 dB, 2.8%, a factor of 10 less than the level of the rectangular window. The Hanning window has been applied to the cosine sequence of Example 3.3 and the resulting $y(n)$ sequence and its DFT are plotted in Figure 3.17. Notice that the windowing has removed the discontinuity between the periodic repetitions of the time waveform and has made $Y(m)$ much more closely resemble its theoretical counterpart. The magnitude spectrum is zero for many of its components where it should be and the leakage is spread to only adjacent frequency components. Consider the peak magnitude of the spectra and realize that it is less than that in Figure 3.14. This additional reduction in amplitude occurs because nonrectangular windowing removes signal energy and is called *process loss*.

3.4.5 Resolution

Examine again the plots of the spectral windows in Figure 3.15. Particularly notice the lowest frequencies at which zero crossings of the frequency axis occur for each window. They are differ-

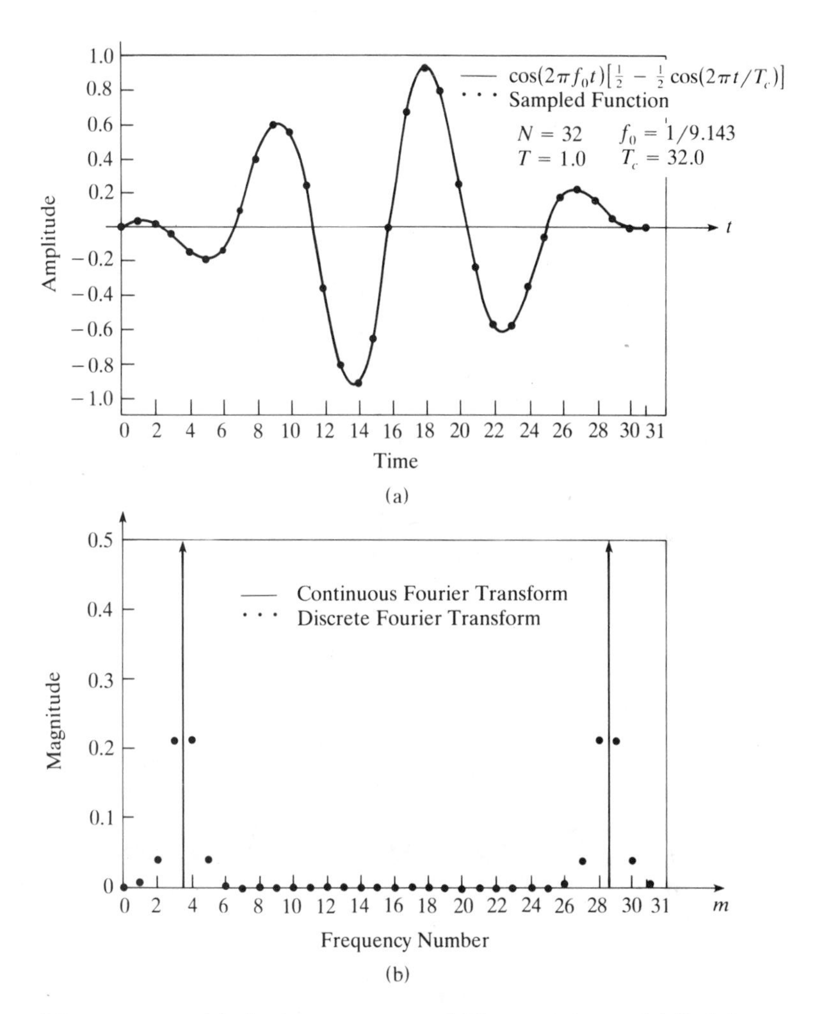

Figure 3.17 (a) Cosine sequence of Figure 3.14 multiplied by a Hanning window and (b) its DFT with frequency numbers. [Adapted from Brigham (1988), Figure 9-9, with permission.]

ent. The *width of the main lobe* is defined as the frequency range of these zero crossings and these are listed in the table in Appendix 3.5. In the previous example, a compromise has occurred. The width of the main lobe of the Hanning window is twice as large as that of the rectangular window. The window width affects the ability to detect the existence of two sinusoids with closely spaced frequencies in a magnitude spectrum. A simple example can illustrate this point. Consider a sequence of two sinusoids with equal amplitudes and frequencies f_1 and $f_1 + \Delta f$,

$$x(n) = \cos(2\pi f_1 nT) + \cos(2\pi(f_1 + \Delta f)nT). \quad (3.33)$$

The DFT of a finite sample with Δf much larger than the main lobe width is plotted in Figure 3.18(a). The presence of the two sinusoids is evident. The shape of the DFT as Δf becomes smaller

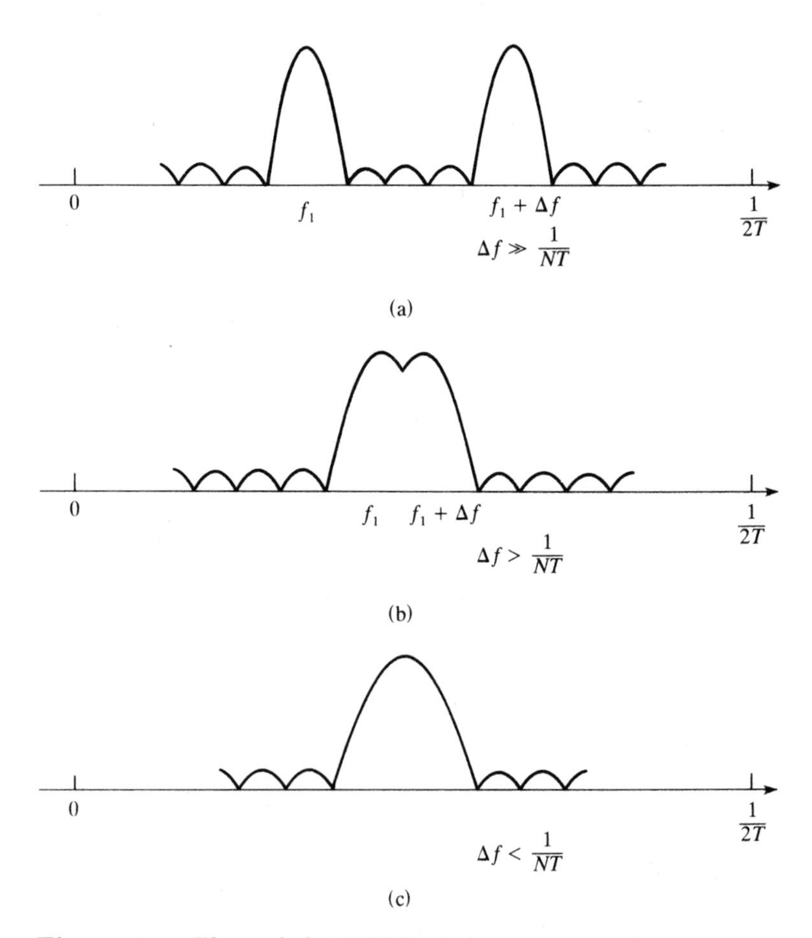

Figure 3.18 Plots of the DTFT of the sequence of two cosine functions whose frequencies differ by Δf Hz: (a) $\Delta f \gg 1/NT$; (b) $\Delta f > 1/NT$; (c) $\Delta f \leq 1/NT$.

is sketched Figure 3.18(b). As one can envision, the spectral peaks are not as easily resolved as Δf approaches 50% of the main lobe width. In fact, for $\Delta f < 1/NT$, the peaks are not distinguishable [Figure 3.18(c)]. For sinusoids with unequal amplitudes, the frequency difference at which they are resolvable is more complex to define. The rule of thumb is that the *resolution* of a data window is one-half of its main lobe width.

Example 3.5

Compute the DFT magnitude spectrum for the function

$$x(t) = e^{-0.063t}\sin(2\pi t) + e^{-0.126t}\sin(2.2\pi t),$$

with $t = nT$ when $T = 0.175$ and $N = 50$, 100, 200, and 400. The

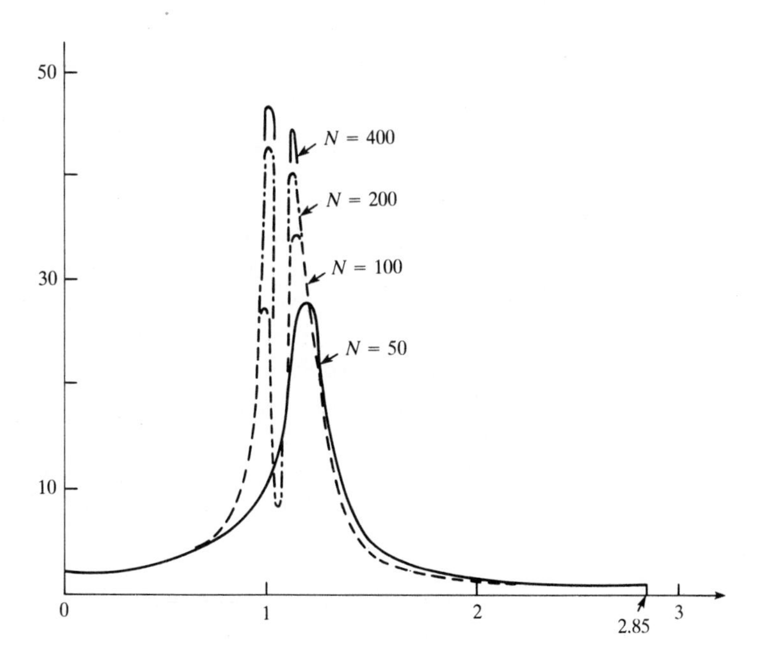

Figure 3.19 The magnitude spectrum of the DFT of the sampled function in Example 3.5 with $T = 0.175$ and $N = 50, 100, 200, 400$. [Adapted from Chen (1979), Figure 4.18, with permission.]

results are plotted in Figure 3.19. The frequency difference, Δf, is 0.1 Hz. For $N = 50$, the main lobe width is 0.228 Hz and the frequency resolution is approximately 0.114 Hz. Notice that the peaks of the two sinusoids are not present in the spectrum. If the number of signal points is increased, the main lobe width will be reduced and the frequency components resolvable. This can be seen for the DFT calculated when $N \geq 100$.

In general, there is a compromise when applying data windows. Study Appendix 3.5. It shows that when the side lobe level is reduced, the width of the main lobe is increased. Thus, when reducing the leakage error, one also reduces the spectral resolution. This is usually not a problem when the spectra are continuous and do not have narrow peaks. Other windows have been designed for special applications such as resolving the frequencies of two simultaneously occurring sinusoids with close frequencies. The most comprehensive source of information is the review article by Harris (1978). It reviews the concepts and purposes of windowing and summarizes the characteristics of over 20 window functions that have been developed.

3.4.6 Detrending

Another aspect similar to windowing is the removal of polynomial time trends and average value. Both of these time functions have energy around zero frequency that is spread into the low frequency ranges of $Y(m)$. This error is easily removed by fitting the time series with a low-order polynomial and then removing the values estimated by it. This process is called *detrending*. A good example of a signal with a linear trend is the record of airline passenger data that is tabulated in Appendix 3.3 and is plotted in Figure 3.20(a). A linear regression curve fitted for this data has the equation

$$f(t) = 198.3 + 3.01t, \tag{3.34}$$

where $f(t)$ is the number of thousands of passengers using air travel at an airport and t is the number of months relative to January 1953. The month-by-month subtraction of $f(t)$ from the

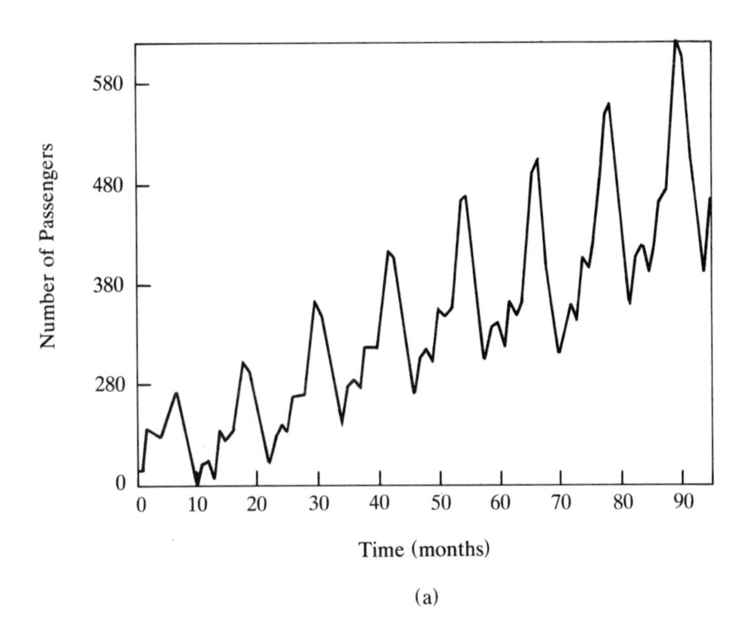

(a)

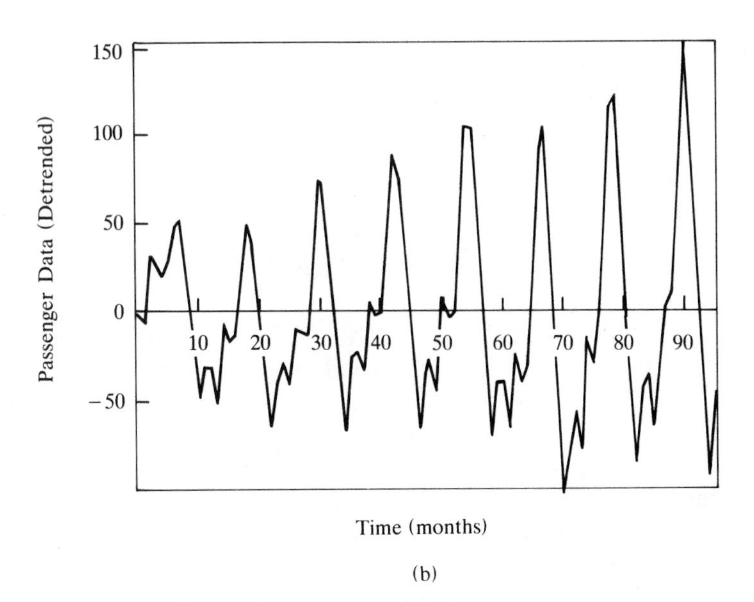

(b)

Figure 3.20 Plot of number of airline passengers (in thousands) for years 1953 through 1960: (a) measured data; (b) detrended data.

measured data produces the detrended data plotted in Figure 3.20(b), with an average value of zero. Most of the situations in engineering require only removing the average value.

3.5 Procedural Summary

In summary, the procedure for calculating the DFT and minimizing the potential sources of error is as follows:

1. detrend the signal;
2. multiply the signal with a suitable data window;
3. zero-pad as necessary to reduce frequency spacing or to implement a fast algorithm;
4. perform the DFT operation on the resulting signal.

Use of these concepts on actual signals will help the review process. The sampled values of several signals are tabulated in the appendix of this chapter. Exercises requiring a computer are suggested in the exercise section. They have been chosen because the approximate periodicities can be appreciated through visual inspection of plots of the signals.

3.6 Selected Applications

Several applications of the DFT are briefly summarized below to show it utility over a broad range of engineering fields.

Food and waste matter are moved through the digestive system in order to nourish living systems. Muscles encircle the tube (tract) of the digestive system and contract to move the food along the tract. The electrical signals from the muscles are measured to study their contraction characteristics. In general, these signals are called electromyograms and, in particular for the digestive system, they are called electrogastrograms. Figure 3.21(a) shows a 4.24-minute recording of the electrogastrogram from the ascending colon of a person. Visual examination of the electrogastrogram

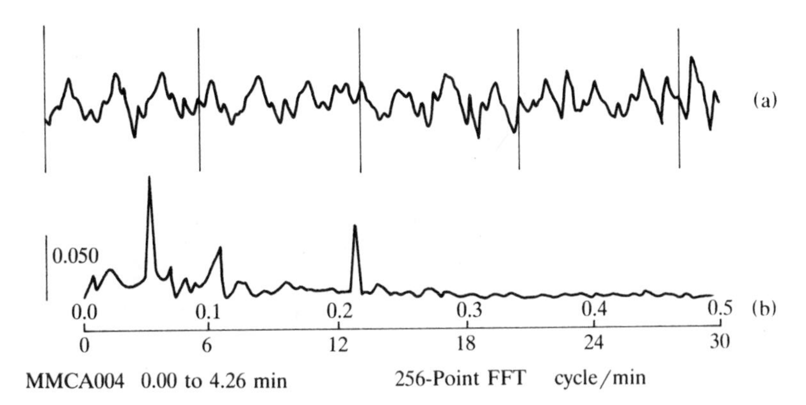

Figure 3.21 (a) The electrogastrogram from the muscle of the ascending colon. Its duration is 4.24 min and the vertical lines demarcate one-second segments. (b) The 256-point FFT of electrogastrogram above. The signal was sampled at 1 Hz and the magnitude bar indicating a 0.05-mV amplitude is plotted on the left side. [From Smallwood et al. (1980), Figure 4, with permission.]

reveals that there are periods of greater and lesser activity, which indicates that the muscle contracts and relaxes approximately three times per minute. The signal was sampled once per second for 4.24 minutes and the DFT was calculated using a radix 2 FFT algorithm. The magnitude spectrum of the DFT in Figure 3.21(b) shows several well-defined peaks. The largest magnitude is 0.1 millivolts and occurs at a frequency of 3.1 cycles per minute. Other major cyclic components oscillate at 6.4 and 12.8 cycles per minute.

The wear and incipient faults in gears of rotating machinery can be monitored with frequency analysis. The method is to measure the vibration in the gearbox and study its magnitude spectrum. Figure 3.22(a) shows a vibration signal and its spectrum for a machine exhibiting uniform gear wear. The tooth-meshing frequency (TM) is the number of teeth meshing per unit time and is 60 per second. The harmonics of TM are always present. As the wear becomes nonuniform, the energy in the harmonics increase with respect to the fundamental frequency. Figure 3.22(b) shows such a situation. This indicates that a fault will occur soon and some corrective steps must be undertaken.

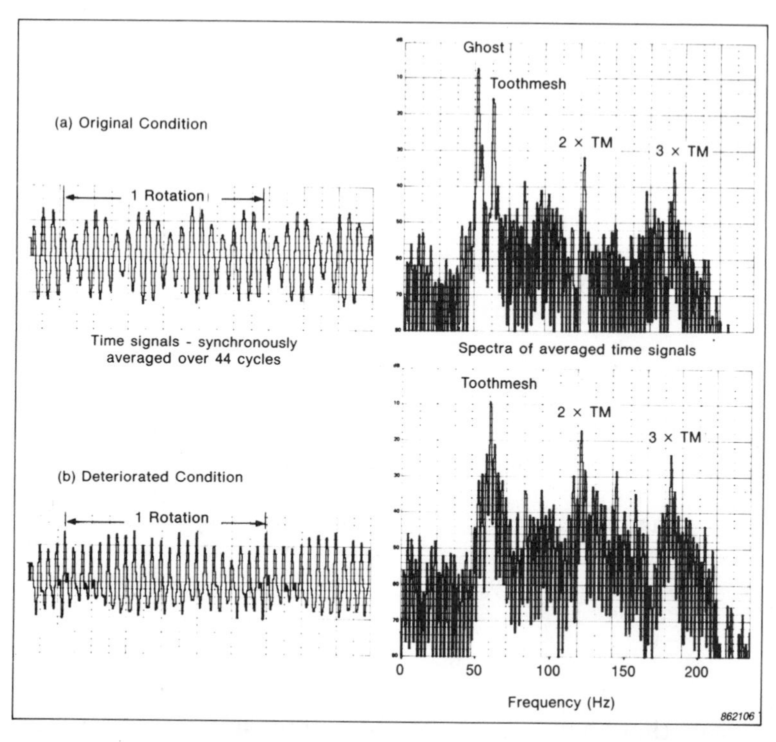

Figure 3.22 (a) The vibration signal and its magnitude spectrum for a rotating gearbox exhibiting normal wear and (b) the vibration signal and its magnitude spectrum for a rotating gearbox exhibiting an incipient fault. [From Angelo (1987), Figure 8, with permission.]

In the production of high-quality audio systems the characteristics of loudspeakers are very important. It is desired that the loudspeakers have a flat frequency response so that all frequencies of sound are faithfully reproduced. If the response is not flat, then electronic equalizers can be designed to make the sound energy produced have a flat spectrum. The testing is done by exciting the loudspeaker with an impulse response and calculating the DFT of the resulting sound. Figures 3.23(a) and 3.23(b) show the impulse response and magnitude spectrum, respectively. Notice that the spectrum has dips at frequencies of 1.1 and 2.2 kHz that will need to be compensated.

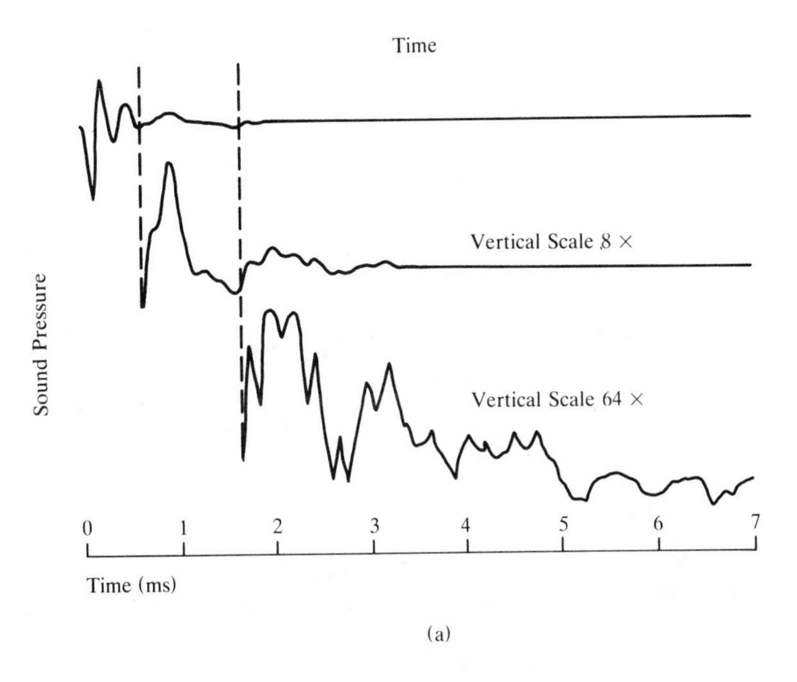

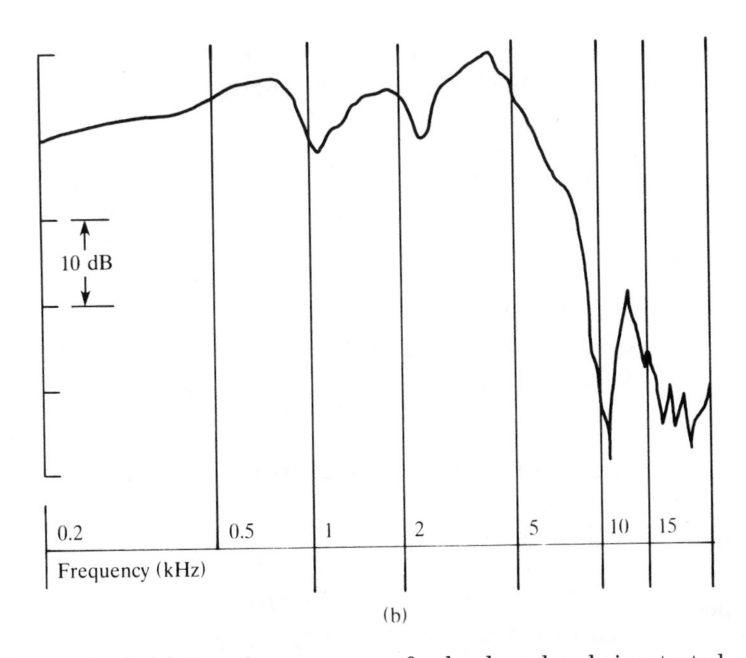

Figure 3.23 (a) Impulse response of a loudspeaker being tested and (b) the magnitude spectrum of its DFT. [From Blesser and Kates (1978), Figures 2-25 and 2-26, with permission.]

References

Discrete Fourier Transforms and Fast Algorithms

Blahut, R. (1985). *Fast Algorithms for Digital Signal Processing*. Addison-Wesley Publishing Co., Reading, MA.

Brigham, E. (1988). *The Fast Fourier Transform and Its Applications*. Prentice-Hall, Englewood Cliffs, NJ.

Burrus, C. and T. Parks (1985). *DFT/FFT and Convolution Algorithms*. John Wiley and Sons, New York.

Cadzow, J. (1987). *Foundations of Digital Signal Processing and Data Analysis*. Macmillan Publishing Co., New York.

Childers, D. and A. Durling (1975). *Digital Filtering and Signal Processing*. West Publishing Co., St. Paul.

Elliot, D. and K. Rao (1982). *Fast Transforms—Algorithms, Analyses, Applications*. Academic Press, New York.

Heideman, M., D. Johnson, and C. Burrus (1984). Gauss and the history of the fast fourier transform. *IEEE Acoust., Speech, and Sig. Proc. Mag.* 1(4):14–21.

IEEE (1979). *Programs for Digital Signal Processing*; Edited by IEEE Acoustics, Speech, and Signal Processing Society Special Committee. IEEE Press, New York, 1979.

Rabiner, L. and B. Gold (1975). *Theory and Application of Digital Signal Processing*. Prentice-Hall, Englewood Cliffs, NJ.

Discrete Fourier Transforms

Chen, C. (1979). *One-Dimensional Signal Processing*. Marcel Dekker, New York.

Harris, F. (1978). On the use of windows for harmonic analysis with the discrete fourier transform. *Proc. IEEE* 66:51–83.

Kuc, R. (1988). *Introduction to Digital Signal Processing*. McGraw-Hill Co., New York.

Marple, S. (1987). *Digital Spectral Analysis with Applications*. Prentice-Hall, Englewood Cliffs, NJ.

Roberts, R. and C. Mullis (1987). *Digital Signal Processing*. Addison-Wesley Publishing Co., Reading, MA.

Stanley, W., G. Dougherty, and R. Dougherty (1984). *Digital Signal Processing*. Reston Publishing Co., Reston, VA.

Williams, C. (1986). *Designing Digital Filters*. Prentice-Hall, Inc., Englewood Cliffs, NJ.

Fourier Series and Transforms

Bloomfield, P. (1976). *Fourier Analysis of Time Series—An Introduction*. John Wiley and Sons, New York.

Bracewell, R. (1978). *The Fourier Transform and its Applications*. McGraw-Hill Book Co., New York.

Hsu, H. (1970). *Fourier Analysis*. Simon and Schuster, New York.

Papoulis, A. (1962). *The Fourier Integral and Its Applications*. McGraw-Hill Book Co., New York.

Papoulis, A. (1977). *Signal Analysis*. McGraw-Hill Book Co., New York.

Zeimer, R., W. Tranter, and D. Fannin (1983). *Signals and Systems—Continuous and Discrete*. Macmillan Publishing Co., New York.

Applications

Angelo, M. (1987). Vibration monitoring of machines. *Bruel & Kjaer Technical Review* No. 1:1–36.

Blesser, B. and J. Kates (1978). Digital processing in audio signals. In *Applications of Digital Signal Processing* (A. Oppenheim, ed.); pp. 29–116. Prentice-Hall, Englewood Cliffs, NJ.

Bodenstein, G. and H. Praetorius (1977). Feature extraction from the electroencephalogram by adaptive segmentation. *Proc. IEEE* 65:642–652.

Licht, T., H. Andersen, and H. Jensen (1987). Recent developments in accelerometer design. *Bruel & Kjaer Technical Review* No. 2:1–22.

Smallwood, R., D. Linkins, H. Kwak, and C. Stoddard (1980). Use of autoregressive-modelling techniques for the analysis of colonic myoelectric activity. *Med. Biol. Engrg. Comput.* 18:591–600.

Whittaker, E. and G. Robinson (1967). *The Calculus of Observations*. Dover Publications, New York.

Exercises

3.1 Prove Parseval's theorem, such as equation (A3.13), for orthogonal function sets.

3.2 Show that the squared error in a truncated Fourier series using $N + 1$ terms is $\sum_{n=N+1}^{\infty} \lambda_n a_n^2$.

3.3 Verify equation (A3.4), $\lambda_n = P$.

3.4 For the time-shifted sawtooth waveform in Figure A3.1:
- a. Form the complex Fourier series.
- b. How do the coefficient magnitudes and phase angle compare to the ones from the unshifted version?
- c. Plot the line spectra for $-10 < n < 10$.
- d. Calculate the time shifts τ_n and plot a time-shift line spectrum.

3.5 Show that the average power of the Fourier series can be expressed in terms of the trigonometric series coefficients as

$$C_0^2 + \sum_{n=1}^{\infty} \tfrac{1}{2} C_n^2.$$

3.6 It is obvious that changing the magnitude of the harmonic coefficients will change the shape of a periodic function. Changes in only the phase angle can do that also without changing the average power. Build a function $g(t)$ with two cosine waveforms of amplitude 1.0, phase angles equal zero, and fundamental frequency of 1.0 Hz; that is,

$$g(t) = C_1 \cos(\omega_0 t) + C_2 \cos(2\omega_0 t).$$

a. Plot $g(t)$.
b. Change C_1 to 2.0 and replot it as $g_1(t)$.
c. Take $g(t)$ and make $\theta_2 = 90°$. Replot it as $g_2(t)$. Notice it is different from $g(t)$ also.
d. Take $g(t)$ and make $\tau_1 = \tau_2 = 0.25$. Notice how the shape has remained the same and $g(t)$ is shifted in time.

3.7 Prove that with the input to the linear system being $\bar{\delta}(t)$ [equation (3.5)], the output is $\bar{h}(t)$ [equation (3.8)].

3.8 In Section 3.2.2 is stated the Poisson sum formula for the frequency domain. Starting with

$$\bar{X}(\omega) = \sum_{n=-\infty}^{\infty} X(\omega + n\omega_1),$$

prove that

$$\bar{X}(\omega) = \frac{2\pi}{\omega_1} \sum_{n=-\infty}^{\infty} x(nT_1) e^{-j\omega n T_1}.$$

3.9 Draw and label the frequency axes over the valid frequency range for the following conditions:
a. $T = 1$ s, $N = 10, 20, 40$;
b. $T = 0.1$ s, $N = 10, 20, 40$;
c. $T = 1$ ms, $N = 10, 20, 40$.

3.10 Using the sequence $x = [1, 1, 0, 0, \cdots]$, do the following:
a. Derive the DTFT.
b. Calculate the DFT for $N = 4$ and $N = 8$.
c. Plot the three magnitude spectra found in parts a and b and compare them.

3.11 Let $f(t) = e^{-t}U(t)$. Find or derive its Fourier transform.
a. Compute the DFT of $f(t)$ with $T = 0.25$ for $N = 4$ and 8.
b. How do the two $X(m)$ compare with $X(f)$?

c. Use zero padding and repeat part a.
d. How do the results of part c compare with the other estimates of $X(f)$?

3.12 Verification for Example 3.1.
a. In Figure 3.9, verify the conjugate symmetry of the real and imaginary parts of the DFT.
b. What would the frequency axis be in Figure 3.10 if the sampling interval were 1 h?
c. In Figure 3.10, verify the values of the magnitude and phase spectra for $f = 0$, 2, and 6 cycle/day.

3.13 For the spectra in Example 3.1, perform the inverse transform using equation (3.17) for $0 \le n \le 24$. Is the signal periodically repeated?

3.14 Repeat Exercise 3.11a with $N = 64$. Plot the spectra.

3.15 Detrend the signal in Exercise 3.11.
a. Plot the detrended signal.
b. Repeat Exercise 3.11c. What values have changed?

3.16 Derive the spectral window [equation (3.30)], of the rectangular data window [equation (3.27)].

3.17 For the cosine sequence in Example 3.3, do the following:
a. prove that the DTFT of the truncated sequence is

$$Y(f) = 0.5D_R(-f) + 0.5D_R(f),$$

where

$$D_R(f) = T\frac{\sin(\pi fNT)}{\sin(\pi fT)}e^{-j\pi fNT}.$$

b. Verify the magnitude values at frequency numbers 3, 4, and 10 when $f_0 = 1/9.143$.

3.18 Repeat Exercise 3.14 after applying a Hanning window. Have the calculated spectra become more like the theoretical spectra?

3.19 Calculate the DFT of the ionospheric signal after detrending, windowing, and padding with 12 zeros. How does it compare with the spectrum in Figure 3.11?

Computer Exercises

3.20 Analyze the star light intensity signal tabulated in Appendix 3.2.

a. Plot at least 25% of the signal.
b. What are the approximate periodicities present?
c. Detrend and window the entire signal; plot the result.
d. Calculate the DFT using an algorithm of your choice.
e. Plot the magnitude and phase spectra.
f. What major frequency components or ranges are present? Do they agree with your initial estimation?
g. Apply another data window and repeat steps c–f. Are there any differences in the two spectra?

3.21 Analyze the passenger signal in Appendix 3.3.
a. Plot the entire signal.
b. What are the approximate harmonics present? At what time of the year do they peak?
c. Detrend and window the entire signal; plot the result. How does the plot compare to Figure 3.20b?
d. Calculate the DFT using an algorithm of your choice.
e. Plot the magnitude and phase spectra.
f. What major frequency components are present? What phase angles and time lags do these components have? Are these results consistent with what is expected from part b?

Appendices

Appendix 3.1 DFT of Ionosphere Data

Frequency cycle/h	Re(f)	Im(f)
0.0000	44.0	0.0
0.0833	−5.7	−67.0
0.1667	27.0	−84.9
0.2500	−15.0	3.0
0.3333	8.0	6.9
0.4167	−21.0	4.0
0.5000	42.0	0.0
0.5833	−21.0	−4.0
0.6667	8.0	−6.9
0.7500	−15.0	−3.0
0.8333	27.0	84.9
0.9167	−5.7	67.0

Appendix 3.2 Brightness Data of a Variable Star

Brightness of a Variable Star at Midnight on 600 Successive Days

Day	Mag.	Day	Mag.	Day	Mag.
1	25	41	10	81	17
2	28	42	7	82	18
3	31	43	5	83	19
4	32	44	3	84	19
5	33	45	3	85	19
6	33	46	3	86	19
7	32	47	4	87	20
8	31	48	5	88	20
9	28	49	7	89	20
10	25	50	10	90	20
11	22	51	13	91	20
12	18	52	16	92	20
13	14	53	19	93	20
14	10	54	22	94	20
15	7	55	24	95	21
16	4	56	26	96	20
17	2	57	27	97	20
18	0	58	28	98	20
19	0	59	29	99	20
20	0	60	28	100	19
21	2	61	27	101	18
22	4	62	25	102	17
23	8	63	24	103	16
24	11	64	21	104	15
25	15	65	19	105	13
26	19	66	17	106	12
27	23	67	15	107	11
28	26	68	13	108	10
29	29	69	12	109	9
30	32	70	11	110	9
31	33	71	11	111	10
32	34	72	10	112	10
33	33	73	10	113	11
34	32	74	11	114	12
35	30	75	12	115	14
36	27	76	12	116	16
37	24	77	13	117	19
38	20	78	14	118	21
39	17	79	15	119	24
40	13	80	16	120	25

Brightness of a Variable Star at Midnight on 600 Successive Days

Day	Mag.	Day	Mag.	Day	Mag.
121	27	161	3	201	25
122	28	162	1	202	26
123	29	163	0	203	27
124	29	164	0	204	27
125	28	165	1	205	26
126	27	166	3	206	25
127	25	167	6	207	24
128	23	168	9	208	22
129	20	169	13	209	20
130	17	170	17	210	18
131	14	171	21	211	17
132	11	172	24	212	15
133	8	173	27	213	14
134	5	174	30	214	13
135	4	175	32	215	13
136	2	176	33	216	12
137	2	177	33	217	12
138	2	178	32	218	12
139	4	179	31	219	13
140	6	180	28	220	13
141	9	181	25	221	13
142	12	182	22	222	14
143	16	183	19	223	14
144	19	184	15	224	15
145	23	185	12	225	15
146	27	186	9	226	16
147	30	187	7	227	17
148	32	188	5	228	17
149	33	189	4	229	17
150	34	190	4	230	17
151	33	191	5	231	18
152	32	192	5	232	18
153	30	193	7	233	19
154	27	194	9	234	19
155	24	195	12	235	20
156	20	196	14	236	20
157	16	197	17	237	21
158	12	193	20	238	21
159	9	199	22	239	22
160	5	200	24	240	22

Brightness of a Variable Star at Midnight on 600 Successive Days

Day	Mag.	Day	Mag.	Day	Mag.
241	22	281	1	321	32
242	22	282	1	322	32
243	22	283	2	323	31
244	21	284	4	324	29
245	20	285	7	325	26
246	19	286	10	326	23
247	17	287	14	327	21
248	16	288	17	328	17
249	14	289	21	329	14
250	12	290	25	330	11
251	11	291	29	331	9
252	9	292	31	332	7
253	8	293	33	333	6
254	7	294	34	334	5
255	8	295	34	335	6
256	8	296	33	336	6
257	9	297	31	337	7
258	10	298	29	338	9
259	12	299	26	339	11
260	14	300	22	340	13
261	17	301	19	341	15
262	20	302	14	342	18
263	23	303	11	343	20
264	25	304	7	344	22
265	27	305	4	345	23
266	29	306	2	346	24
267	30	307	1	347	25
268	30	308	0	348	25
269	30	309	1	349	25
270	29	310	2	350	24
271	27	311	5	351	24
272	25	312	7	352	22
273	22	313	11	353	21
274	19	314	15	354	19
275	16	315	19	355	18
276	12	316	22	356	17
277	9	317	25	357	16
278	6	318	28	358	15
279	4	319	30	359	15
280	2	320	32	360	14

Brightness of a Variable Star at Midnight on 600 Successive Days

Day	Mag.	Day	Mag.	Day	Mag.
361	14	401	7	441	32
362	14	402	8	442	30
363	14	403	10	443	28
364	14	404	12	444	24
365	14	405	15	445	20
366	14	406	18	446	16
367	14	407	22	447	13
368	14	408	24	448	9
369	14	409	27	449	6
370	14	410	29	450	3
371	15	411	31	451	2
372	15	412	31	452	1
373	15	413	31	453	1
374	15	414	31	454	2
375	16	415	29	455	4
376	16	416	27	456	6
377	17	417	24	457	9
378	18	418	21	458	13
379	19	419	18	459	17
380	20	420	14	460	20
381	21	421	10	461	23
382	22	422	7	462	26
383	23	423	5	463	28
384	23	424	2	464	30
385	24	425	1	465	31
386	24	426	0	466	31
387	24	427	1	467	31
388	23	428	2	468	29
389	22	429	5	469	27
390	21	430	8	470	24
391	19	431	12	471	22
392	17	432	15	472	19
393	15	433	19	473	16
394	13	434	23	474	13
395	11	435	27	475	11
396	9	436	30	476	9
397	7	437	32	477	8
398	6	438	34	478	7
399	6	439	34	479	7
400	6	440	34	480	7

Brightness of a Variable Star at Midnight on 600 Successive Days

Day	Mag.	Day	Mag.	Day	Mag.
481	8	521	15	561	26
482	9	522	16	562	23
483	11	523	18	563	20
484	12	524	19	564	16
485	14	525	21	565	12
486	16	526	22	566	8
487	18	527	24	567	6
488	20	528	24	568	3
489	21	529	25	569	1
490	22	530	26	570	0
491	23	531	26	571	0
492	23	532	25	572	1
493	23	533	24	573	3
494	23	534	23	574	6
495	23	535	21	575	10
496	22	536	19	576	13
497	21	537	16	577	17
498	20	538	14	578	21
499	19	539	12	579	25
500	18	540	9	580	28
501	18	541	7	581	31
502	17	542	5	582	33
503	17	543	5	583	34
504	16	544	4	584	34
505	16	545	5	585	33
506	16	546	6	586	31
507	16	547	8	587	29
508	15	548	10	588	26
509	15	549	13	589	22
510	15	550	16	590	18
511	14	551	20	591	15
512	14	552	23	592	11
513	13	553	26	593	8
514	13	554	29	594	5
515	13	555	31	595	3
516	13	556	32	596	2
517	13	557	32	597	2
518	13	558	32	598	2
519	14	559	31	599	4
520	14	560	29	600	5

Appendix 3.3 International Airline Passenger Data

Monthly Total (in 1000's) of International Airline Passengers

	1953	1954	1955	1956	1957	1958	1959	1960
Jan.	196	204	242	284	315	340	360	417
Feb.	196	188	233	277	301	318	342	391
Mar.	236	235	267	317	356	362	406	419
Apr.	235	227	269	313	348	348	396	461
May	229	234	270	318	355	363	420	472
June	243	264	315	374	422	435	472	535
July	264	302	364	413	465	491	548	622
Aug.	272	293	347	405	467	505	559	606
Sept.	237	259	312	355	404	404	463	508
Oct.	211	229	274	306	347	359	407	461
Nov.	180	203	237	271	305	310	362	390
Dec.	201	229	278	306	336	337	405	432

Appendix 3.4 Review of Fourier Series and Transform

A3.4.1 Fourier Series

A3.4.1.1 Definition

This treatment of Fourier series is intended to be strictly a review of the concepts of harmonics, complex numbers, phase shifts, and line spectra. Comprehensive treatments can be found in the books listed in the reference section.

For periodic finite power signals, the trigonometric functions form a very powerful orthogonal function set. Another definition and one that is more useful for signal analysis is composed of complex exponentials. The orthogonal function set is

$$\Phi_n(t) = e^{jn\omega_0 t} \quad \text{for } -\infty \le n \le \infty \text{ and } \omega_0 = \frac{2\pi}{P} = 2\pi f_0, \quad \text{(A3.1)}$$

where P is the period and ω_0 is the fundamental frequency in radians per unit time. Figure A3.1 shows two periodic sawtooth waveforms with a period of four seconds. Thus, $P = 4$ s, $\omega_0 = 0.5\pi$ rad/s, and $f_0 = 0.25$ Hz. The orthogonality condition and the

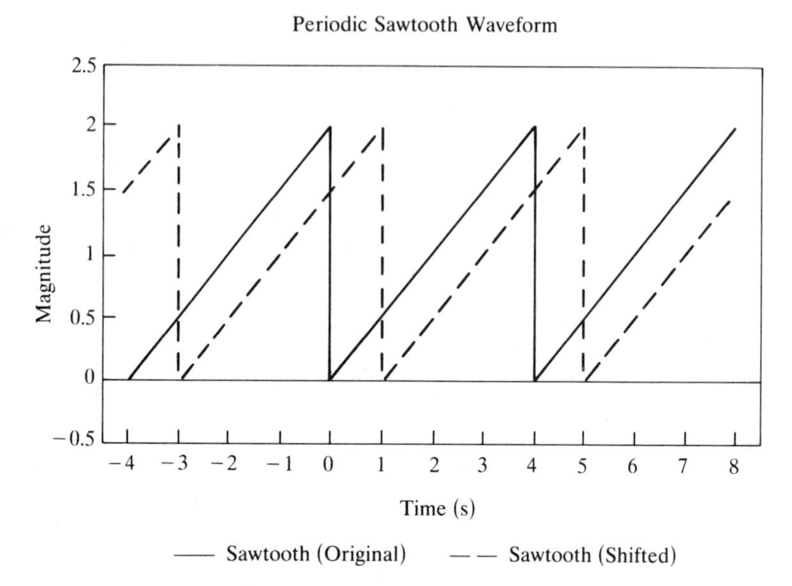

Figure A3.1 Periodic sawtooth waveforms with period of 4 s and amplitude of 2 units.

coefficient evaluation equation change slightly for complex function sets in that one term in the defining integral is a complex conjugate term (indicated by an asterisk). For this orthogonality condition, equation (A2.2) becomes

$$\int_0^P \Phi_m(t)\Phi_n^*(t)\,dt = \begin{cases} \lambda_n & \text{for } m = n, \\ 0 & \text{for } m \neq n, \end{cases} \tag{A3.2}$$

where the coefficients are evaluated from

$$z_n = \frac{1}{\lambda_n}\int_0^P f(t)\Phi_n^*(t)\,dt$$

$$= \frac{1}{\lambda_n}\int_0^P f(t)e^{-jn\omega_0 t}\,dt, \qquad -\infty \leq n \leq \infty. \tag{A3.3}$$

It is easily verified that

$$\lambda_n = P. \tag{A3.4}$$

The series expansion is now

$$f(t) = \sum_{n=-\infty}^{\infty} z_n \Phi_n(t)$$

$$= \sum_{n=-\infty}^{\infty} z_n e^{jn\omega_0 t}$$

$$= \cdots + z_{-n}\Phi_{-n}(t) + \cdots + z_{-1}\Phi_{-1}(t) + z_0\Phi_0(t)$$

$$+ z_1\Phi_1(t) + \cdots + z_n\Phi_n(t) + \cdots$$

$$= \cdots + z_{-n}e^{-jn\omega_0 t} + \cdots + z_{-1}e^{-j\omega_0 t} + z_0 + z_1 e^{j\omega_0 t} + \cdots$$

$$+ z_n e^{jn\omega_0 t} + \cdots. \qquad \text{(A3.5)}$$

The coefficients and the terms of this equation have some conjugate symmetries. Expand equation (A3.5) with Euler's formula for the complex exponential,

$$e^{-jn\omega_0 t} = \cos(n\omega_0 t) - j\sin(n\omega_0 t),$$

then

$$z_n = \frac{1}{P}\int_0^P f(t)(\cos(n\omega_0 t) - j\sin(n\omega_0 t))\,dt. \qquad \text{(A3.6)}$$

Thus, z_n is a complex number and is a conjugate of the z_{-n} coefficient when $f(t)$ is a real signal.

The exponential form is directly related to the trigonometric form. The conversion is most directly implemented by expressing the complex coefficients in polar form and using the conjugate symmetry properties,

$$z_n = Z_n e^{j\theta_n} \quad \text{and} \quad Z_{-n} = Z_n, \qquad \theta_{-n} = -\theta_n. \qquad \text{(A3.7)}$$

Inserting the relations into equations (A3.7) into equation (A3.5) yields

$$f(t) = \sum_{n=-\infty}^{-1} Z_n e^{j\theta_n} e^{jn\omega_0 t} + z_0 + \sum_{n=1}^{\infty} Z_n e^{j\theta_n} e^{jn\omega_0 t}. \qquad \text{(A3.8)}$$

Knowing to use Euler's formula for the cosine function equation (A3.8) becomes

$$f(t) = z_0 + \sum_{n=1}^{\infty} Z_n(e^{j\theta_n}e^{jn\omega_0 t} + e^{-j\theta_n}e^{-jn\omega_0 t})$$

$$= z_0 + \sum_{n=1}^{\infty} 2Z_n \cos(n\omega_0 t + \theta_n)$$

$$= C_0 + \sum_{n=1}^{\infty} C_n \cos(n\omega_0 t + \theta_n). \qquad \text{(A3.9)}$$

Thus one can see that there are direct correspondences between the two forms. The average terms have the same magnitude, whereas the magnitudes in the complex coefficients have one-half the magnitude of the trigonometric form. The second equivalence is in the phase angle. The phase angle of the complex coefficients for positive n are equal to the phase shift of the cosine terms. Remember that

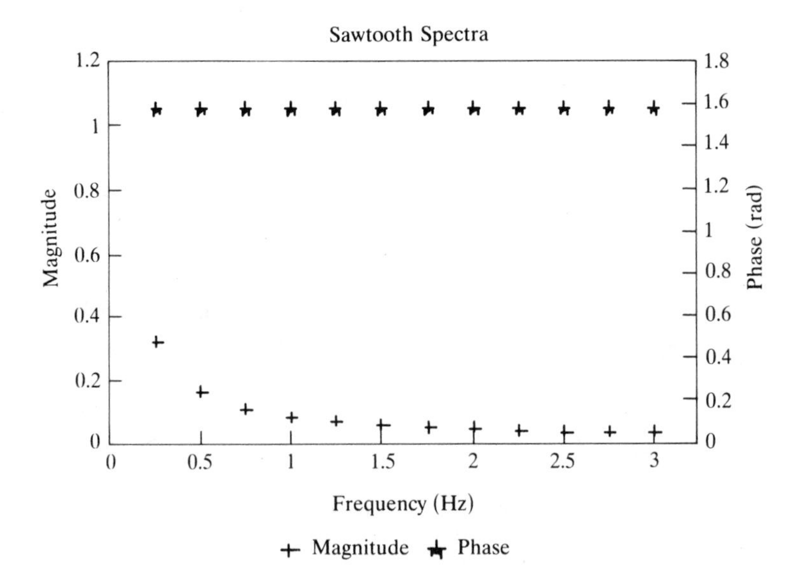

Figure A3.2 Magnitude and phase line spectra for polar form of periodic sawtooth function indicated by the solid line in Figure A3.1.

phase shifts translate into time shifts τ_n of the cosine waves through the relationship

$$\theta_n = n\omega_0\tau_n. \qquad \text{(A3.10)}$$

The information contained in the magnitudes of the coefficients and in the phase angles is very important in frequency analysis. Their values as a function of frequency are called the *magnitude and phase spectra*. Because these values exist only at harmonic frequencies, the spectra are *line spectra*. They are plotted in Figure A3.2 for the sawtooth waveform.

A3.4.1.2 Convergence

The Fourier series will converge to a periodically defined function in almost all conceivable practical situations. The function simply has to satisfy the Dirichlet conditions:

1. $f(t)$ must have a finite number of discontinuities over the period;
2. $f(t)$ must have a finite number of maxima and minima over the period;
3. $f(t)$ must be bounded or absolutely integrable, that is,

$$\int_0^P |f(t)|\,dt < \infty.$$

In addition, the Fourier series exists as long as the range of integration is over one full period; that is, the initial time is not important. In general,

$$z_n = \frac{1}{P}\int_{t_1}^{t_1+P} f(t)e^{-jn\omega_0 t}\,dt \quad \text{for } -\infty \le t_1 \le \infty. \qquad \text{(A3.11)}$$

At any points of discontinuity in $f(t)$ the series will converge to a value that is the average of the upper and lower values at the point of discontinuity. That is, if $f(t)$ is discontinuous at time t_1, the series value is

$$f(t_1) = \tfrac{1}{2}(f(t_1^-) + f(t_1^+)). \qquad \text{(A3.12)}$$

The total energy in one period is simply expressed as

$$E_{\text{tot}} = \sum_{n=-\infty}^{\infty} a_n^2 \lambda_n = \sum_{n=-\infty}^{\infty} P Z_n^2. \qquad \text{(A3.13)}$$

The average power is

$$P_{\text{av}} = \frac{1}{P} \sum_{n=-\infty}^{\infty} Z_n^2. \qquad \text{(A3.14)}$$

A3.4.2 The Fourier Transform

The Fourier transform is suitable for finding the frequency content of aperiodic signals. Conceptually and mathematically, the Fourier transform can be derived from the Fourier series relationships by a limiting operation. Consider the waveform from one cycle of a periodic process, let it stay unchanged and let the period approach an infinite duration. Figure A3.3 shows the sawtooth

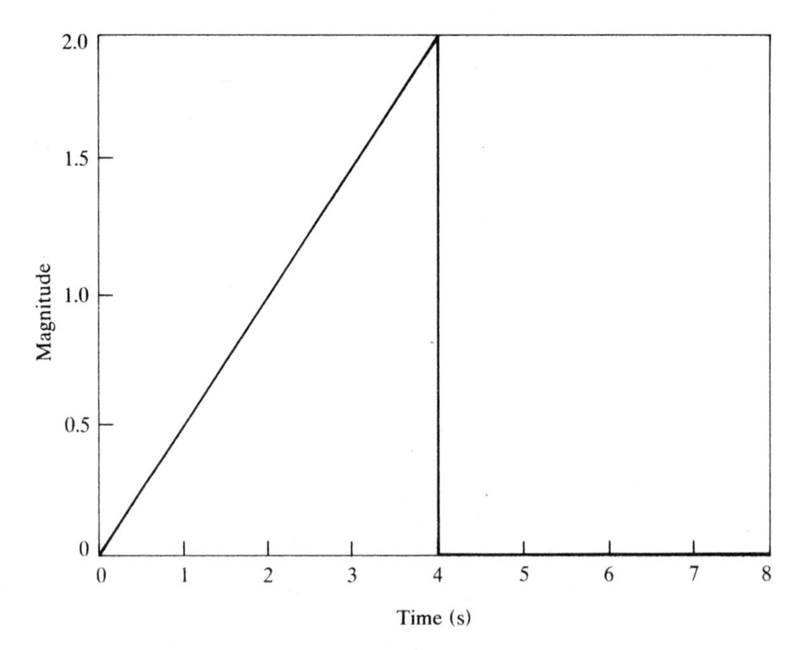

Figure A3.3 The sawtooth waveform as an aperiodic waveform.

waveform after such a manipulation. Several other properties change concurrently. First, the signal becomes an energy signal. Second, the spectra becomes continuous. This is understood by considering the frequency difference between harmonics of the line spectra during the limiting operation. This frequency difference is the fundamental frequency. So

$$\Delta f = f_0 = \lim_{P \to \infty} \frac{1}{P} = 0 \quad \text{(A3.15)}$$

and the line spectra approach continuous spectra, and the summation operation of equation (A3.5) becomes an integration operation. The resulting relationships for the time function become

$$x(t) = \int_{-\infty}^{\infty} X(f) e^{j2\pi ft}\, df = \frac{1}{2\pi} \int_{-\infty}^{\infty} X(\omega) e^{j\omega t}\, d\omega \quad \text{(A3.16)}$$

For the frequency transformation in cycles or radians per unit time, the interval of integration becomes infinite, or equation (A3.11) becomes

$$X(f) = \int_{-\infty}^{\infty} x(t) e^{-j2\pi ft}\, dt \quad \text{or} \quad X(\omega) = \int_{-\infty}^{\infty} x(t) e^{-j\omega t}\, dt. \quad \text{(A3.17)}$$

The derivation of these Fourier transform pairs is presented in great detail in all of the cited references. The properties of the Fourier transform are the same as those of the complex coefficients of the Fourier series. Writing equation (A3.17) by expanding the complex coefficient, it becomes

$$X(f) = \int_{-\infty}^{\infty} x(t)(\cos(2\pi ft) - j\sin(2\pi ft))\, dt. \quad \text{(A3.18)}$$

Thus, $X(f)$ is a complex function. The fact that its real part is an even function and its imaginary part is an odd function can be easily proved for real signals by inserting $-f$ for every f in the integrand. Thus,

$$\mathrm{Re}(f) = \Re[X(f)] = \Re[X(-f)],$$

$$\mathrm{Im}(f) = \Im[X(f)] = -\Im[X(-f)] \tag{A3.19}$$

In polar form this becomes

$$X(f) = |X(f)|\exp(\theta(f)),$$

where

$$|X(f)| = \sqrt{(\mathrm{Re}^2(f) + \mathrm{Im}^2(f))},$$

$$\theta(f) = -a\tan\frac{\mathrm{Im}(f)}{\mathrm{Re}(f)}. \tag{A3.20}$$

Because of the even and odd properties of $\mathrm{Re}(f)$ and $\mathrm{Im}(f)$, it is easily seen that $|X(f)|$ is an even function of frequency and that $\theta(f)$ is an odd function of frequency. The transforms for many waveforms can be found in the references. The spectra for the sawtooth waveform in Figure A3.3 are plotted in Figure A3.4.

In aperiodic signals energy is defined as

$$E_{\mathrm{tot}} = \int_{-\infty}^{\infty} x^2(t)\, dt = \int_{-\infty}^{\infty} x(t)\left(\int_{-\infty}^{\infty} X(f)e^{j2\pi ft}\, df\right) dt,$$

with one expression of the signals described in its transform definition. Now, reorder the order of integration and the energy expression becomes

$$E_{\mathrm{tot}} = \int_{-\infty}^{\infty} X(f)\left(\int_{-\infty}^{\infty} x(t)e^{j2\pi ft}\, dt\right) df. \tag{A3.21}$$

The expression within the brackets is the Fourier transform of $x(t)$ with the substitution of $-f$ for f, or it is equivalent to $X(-f)$.

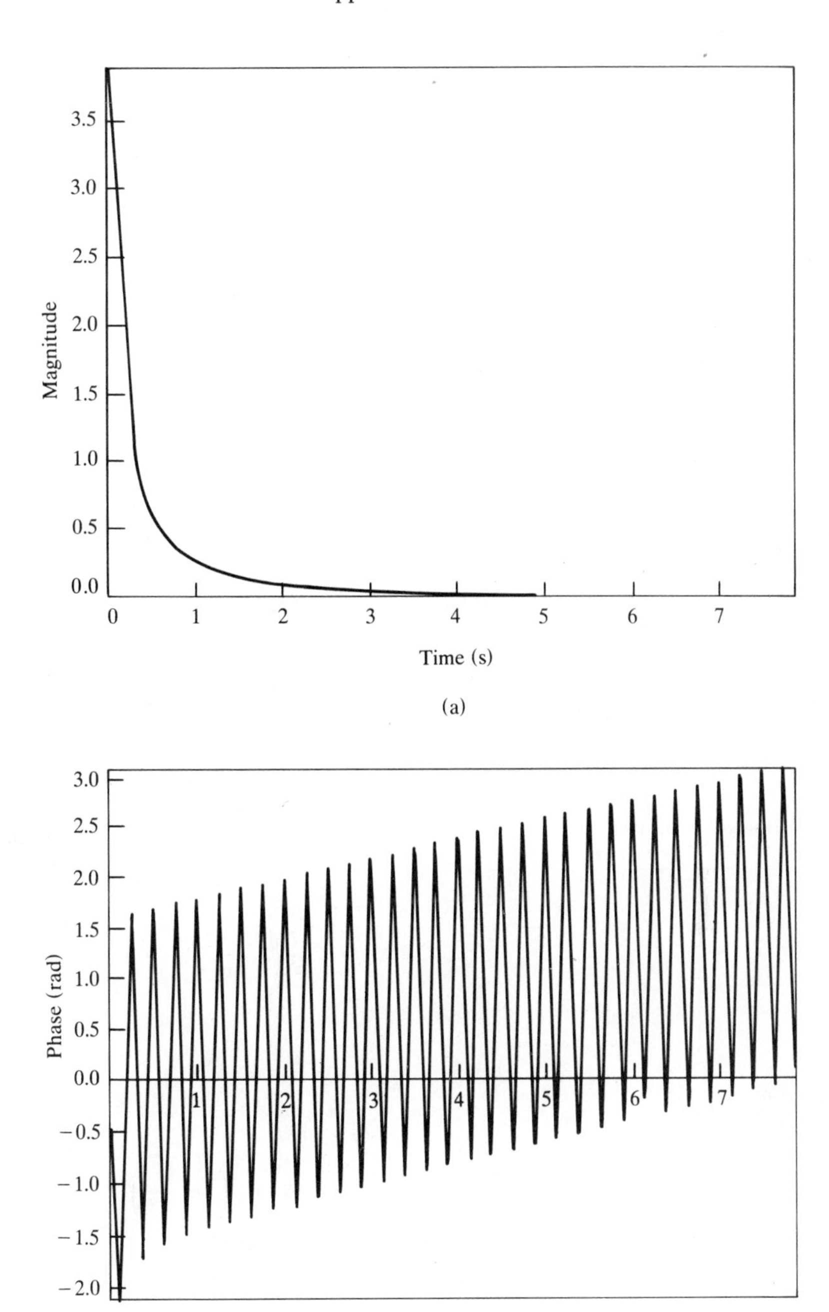

Figure A3.4 The (a) magnitude and (b) phase spectra for the sawtooth waveform in Figure A3.3.

Thus,

$$E_{\text{tot}} = \int_{-\infty}^{\infty} X(f).X(f)\,df = \int_{-\infty}^{\infty} X(f)X^*(f)\,df = \int_{-\infty}^{\infty} |X(f)|^2\,df. \tag{A3.22}$$

In other words, the energy equals the area under the squared magnitude spectrum. The function $|X(f)|^2$ is the *energy per unit frequency*, or *the energy density spectrum*.

Appendix 3.5 Data and Spectral Windows

Data and Spectral Windows — Characteristics and Performance

Data Window*	Spectral Window	Main-Lobe Width	Side-Lobe Level
Rectangular			
$d_R(n) = 1, 0 \le n \le N-1$	$D_R(f) = T\dfrac{\sin(\pi fNT)}{\sin(\pi fT)} \times e^{-j\pi fT(N-1)}$	$\dfrac{2}{NT}$	22.4%
Triangular–Bartlett			
$d_B(n) = \begin{cases} \dfrac{2n}{N-1}, & 0 \le n \le \dfrac{N-1}{2} \\ 2 - \dfrac{2n}{N-1}, & \dfrac{N-1}{2} \le n \le N-1 \end{cases}$	$D_B(f) = \dfrac{2T}{N}\left(\dfrac{\sin(\pi fNT/2)}{\sin(\pi fT)}\right)^2 \times e^{-j\pi fT(N-1)}$	$\dfrac{4}{NT}$	4.5%
Hanning–Tukey			
$d_T(n) = \dfrac{1}{2}\left(1 - \cos\left(\dfrac{2\pi n}{N-1}\right)\right)$	$D_T(f) = 0.5D_R(f) + 0.25D_R\left(f + \dfrac{1}{NT}\right) + 0.25D_R\left(f - \dfrac{1}{NT}\right)$	$\dfrac{4}{NT}$	2.8%
Hamming			
$d_H(n) = 0.54 - 0.46\cos\left(\dfrac{2\pi n}{N-1}\right)$	$D_H(f) = 0.54D_R(f) + 0.23D_R\left(f + \dfrac{1}{NT}\right) + 0.23D_R\left(f - \dfrac{1}{NT}\right)$	$\dfrac{4}{NT}$	0.9%

Data and Spectral Windows — Characteristics and Performance

Data Window*	Spectral Window	Main-Lobe Width	Side-Lobe Level
Blackman			
$d_{BL}(n) = 0.42 - 0.5\cos\left(\frac{2\pi n}{N-1}\right) + 0.08\cos\left(\frac{4\pi n}{N-1}\right)$	$D_{BL}(f) = 0.42D_R(f) + 0.25D_R\left(f + \frac{1}{NT}\right) + 0.25D_R\left(f - \frac{1}{NT}\right) + 0.04D_R\left(f + \frac{2}{NT}\right) + 0.04D_R\left(f - \frac{2}{NT}\right)$	$\frac{6}{NT}$	0.1%
Parzen			
$d_P(n) = \begin{cases} 1 - 6\left(1 - 2\frac{n}{N}\right)^2 + 6\left(\left\lvert 1 - 2\frac{n}{N}\right\rvert\right)^3 & \frac{N}{4} \le n \le 3\frac{N}{4} \\ 2\left(1 - \left\lvert 1 - 2\frac{n}{N}\right\rvert\right)^3 & 0 \le n < \frac{N}{4}, 3\frac{N}{4} < n \le N \end{cases}$	$D_P(f) = \frac{64T}{N^3}\left(\frac{3}{2}\frac{\sin^4(\pi fNT/4)}{\sin^4\pi fT} - \frac{\sin^4(\pi fNT/4)}{\sin^2(\pi fT)}\right) \times e^{-j\pi fT(N-1)}$	$\frac{6}{NT}$	0.22%

*All data windows have $d(n) = 0$ for $n < 0$ and $n \ge N$.

Chapter 4 Concepts of Probability and Estimation

4.1 Introduction

Many situations occur that involve nondeterministic or random phenomena. Some common ones are the effects of wind gusts on the position of a television antenna, air turbulence on the bending on an airplane's wing, and rough terrain on the motion of a wheeled vehicle. Similarly, there are many single and time-series measurements with random characteristics. Their behavior is not predictable with certainty for various reasons; it is too complex to model, knowledge is insufficient for modeling, or, as with some noise processes, it is essentially indeterminate. In order to analyze and understand these phenomena, a probabilistic approach must be used. The concepts and theory of probability and estimation provide a fundamental mathematical framework for the techniques of analyzing random signals. It is assumed that the reader has had an introduction to probability and statistics. Hence, this chapter will provide a brief summary of the concepts of probability and random variables before introducing some concepts of estimation that are essential for signal analysis. If one desires a comprehensive treatment of probability and random variables from an engineering and scientific viewpoint, the books by Papoulis (1984) and Davenport (1970) are excellent; less comprehensive but also a good source is the book by Stark and Woods (1986). If one desires a review, any introductory textbook is suitable. A few are listed in the reference section [Maisle (1971), O'Flynn (1982), and Peebles (1987)].

4.2 Introduction to Random Variables

Signals and data that possess random characteristics arise in a broad array of technical, scientific, and economic fields. Consider the record of voltage deviations in a turboalternator, shown in Figure 4.1. The average value over time seems constant and there are many undulations. Notice that these undulations do not have a consistent period. The frequency analysis methodologies treated in Chapter 3 would not be applicable for analyzing the frequency content of this signal. Noise in electrical circuits and measurements is another common random signal. In the field of reliability and quality control, one is concerned with the characteristics of the output of a production line or a device. Figure 4.2 presents an example showing the size of paint droplets produced by a new design of a spray-painting nozzle. In this case, the concern is the distribution of droplet sizes, and a histogram description is appropriate. These two examples concern continuous random variables but there are many applications that concern discrete random variables. These are mostly counting processes. For instance, Figure 4.3 shows the daily census over time of the number of patients in a hospital. Other counting processes include numbers of parti-

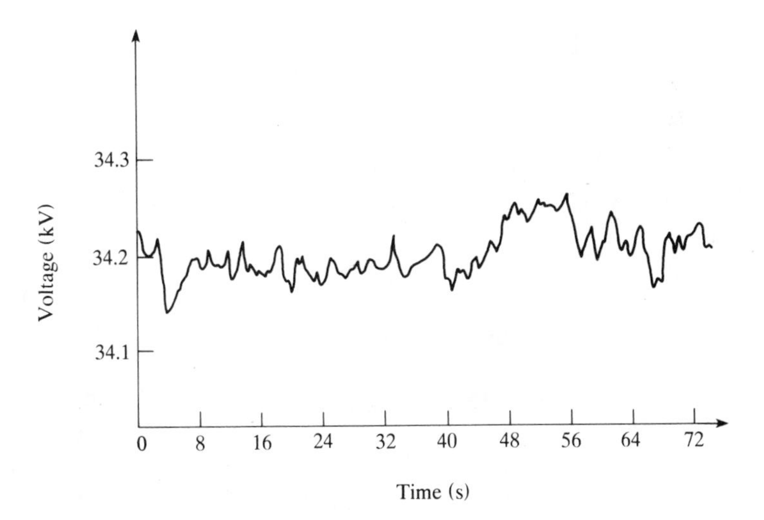

Figure 4.1 Voltage deviations from the stator terminals of a 50-megawatt turboalternator. [Adapted from Jenkins and Watts (1968), Figure 1.1, with permission.]

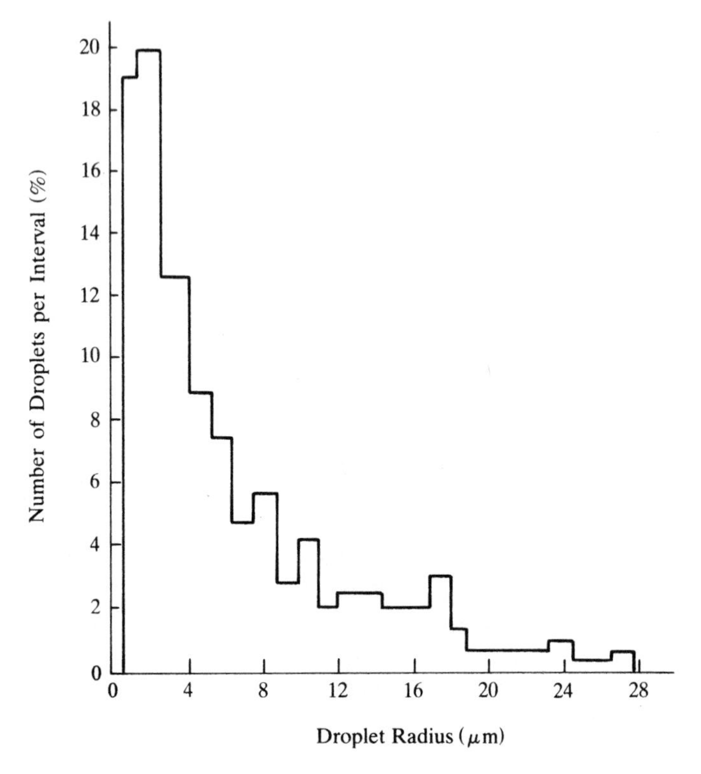

Figure 4.2 Paint droplet produced by a spray gun: percentage of droplets with different radii (in microns).

cles produced during radioactive decay, numbers of calls in a telephone switching and routing unit, and the frequency of heart-beats.

4.2.1 Probability Descriptors

4.2.1.1 Sample Space and Axioms of Probability

In the theory of probability, one is concerned with the *sample space of events*, that is, the set of all possible outcomes of experiments. These events are mapped numerically to values on the real line; these values are the *random variable*. In engineering, almost all measurements or procedures produce some numerical

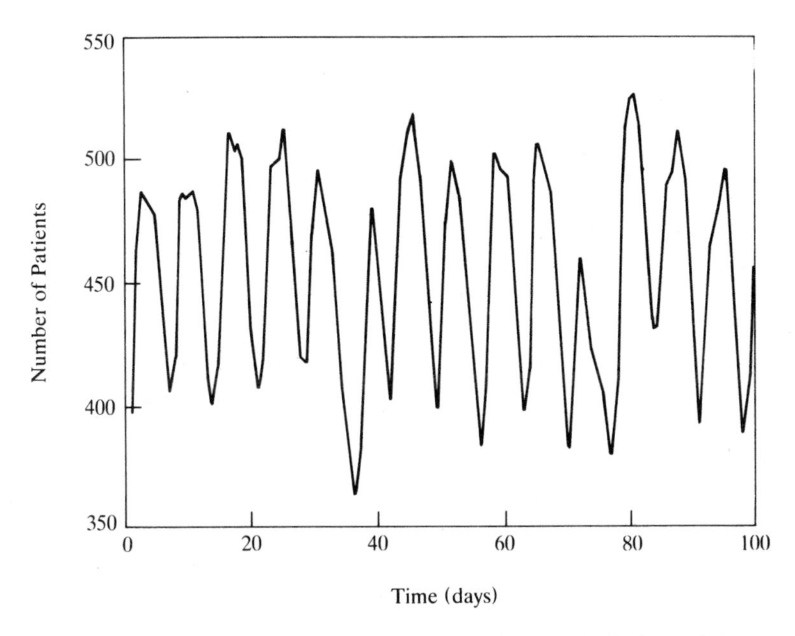

Figure 4.3 Daily inpatient census in a hospital. [Adapted from Pandit and Wu (1983), Figure 9.12, with permission.]

result such that a numerical value or sequence is inherently assigned to the outcome. For example, the production of a paint droplet is an outcome of an experiment and its size a random variable. Some outcomes are *nominal*, like the colors produced by a flame or gray levels in an image from an x-ray. These nominal variables can be mapped to a numeric field so that, in general, the values of random variables will be defined over some real number field. Continuous and discrete random variables obey the same general laws and have the same general descriptors. The range of all possible values of a random variable comprises its *sample space* S. For the paint droplets of Figure 4.2, this is the range of lengths continuous from 0 to 28 μm (microns), and for a die, it is the integers from 1 to 6 inclusive. A collection of values is called a set and is usually labeled with a capital letter. All the values in a sample space are associated with probabilities of occurrence and obey the *axioms of probability*. Each value in a discrete sample space and any set A of values in a continuous or discrete sample space is associated with a probability $P[A]$. The axioms of probability are as follows:

1. $0 \leq P[A] \leq 1$.
2. $P[S] = 1$.
3. If sets A and B are mutually exclusive, that is, they have no common values,

$$P[A \cup B] = P[A] + P[B].$$

An empty set contains none of the values in the sample space, is called a null set, and has a probability of zero. The easiest interpretation of probability is the concept of *relative frequency* that is developed from the *law of large numbers* [Papoulis (1984)]. Consider the situation of removing a black marble B from a box of marbles with an assortment of colors. If there are N marbles and N_B black marbles, the probability $P[B]$ of choosing a black marble is the relative frequency of black marbles, N_B/N.

4.2.1.2 Probability Distribution and Density Functions

The frequency of occurrence of values in a sample space is described by a pair of complementary functions called probability distribution and density functions. Their basic properties satisfy the axioms of probability. The *probability distribution function* (PDF) is defined as follows: For either a continuous or a discrete random variable x, the probability that it has a magnitude equal to or less than a specific value α is

$$P[x \leq \alpha] = F_x(\alpha). \tag{4.1}$$

The properties of a PDF $F_x(\alpha)$ are

$$0 \leq F_x(\alpha) \leq 1, \qquad -\infty \leq \alpha \leq -\infty, \tag{4.2a}$$

$$F_x(-\infty) = 0, \qquad F_x(\infty) = 1, \tag{4.2b}$$

$$F_x(\alpha) \text{ is nondecreasing with } \alpha, \tag{4.2c}$$

$$P[\alpha_1 \leq x \leq \alpha_2] = F_x(\alpha_2) - F_x(\alpha_1). \tag{4.2d}$$

For a continuous random variable, the *probability density function* (pdf), $f_x(\alpha)$, is essentially a derivative of $F_x(\alpha)$. Its properties are

$$f_x(\alpha) \geq 0, \qquad -\infty \leq \alpha \leq -\infty, \tag{4.3a}$$

$$\int_{-\infty}^{\infty} f_x(u)\, du = 1, \tag{4.3b}$$

$$F_x(\alpha) = \int_{-\infty}^{\alpha} f_x(u)\, du, \tag{4.3c}$$

$$P[\alpha_1 \leq x \leq \alpha_2] = \int_{\alpha_1}^{\alpha_2} f_x(u)\, du, \tag{4.3d}$$

For discrete random variables, the PDF is discontinuous and $f_x(\alpha) = P[x = \alpha]$. Equations (4.3a)–(4.3d) are also correct if the delta function is utilized in the definition of the probability density function. However, discrete random variables are not a concern; for more information refer to O'Flynn (1982) or Peebles (1987). Probabilities are most often calculated using equation (4.3d). Probability density functions often can be represented with formulas as well as graphs. The *uniform pdf* describes values that are equally likely to occur over a finite range,

$$f_x(\alpha) = \begin{cases} \dfrac{1}{b-a}, & a \leq \alpha \leq b, \\ 0, & \text{for } x \text{ elsewhere.} \end{cases} \tag{4.4}$$

The *exponential random variable* has a semiinfinite range with the pdf

$$f_x(\alpha) = \begin{cases} \dfrac{1}{b} e^{-(\alpha-a)/b}, & a \leq \alpha \leq \infty, \\ 0, & \alpha < a. \end{cases} \tag{4.5}$$

This is commonly used to describe failure times in equipment and times between calls entering a telephone exchange.

At this point it is important to state that there is a diversity in notation for PDFs and pdfs. When there is no confusion concerning the random variable and its specific values, a simpler notation is often utilized [Papoulis (1984)]:

$$f(x) = f_x(\alpha) \quad \text{and} \quad F(x) = F_x(\alpha). \tag{4.6}$$

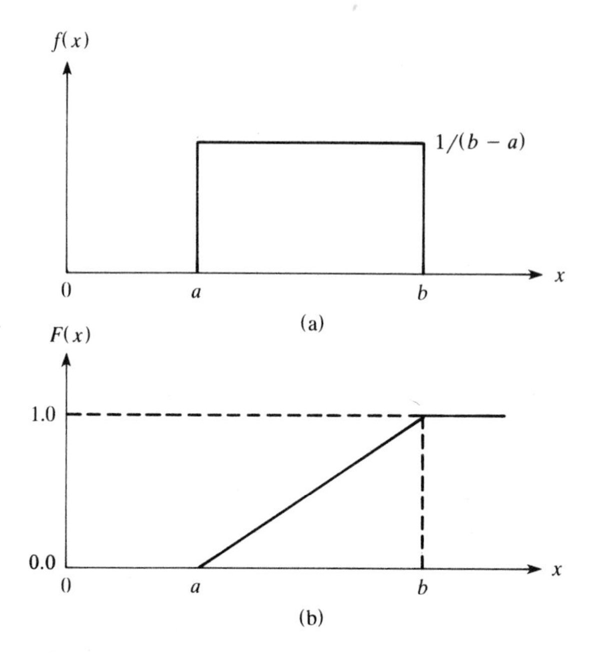

Figure 4.4 The (a) probability density and (b) probability distribution functions for the uniform random variable. [Adapted from Peebles (1987), Figure 2.5-2, with permission.]

This simpler notation will be utilized when the context of the discussion is clear.

Another pdf is the *Gaussian, or normal*, pdf. It has the formula

$$f(x) = \frac{1}{\sqrt{2\pi}\sigma} \exp\left(-\frac{(x-m)^2}{2\sigma^2} \right), \qquad -\infty \leq x \leq -\infty \quad (4.7)$$

(the parameters m and σ have direct meanings that are explained in the next section). It is useful for describing many phenomena, such as random noise and biological variations.

The probability distribution and density functions for uniform, exponential, and normal random variables are plotted in Figures 4.4–4.6, respectively. There are many pdfs, and several others are presented in Appendix 4.1. A comprehensive treatment of the many types of probability function can be found in Larson and Shubert (1979).

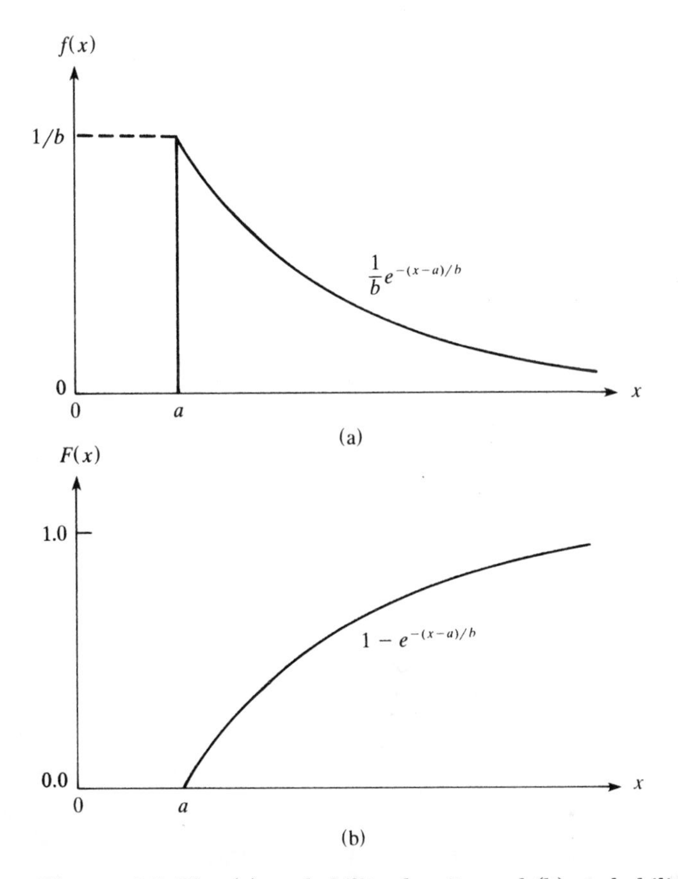

Figure 4.5 The (a) probability density and (b) probability distribution functions for the exponential random variable. [Adapted from Peebles (1987), Figure 2.5-3, with permission.]

It is often desired to calculate probabilities from distribution functions whose formulae have been encoded within computer programs. The derivation of the PDF for some models is easy and the PDFs have closed-form solutions. The software to produce some of these PDFs is found in the Scientific Subroutine Library by Peerless Engineering Service (1984). The PDFs without closed solutions can be calculated with infinite or finite asymptotic series. These are complex, and numerical precision is important. The handbook by Abramowitz and Stegun (1965) presents several solutions for important distribution functions.

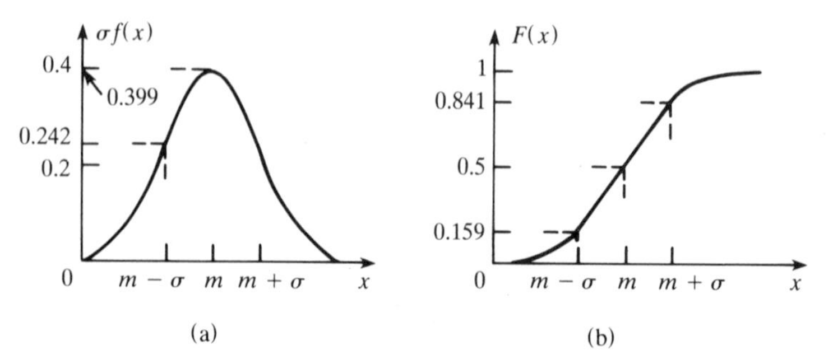

Figure 4.6 The (a) probability density and (b) probability distribution functions for the normal random variable. [Adapted from Papoulis (1984), Figure 4.7, with permission.]

Example 4.1

The failure time for a certain kind of light bulb has an exponential pdf with $b = 3$ and $a = 0$ months. What is the probability that one will fail between 2 and 5 months?

$$P[2 \le x \le 5] = \int_2^5 \tfrac{1}{3} e^{-x/3}\, dx = -(e^{-5/3} - e^{-2/3}) = 0.324.$$

4.2.2 Moments of Random Variables

Not only are the forms of the pdfs very important, but also some of their *moments*. The moments not only quantify useful properties but in many situations only the moments are available. The general moment of a function $g(x)$ of the random variable x is symbolized by the *expectation operator*,

$$E[g(x)] = \int_{-\infty}^{\infty} g(x)\, f(x)\, dx. \tag{4.8}$$

The function $g(x)$ can take any form but it is usually a polynomial, most often of the first or second order. In addition to using the expectation operator, some moments are given special symbols. The *mean*, or average value, is

$$E[x] = m_1 = m = \int_{-\infty}^{\infty} x f(x)\, dx. \tag{4.9}$$

The mean squared value is similarly defined as

$$E[x^2] = m_2 = \int_{-\infty}^{\infty} x^2 f(x)\, dx. \tag{4.10}$$

The higher-order moments of a random variable are used in advanced statistics. Similarly, *central moments* (moments about the mean) are defined for $g(x) = (x - m)^k$ and symbolized with the notation μ_k. The most-used central moments is for $k = 2$ and is the *variance*, with the symbol σ^2 and the definition

$$E\left[(x - m)^2\right] = \sigma^2 = \int_{-\infty}^{\infty} (x - m)^2 f(x)\, dx. \tag{4.11}$$

The mean, the mean square, and the variance are interrelated by the equation

$$\sigma^2 = m_2 - m^2. \tag{4.12}$$

The variance is an important parameter because it indicates the spread of magnitudes of a random variable. Its square root, σ—the *standard deviation*, is used synonymously. This indication can be easily seen in Figure 4.7, which shows two uniform density func-

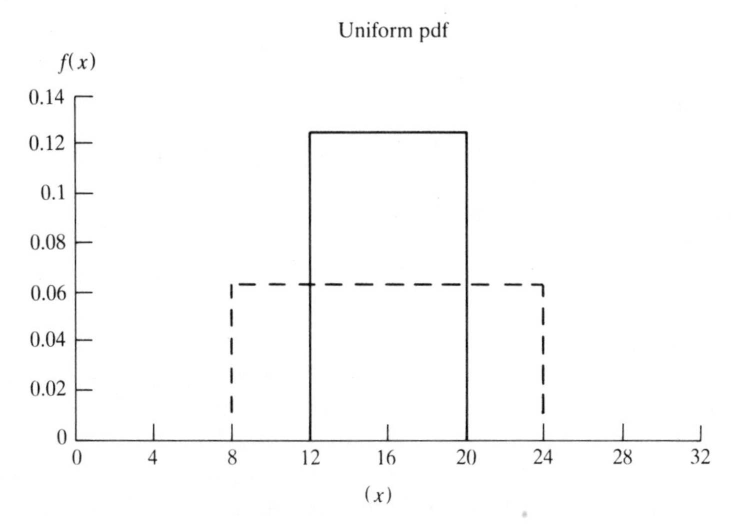

Figure 4.7 The plots of two uniform density functions: (—) $m = 16$ and $\sigma^2 = 5.33$; (--) $m = 16$ and $\sigma^2 = 21.33$.

tions with the same mean and different ranges of values. The variance for the density function with the broader spread is larger.

Example 4.2

Find the mean and variance of the uniform density function in Figure 4.4 with $a = 2$ and $b = 7$.

$$m = \int_{-\infty}^{\infty} xf(x)\,dx = \int_{2}^{7} x \cdot 0.2\,dx = 0.1\,(49 - 4) = 4.5,$$

$$\sigma^2 = \int_{-\infty}^{\infty} (x - m)^2 f(x)\,dx = \int_{2}^{7} (x - 4.5)^2 0.2\,dx = 2.0833.$$

Example 4.3

Find the mean of the Rayleigh density function shown in Appendix 4.1.

$$m = \int_{-\infty}^{\infty} xf(x)\,dx = \int_{0}^{\infty} x\frac{x}{\alpha^2}\exp\left(\frac{-x^2}{2\alpha^2}\right)dx = \frac{\alpha}{\sqrt{2}}\Gamma(0.5) = \frac{\alpha\sqrt{\pi}}{\sqrt{2}},$$

where $\Gamma(y)$ is the gamma function.

4.2.3 Gaussian Random Variable

The Gaussian, or normal, density function is extremely important because of its many applications and its tractability. The *central limit theorem* makes it important because under certain conditions many processes formed by summing together several random variables with finite variances can be described in the limit by a normal pdf. Thus, it is a rational representation for many noise and biological processes that naturally consist of a summation of many random events. An efficient aspect of the normal pdf is that its mean m and variance σ^2 are written directly as parameters of the function. Reexamine equation (4.7). It is often abbreviated as $N(m, \sigma^2)$. Derivations of these moments can be found in Fante (1988). Calculating probabilities for this pdf is not simple because there is no closed-form solution for integrating over a finite range. To assist in this problem, a variable transformation called *standardization* is performed. The transformation is linear and is

$$y = \frac{x - m}{\sigma}. \tag{4.13}$$

Performing this transformation on equation (4.7) produces

$$f(y) = \frac{1}{\sqrt{2\pi}} \exp\left(-\frac{y^2}{2}\right), \qquad -\infty \leq y \leq -\infty. \tag{4.14}$$

This is a Gaussian pdf with a mean of 0 and a variance of 1 and is called the *unit normal pdf*, $N(0, 1)$. Probabilities over a finite range of values of y are defined mathematically as

$$P[0 \leq y \leq Y] = \operatorname{erf}(Y) = \int_0^Y f(y)\, dy. \tag{4.15}$$

The solution of this integral is an infinite series and the function is called the *error function* (erf). The tabulation of $\operatorname{erf}(Y)$ versus Y can be found in most basic statistics textbooks and in books of tables. Rational polynomial approximations for calculating values of $N(0, 1)$ and the error function can be found in Abramowitz and Stegun (1964). Because $f(y)$ is an even function, for a positive number Y

$$P[-Y \leq y \leq 0] = P[0 \leq y \leq Y] = \operatorname{erf}(Y). \tag{4.16}$$

The error function is plotted in Figure 4.8 and is tabulated in Appendix 4.2.

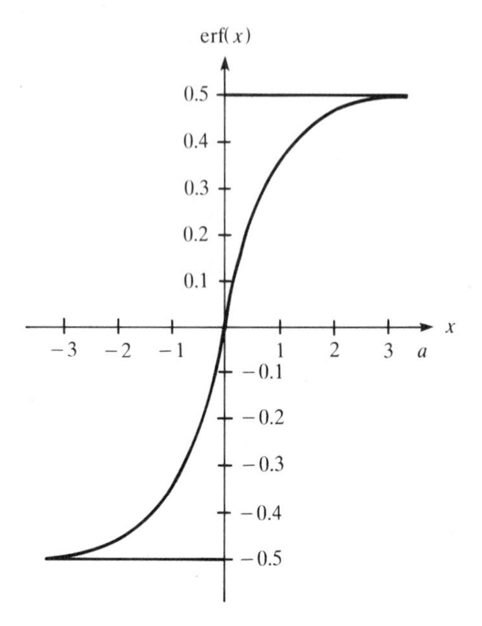

Figure 4.8 The error function of x is plotted. [Adapted from Stark and Woods (1986), Figure 2.4-2, with permission.]

Example 4.4

For $f(x) = N(3, 4)$ find $P[x \leq 5.5]$. In detail this is

$$P[x \leq 5.5] = \int_{-\infty}^{5.5} \frac{1}{\sqrt{2\pi}\,2} \exp\left(-\frac{(x-3)^2}{2 \cdot 4}\right).$$

Standardizing using equation (4.13), the transformation is $y = (x - 3)/2$ and the probabilities become

$$\begin{aligned} P[x \leq 5.5] &= P[y \leq 1.25] \\ &= \int_{-\infty}^{1.25} N(0,1)\,dy = \int_{-\infty}^{0} N(0,1)\,dy + \int_{0}^{1.25} N(0,1)\,dy \\ &= 0.5 + \mathrm{erf}(1.25) = 0.8944. \end{aligned}$$

Example 4.5

A manipulation more commonly encountered for finding probabilities occurs when both bounds are finite. For $f(x) = N(2,4)$ find $P[1 \le x \le 4]$. The standardization is $y = (x - 2)/2$. Therefore,

$$P[1 \le x \le 4] \doteq P[-0.5 \le y \le 1]$$

$$= \int_{-0.5}^{0} N(0,1)\, dy + \int_{0}^{1} N(0,1)\, dy$$

$$= \text{erf}(0.5) + \text{erf}(1) = 0.53$$

4.3 Joint Probability

4.3.1 Bivariate Distributions

The concept of *joint probability* is a very important one in signal analysis. Very often the values of two variables from the same or different sets of measurements are being compared or studied and a *two-dimensional sample space* exists. The joint, or bivariate, probability density function and its moments are the basis for describing any interrelationships or dependencies between the two variables. A simple example is the selection of a resistor from a box of resistors. The random variables are the resistance r and wattage w. If a resistor is selected, it is desired to know the probabilities associated with ranges of resistance and wattage values, that is,

$$P[(r \le R) \text{ and } (w \le W)] = P[r \le R, w \le W], \qquad (4.17)$$

where R and W are particular values of resistance and wattage, respectively. For signals this concept is extended to describe the relationship between values of a process $x(t)$ at two different times, t_1 and t_2, and between values of two continuous processes, $x(t)$ and

$y(t)$, at different times. These probabilities are written

$$P[x(t_1) \leq \alpha_1, x(t_2) \leq \alpha_2] \quad \text{and} \quad P[x(t_1) \leq \alpha_1, y(t_2) \leq \beta_2], \tag{4.18}$$

respectively. The joint probabilities are functionally described by *bivariate probability distribution and density functions*. The bivariate PDF is

$$F_{xy}(\alpha, \beta) = P[x \leq \alpha, y \leq \beta] \tag{4.19}$$

and is related to the bivariate pdf through the double integration

$$F_{xy}(\alpha, \beta) = \int_{-\infty}^{\beta} \int_{-\infty}^{\alpha} f_{xy}(u, v)\, du\, dv. \tag{4.20}$$

These functions have some important properties:

$$F_{xy}(\alpha, \infty) = F_x(\alpha), \quad \text{the marginal PDF for variable } x; \tag{4.21a}$$

$$F_{xy}(\infty, \infty) = 1; \tag{4.21b}$$

$$f_{xy}(\alpha, \beta) \geq 0; \tag{4.21c}$$

$$f_x(\alpha) = \int_{-\infty}^{\infty} f_{xy}(\alpha, v)\, dv, \quad \text{the marginal pdf for variable } x. \tag{4.21d}$$

Again, when there is no confusion concerning the random variable and its specific values, a simpler notation is often utilized:

$$f(x, y) = f_{xy}(\alpha, \beta) \quad \text{and} \quad F(x, y) = F_{xy}(\alpha, \beta). \tag{4.22}$$

Related to this is the notion of *conditional probability*; that is, given the knowledge of one variable, what are the probability characteristics of the other variable? A conditional pdf for resistance, knowing that a resistor with a particular value of wattage has been selected, is written $f(r|w)$. The conditional, *marginal*, and joint pdfs are related through *Bayes' rule* by

$$f(r|w) = \frac{f(w|r)\,f(r)}{f(w)} = \frac{f(r,w)}{f(w)}. \qquad (4.23)$$

Conditional density functions have the same properties as marginal density functions. Bayes' rule and conditional probability provide a foundation for the concept of *independence* of random variables. If knowledge of wattage does not contain any information about resistance, then

$$f(r|w) = f(r), \qquad (4.24)$$

or the conditional density function equals the marginal density function. Using equation (4.24) with equation (4.23) gives

$$f(r,w) = f(r)\,f(w), \qquad (4.25)$$

or the bivariate density function is the product of the marginal density functions. Two random variables r and w are independent if either equation (4.24) or equation (4.25) is true. Comprehensive treatments of joint probability relationships can be found in books such as Davenport (1970) and O'Flynn (1982).

4.3.2 Moments of Bivariate Distributions

Moments of bivariate distributions are defined with a function $g(x,y)$. The expectation is

$$E[g(x,y)] = \int_{-\infty}^{\infty}\int_{-\infty}^{\infty} g(x,y)\,f(x,y)\,dx\,dy. \qquad (4.26)$$

Two moments are extremely important: the *mean product* $E[xy]$, with $g(x,y) = xy$, and the *first-order central moment* σ_{xy}, with $g(x,y) = (x - m_x)(y - m_y)$. The moment σ_{xy} is also called the *covariance*. They are all related by

$$E[xy] = \sigma_{xy} + m_x m_y. \qquad (4.27)$$

The covariance is an indicator of the strength of linear relationship between two variables and defines the state of *correlation*. If

$\sigma_{xy} = 0$, then x and y are *uncorrelated* and $E[xy] = m_x m_y$. If variables x and y are independent, then also $E[xy] = m_x m_y$, and the covariance is zero. However, the converse is not true. In other words, independent random variables are uncorrelated, but uncorrelated random variables can still be dependent. This linear relationship will be explained in the section on estimation.

When $\sigma_{xy} \neq 0$, the random variables are correlated. However, the strength of the relationship is not quantified because the magnitude of the covariance depends upon the units of the variables. For instance, the use of watts or milliwatts as units for w will make a factor of 10^3 difference in the magnitude of the covariance σ_{rw}. This situation is solved by using a unitless measure of *linear dependence* called the *correlation coefficient* ρ, where

$$\rho = \frac{\sigma_{rw}}{\sigma_r \sigma_w} \tag{4.28}$$

and

$$-1 \leq \rho \leq 1. \tag{4.29}$$

The measure is linear because if $r = k_1 w + k_2$, where k_1 and k_2 are constants, then $\rho = \pm 1$. For uncorrelated random variables $\rho = 0$.

Example 4.6

An example of a two-dimensional pdf is

$$f(x,y) = \begin{cases} abe^{-(ax+by)} & \text{for } x \geq 0, y \geq 0, \\ 0 & \text{elsewhere.} \end{cases}$$

It can be seen that the pdf is separable into two functions,

$$f(x,y) = ae^{-(ax)} \cdot be^{-(by)} = f(x) \cdot f(y).$$

Thus the variables x and y are independent and also $\rho = 0$.

Example 4.7

Another bivariate pdf is

$$f(x,y) = \begin{cases} xe^{-x(y+1)} & \text{for } x \geq 0,\ y \geq 0, \\ 0 & \text{elsewhere.} \end{cases}$$

These variables are dependent because the pdf $f(x, y)$ cannot be separated into two marginal pdfs. The marginal density function for x is

$$f(x) = \int_0^\infty f(x,y)\,dy = xe^{-x}\int_0^\infty e^{-xy}\,dy = e^{-x} \quad \text{for } x \geq 0.$$

The conditional density function for y is

$$f(y|x) = \frac{f(x,y)}{f(x)} = \frac{xe^{-x(y+1)}}{e^{-x}} = \begin{cases} xe^{-xy} & \text{for } x \geq 0,\ y \geq 0, \\ 0 & \text{elsewhere.} \end{cases}$$

Example 4.8

An extremely important two-dimensional pdf is the bivariate normal distribution. It has the form

$$f(x,y) = \frac{1}{2\pi\sigma_x\sigma_y\sqrt{1-\rho^2}}e^{-a/2},$$

with

$$a = \frac{1}{1-\rho^2}\left(\frac{(x-m_x)^2}{\sigma_x^2} - 2\rho\frac{(x-m_x)}{\sigma_x}\frac{(y-m_y)}{\sigma_y} + \frac{(y-m_y)^2}{\sigma_y^2}\right).$$

The means and variances for the random variables x and y and their correlation coefficient are explicitly part of the pdf. Sketches of the surface for two different values of ρ are shown in Figure 4.9.

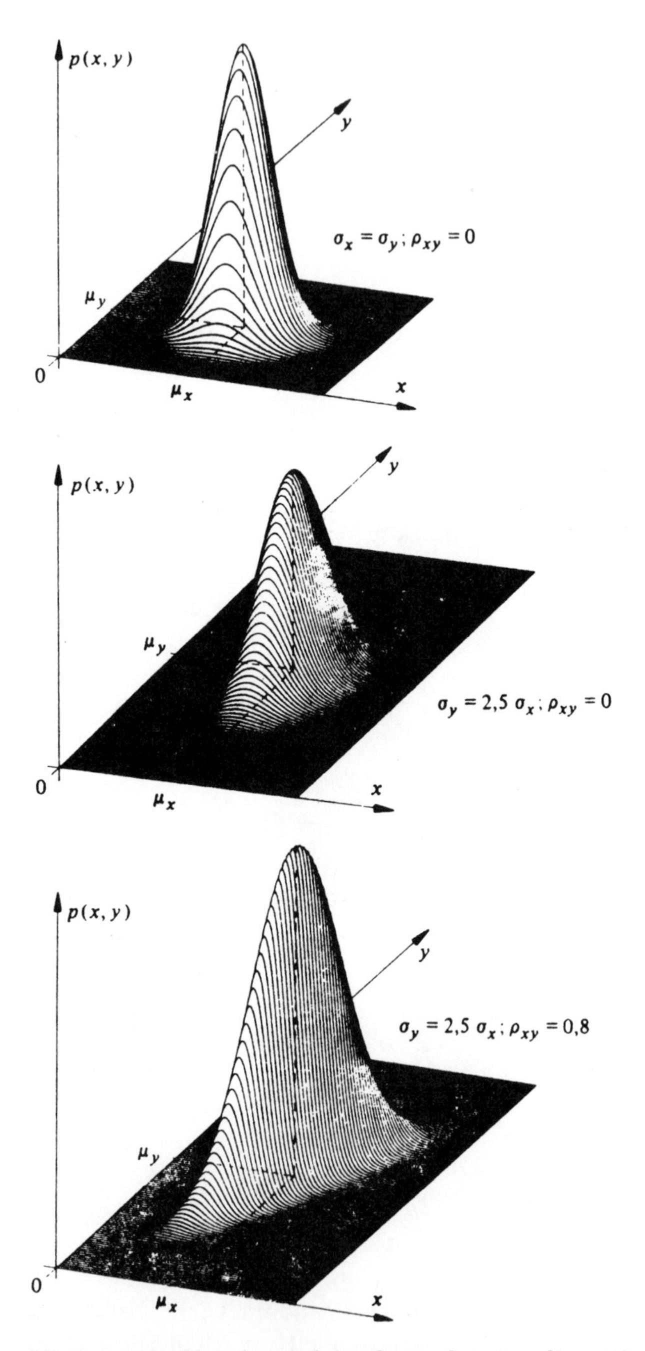

Figure 4.9 Sketches of surfaces for two-dimensional normal probability density functions with two different values of ρ are shown. [From de Coulon (1986), Figure 5.26, with permission.]

If $\rho = 0$, then the middle term in the exponent is also zero and $f(x, y) = f(x)f(y)$. Thus, uncorrelated normal random variables are also independent. This is not true in general for other two-dimensional pdfs.

4.4 Concept of Sampling and Estimation

The study or investigation of a random phenomenon usually requires knowledge of its statistical properties, that is, its probabilistic description. Most often this description is not available and it must be discovered from experimental measurements. The measurements produce a *sample* of N data values $\{x_1, x_2, \ldots, x_N\}$. Mathematical operations are performed on the sample to determine the statistical properties. These operations are called *estimators* and this entire process is called *estimation*. An important aspect of estimation is its accuracy, which can be quite difficult to determine. This section will introduce the notion of *sample moments* and some general characteristics of estimation. All of this will be exemplified by focusing on estimating the mean and variance of a data sample.

4.4.1 Sample Moments

The procedure for estimating many properties is often obtained by directly translating the mathematical definition of its theoretical counterpart. Consider approximating the general expectation operator of equation (4.8) with an infinite sum. Divide the real line into intervals of length Δx with boundary points d_i. Then

$$E[g(x)] = \int_{-\infty}^{\infty} g(x) f(x)\, dx \simeq \sum_{i=-\infty}^{\infty} g(d_i) f(d_i)\, \Delta x.$$

Notice that a very important assumption is made: All of the sample values of x have the same pdf, that is, they are *identically distributed*. Because data points may equal boundary values, the intervals are half-open and

$$f(d_i)\, \Delta x \simeq P[d_i \le x < d_i + \Delta x].$$

Using the interpretation of relative frequency, then

$$E[g(x)] \simeq \sum_{i=-\infty}^{\infty} g(d_i) P[d_i \le x < d_i + \Delta x] = \sum_{i=-\infty}^{\infty} g(d_i) \frac{N_i}{N},$$

where N_i is the number of measurements in the interval $[d_i \le x < d_{i+1}]$. Because $g(d_i)N_i$ approximates the sum of values of $g(x)$ for points within interval i, then

$$E[g(x)] \simeq \frac{1}{N} \sum_{j=1}^{N} g(x_j), \tag{4.30}$$

where x_j is the jth data point. Equation (4.30) is the *estimator* for the average value of $g(x)$. Because no criterion was optimized for the derivation of the estimator, it can be considered an empirical *sample moment*. Notice that this is a commonly used estimator. It is obvious if $g(x) = x$, or

$$E[x] \simeq \hat{m} = \frac{1}{N} \sum_{j=1}^{N} x_j. \tag{4.31}$$

This is the estimator for the mean value. The circumflex is used to denote an estimator of the theoretical function. Similarly, an estimator for the variance is

$$\hat{\sigma}^2 = \frac{1}{N} \sum_{j=1}^{N} (x_j - m)^2.$$

Typically, the mean is not known, and in this case the variance estimator becomes

$$\hat{\sigma}^2 = \frac{1}{N-1} \sum_{j=1}^{N} (x_j - \hat{m})^2. \tag{4.32}$$

Notice that the coefficient in equation (4.32) is $1/(N-1)$ instead of $1/N$. The reason will be given in Section 4.7. The actual values that are calculated using equations (4.31) and (4.32) are *estimates* of the mean and variance, respectively.

There are other methods for developing estimators. For instance in Exercise 2.1, the estimator for the sample mean is derived using a mean-square criterion. The result is the same as equation (4.31). This is not usually the situation with other estimators; different criteria can produce different estimators for the same parameter.

The estimate of a moment of a random variable is also a random variable. Consider estimating the mean value of the daily river flow plotted in Figure 4.10 from the table in Appendix 4.3. Assume that only the first 10 values were available. Then $\hat{m} = 2368$. If only the last 10 values were available, then $\hat{m} = 1791$. Thus, $\hat{m}$ itself is a random variable with a pdf $f(\hat{m})$ and it is necessary to know its relationship to m. In general, it is necessary to know the relationship between a sample estimate of a moment or function and its true value. There is an elegant branch of mathematical statistics, called *estimation and sampling theory*, that has as its definition the development of these relationships [Fisz (1963)]. The premises for most of the developments are that the measure-

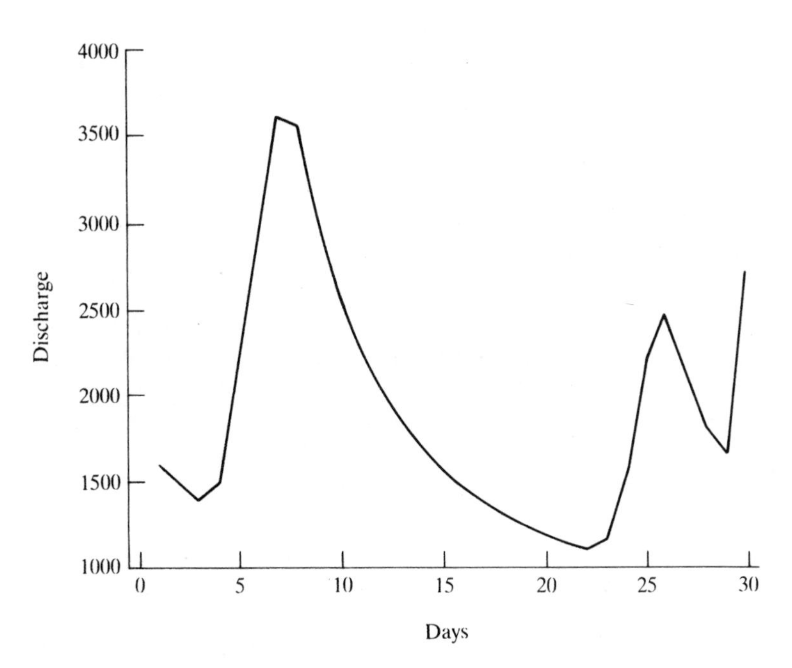

Figure 4.10 Daily river flow during March 1981 of the Duck River.

ments arise from the same underlying distribution and that they are independent. The latter condition can sometimes be relaxed if the number of measurements is large enough to represent the sample space. For many probabilistic parameters the distribution of values produced by an estimator, the *sampling distribution*, has been derived and it relates the estimate to the true value. The use of sampling distributions is essential in signal analysis and this will be illustrated in detail for the sample mean.

4.4.2 Significance of The Estimate

For the sample mean estimator, the *Student's t distribution* relates m and $\hat{m}$ through the t variable if either the random variable x has a normal distribution or N is large. The t variable is a standardized error and is defined as

$$t = \frac{m - \hat{m}}{\hat{\sigma}/\sqrt{N}}, \tag{4.33}$$

where $\hat{\sigma}$ is the *sample standard deviation* $\sqrt{\hat{\sigma}^2}$. The pdf of the Student's t variable has a complicated form, which is

$$f(t) = \frac{\Gamma(\frac{1}{2})}{\sqrt{N-1}\,\Gamma(\frac{1}{2})\Gamma(\frac{1}{2}(N-1))} \frac{1}{\left[1 + t^2/(N-1)\right]^{N/2}}, \tag{4.34}$$

where $\Gamma(u)$ is the gamma function; $f(t)$ is plotted in Figure 4.11. Notice that $f(t)$ asymptotically approaches zero as the absolute value of t approaches infinity. This means that the error in estimating the mean can be infinitely large; however, judging from the magnitude of $f(t)$, the likelihood is very small. This is formalized in the following manner. A bound t_c is set for t such that the probability that the error is larger than this bound is very small, that is,

$$P\left[|t| \geq t_c\right] = 0.05 = \alpha. \tag{4.35}$$

The range of values of t satisfying equation (4.35) is called the *critical region* or *significance region*. The remaining values of t comprise the *confidence region*. These regions are illustrated also in Figure 4.11. The probability α is the *critical level* and t_c is the

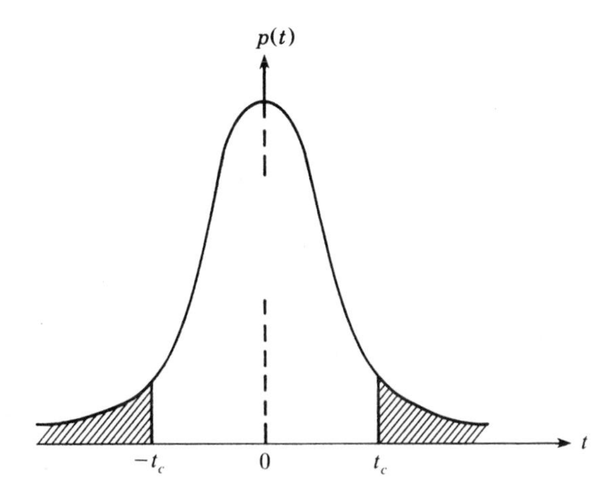

Figure 4.11 Student's t pdf with critical regions shaded.

critical value. (The words critical and significance are used interchangeably.) The utility of this construction is that the critical value can be easily translated into useful bounds for the real value of the mean via equation (4.33). Rewriting this equation for relating m to its estimate $\hat{m}$ gives

$$m = \hat{m} - t\frac{\hat{\sigma}}{\sqrt{N}}. \tag{4.36}$$

This is effectively a variable transformation from t space to m space. Inserting the critical values of t ($\pm t_c$) into this equation produces the α-level *lower* (m_l) and *upper* (m_u) *bounds* for the mean, respectively, as

$$m_l = \hat{m} - t_c\frac{\hat{\sigma}}{\sqrt{N}}, \tag{4.37}$$

$$m_u = \hat{m} + t_c\frac{\hat{\sigma}}{\sqrt{N}}. \tag{4.38}$$

Thus, given the sample mean the confidence interval at level α for the true mean is

$$P[m_l \leq m \leq m_u] = 1 - \alpha = 0.95. \tag{4.39}$$

The equation for $f(t)$ is complicated to evaluate, so the critical values for various significance levels are tabulated in Appendix 4.4. There is one more aspect to the Student's t distribution; notice in equation (4.34) that $f(t)$ depends upon the factor $(N - 1)$, this is called the *degrees of freedom* ν. One simply designates a significance level, α, and finds the value of t_c for the needed degrees of freedom in the table. The notion of degrees of freedom relates to the number of independent variables that are being summed. The estimate $\hat{m}$ is a sum of N independent measurements. However, the t variable is not; it is a sum of terms $\{x_i/\hat{\sigma}; 1 \leq i \leq N\}$, where $\hat{\sigma}$ is also a function of x_i. A sum of any $N - 1$ terms $x_i/\hat{\sigma}$ is independent. The Nth term is constrained in value because $\hat{\sigma}$ is calculated. Hence, the t variable has only $N - 1$ independent terms, that is, only $N - 1$ degrees of freedom. This concept of the number of independent variables in a summation is used with all estimators.

Example 4.9

For the river flow data plotted in Figure 4.10, estimate the mean monthly river flow. The data are tabulated in Appendix 4.3 for the month of March. Use a critical level of 0.05. The degrees of freedom is $\nu = N - 1 = 29$. The critical value is $t_c = 2.045$. Next, the sample mean and variance must be calculated.

$$\hat{m} = \tfrac{1}{30} \sum_{i=1}^{30} x_i = 1913,$$

$$\hat{\sigma} = \sqrt{\tfrac{1}{29} \sum_{i=1}^{30} (x_i - \hat{m})^2} = 700.8.$$

The bounds for the mean are

$$m_l = \hat{m} - t_c \frac{\hat{\sigma}}{\sqrt{N}} = 1913 - 2.045 \frac{700.8}{\sqrt{30}} = 1651.3,$$

$$m_u = \hat{m} + t_c \frac{\hat{\sigma}}{\sqrt{N}} = 1913 + 2.045 \frac{700.8}{\sqrt{30}} = 2174.7.$$

The confidence interval for the true mean is

$$P[1651.3 \le m \le 2174.7] = 0.95.$$

Thus, it can be said that, based on the 30 observations, the mean river flow is within the range stated with a confidence of 0.95.

An alternative aspect of the procedures used for estimation is called *hypothesis testing*. In this situation a statement is made about the value of some parameter or moment of a random variable. This is the *null hypothesis*. The null hypothesis is tested for feasibility using the *sampling distribution* of the parameter or moment. The sampling distribution relates the true value of a parameter or moment to a sampled value. The general procedure is similar to the procedure just explained in the previous example. For instance, a hypothesis is made about the true mean value of the distribution from which a set of data has been measured; that is, the true mean has a value m_0. This is symbolized by H_0. The alternative to H_0 is: Given the estimate of the mean, the true mean value is other than m_0. This is given the symbol H_1. Mathematically, this is concisely stated as

$$\begin{aligned} H_0&: m = m_0, \\ H_1&: m \neq m_0. \end{aligned} \tag{4.40}$$

In the previous example the sampling distribution is the Student's t and bounds for a confidence interval are defined in equation (4.39) given $\hat{m}$, $\hat{\sigma}$, t_c, and α. If m_0 lies within the confidence region, then H_0 is accepted as true, otherwise it is rejected and H_1 is accepted as true. Hypothesis testing is covered in detail in textbooks on introductory statistics.

4.5 Estimate of Correlation

It is common to need to assess the dependence between two variables or the strength of some cause and effect relationship using the correlation coefficient. The correlation measure is used to compare the rainfall patterns in cities, the similarity between electrocardiographic waveforms, the incidence of diseases with pollution, and so forth. The estimator for the correlation coefficient ρ is a direct translation of the theoretical definition given in Section 4.3.2. The sample covariance is

$$\hat{\sigma}_{xy} = \frac{1}{N-1} \sum_{i=1}^{N} (x_i - \hat{m}_x)(y_i - \hat{m}_y). \tag{4.41}$$

Using the estimators for the sample variance, the sample correlation coefficient is defined as

$$\hat{\rho} = \frac{\hat{\sigma}_{xy}}{\hat{\sigma}_x \hat{\sigma}_y}. \tag{4.42}$$

If both variables, x and y, have normal distributions, this estimator is also a maximum likelihood estimator. The sample correlation is calculated to find the strength of a relationship, therefore it is necessary to test it with some hypotheses. The sampling distribution of $\hat{\rho}$ is quite asymmetric and a transformation that creates an approximately normal random variable z is implemented,

$$z = \frac{1}{2} \ln\left(\frac{1+\hat{\rho}}{1-\hat{\rho}} \right). \tag{4.43}$$

Its mean and variance are, respectively [Fisz (1963)],

$$m_z = \frac{1}{2} \ln\left(\frac{1+\rho}{1-\rho} \right), \qquad \sigma_z^2 = \frac{1}{N-3}. \tag{4.44}$$

Example 4.10

For hospital patients suffering circulatory shock it is desired to know (a) if there is a correlation between the blood pH in the venous system (x) and the arterial system (y) and (b) if so, what is the confidence interval? The pH values are plotted in Figure 4.12.

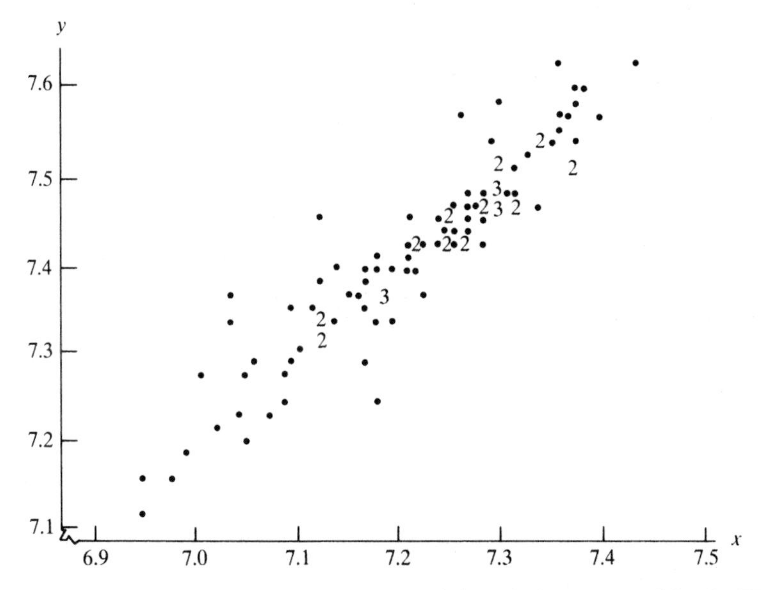

Figure 4.12 The arterial blood pH (y) and the venous blood pH (x) for critically ill patients. The numbers indicate multiple points at that coordinate. [Adapted from Afifi and Azen (1979), Figure 3.1.2, with permission.]

The sample moments and correlation coefficient calculated from measurements on 108 patients are

$$\hat{m}_x = 7.373, \qquad \hat{m}_y = 7.413, \qquad \hat{\sigma}_x^2 = 0.1253, \qquad \hat{\sigma}_y^2 = 0.1184,$$

$$\hat{\sigma}_{xy} = 0.1101, \qquad \hat{\rho} = 0.9039.$$

The first hypothesis to be tested is whether the variables are correlated. The null hypothesis is $\rho = 0$, Using the z transformation, $m_z = 0.0$ and $\sigma_z^2 = 1/(N-3) = 0.00952$ and the 95% confidence interval is $\pm 1.96\sigma_z$. That is, if

$$-0.1913 \leq z \leq 0.1913,$$

then $\rho = 0$. Using equation (4.43),

$$z_{\hat{\rho}} = \frac{1}{2}\ln\left(\frac{1+\hat{\rho}}{1-\hat{\rho}}\right) = \frac{1}{2}\ln\left(\frac{1.9039}{0.0961}\right) = 1.493,$$

and the variables are correlated. Now the confidence interval for ρ must be established. The bounds for the confidence interval of z are $z_{\hat{\rho}} \pm 1.96\sigma_z$ or

$$1.3017 \leq z \leq 1.6843.$$

Using equation (4.43), the inverse transformation from z to ρ is

$$\rho = \frac{e^{2z} - 1}{e^{2z} + 1}$$

and substituting the upper and lower bounds for z yields

$$0.8622 \leq \rho \leq 0.9334$$

and the two pH values are highly correlated.

An important relationship between correlation and linear regression is apparent when the conditional pdf $f(y|x)$ of the normal distribution is considered. This is directly derived from Bayes' rule, equation (4.21), and is reserved as an exercise. The conditional mean $m_{y|x}$ and standard deviation $\sigma_{y|x}$ are

$$m_{y|x} = m_y + \rho\frac{\sigma_y}{\sigma_x}(x - m_x) = m_y - \frac{\sigma_{xy}}{\sigma_x^2}m_x + \frac{\sigma_{xy}}{\sigma_x^2}x, \quad (4.45)$$

$$\sigma_{y|x} = \sigma_y\sqrt{1 - \rho^2}. \quad (4.46)$$

Now return to the solution of the linear model relating two sets of measurements $\{y_i: 1 \leq i \leq N\}$ and $\{x_i: 1 \leq i \leq N\}$. The model is

$$\hat{y} = a + bx. \quad (2.2)$$

Solving equations (2.6) and (2.7) explicitly for the coefficients yields

$$a = \hat{m}_y - \frac{\hat{\sigma}_{xy}}{s_x^2}\hat{m}_x; \qquad b = \frac{\hat{\sigma}_{xy}}{s_x^2}. \quad (4.47)$$

The solution for these coefficients contains the estimates of the moments in equation (4.45) in the same algebraic form. Hence, this shows that the correlation coefficient is a measure of *linear dependence* between two samples or time series.

4.6 Density Function Estimation

4.6.1 General Principle

There are situations when knowing the statistical parameters of the data is not sufficient and it is desired to discover or model the data's probability distribution. In quality control, for example, as in Example 4.1, does an exponential pdf portray the distribution of failure times accurately, or is another model necessary? This hypothesis can be tested using *Pearson's* χ^2 *statistic* [Fisz (1963) and Otnes and Enochson (1972)]. Let $F(x)$ be the hypothesized PDF for the data. Divide the range of x into N_b disjoint intervals S_j such that

$$P[S_j] = P[d_{j-1} \leq x < d_j], \qquad 1 \leq j \leq N_b, \tag{4.48}$$

are the theoretical probabilities of occurrence. The test depends on comparing the observed number of occurrences of values in a set of samples to the number expected from the proposed distribution. Let o_j and e_j represent the number of observed and expected occurrences, respectively, in S_j. If N equals the total number of data points, $e_j = NP[S_j]$. The metric for this comparison, the chi-square statistic, is

$$\chi^2 = \sum_{j=1}^{N_b} \frac{(o_j - e_j)^2}{e_j}, \tag{4.49}$$

where

$$\sum_{j=1}^{N_b} o_j = N. \tag{4.50}$$

The e_j are calculated from the proposed model. The chi-square statistic in equation (4.49) has been developed upon the hypothesis that the sample has the proposed PDF and the density function for χ^2 is

$$f(u) = \begin{cases} \dfrac{1}{2^{\nu/2}\Gamma(\nu/2)} e^{-u/2} u^{\nu/2-1}, & u \ge 0, \\ 0, & u < 0, \end{cases} \tag{4.51}$$

where $u = \chi^2$ and ν = the degree of freedom = $N_b - 1$. This density function is plotted in Figure 4.13. Obviously χ^2 is positive so that an α-level significance region is established only over one tail of the distribution such that

$$P\left[\chi^2 \ge \chi^2_{\nu,\alpha}\right] = \alpha. \tag{4.52}$$

The value of $\chi^2_{\nu,\alpha}$ corresponding to various significance levels is tabulated in Appendix 4.5. If the calculated value of χ^2 lies within the significance region, then the proposed model is rejected as being appropriate and another one must be selected. A simple example will illustrate this procedure.

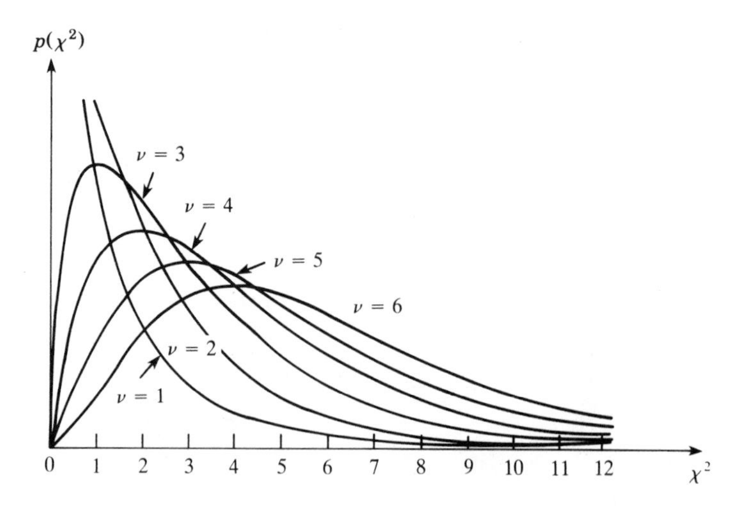

Figure 4.13 Chi-square density function with various degrees of freedom ν.

Example 4.11

Cards are drawn at random 20 times from a full deck. The result is 8 clubs, 3 diamonds, 5 hearts, and 4 spades. The hypothesis being tested is the honesty of the deck of cards. Thus, the probability for drawing a card of a desired suit is $\frac{1}{4}$ and this is the proposed model. The mapping from the nominal variables to a magnitude or selection level is clubs = 1, diamonds = 2, hearts = 3, spades = 4. The expected number for each level is 5; therefore, $e_1 = e_2 = e_3 = e_4 = 5$. The observed numbers are $o_1 = 8$, $o_2 = 3$, $o_3 = 5$, $o_4 = 4$. Calculating χ^2 from equation (4.49) produces

$$\chi^2 = \sum_{j=1}^{N_b} \frac{(o_j - e_j)^2}{e_j} = \frac{9}{5} + \frac{4}{5} + \frac{1}{5} + 0 = 2.8.$$

The confidence limit is based on a 95% confidence level and $\nu = 4 - 1 = 3$ degrees of freedom. The appropriate entry in the table in Appendix 4.5 shows that $\chi^2_{3,0.05} = 7.815$. Thus, the inequality $\chi^2 \geq \chi^2_{\nu,\alpha}$ is not satisfied and the proposed probability model is accepted as true, that is, the deck of cards is honest.

4.6.2 Detailed Procedure

In practice, one does not have well-defined categories as in the previous example and one needs to produce a histogram. Consider the collector currents in a sample of 100 manufactured transistors. The current is measured with an accuracy of 0.05 μA (microamperes) and the number with different values is plotted in Figure 4.14(a). The production of the histogram requires defining the number and boundaries of these categories. Define the range of interest for the random variable, $a \leq x \leq b$. Then divide this range into k subintervals or class intervals. The width W of these class intervals is $W = (b - a)/k$. To account for the entire sample space, define two classes for $x < a$ and $x \geq b$. These are $k + 2$ classes or bins and number them consecutively 0 to $k + 1$. Let j indicate the bin number and N_j the number in each bin. The histogram is

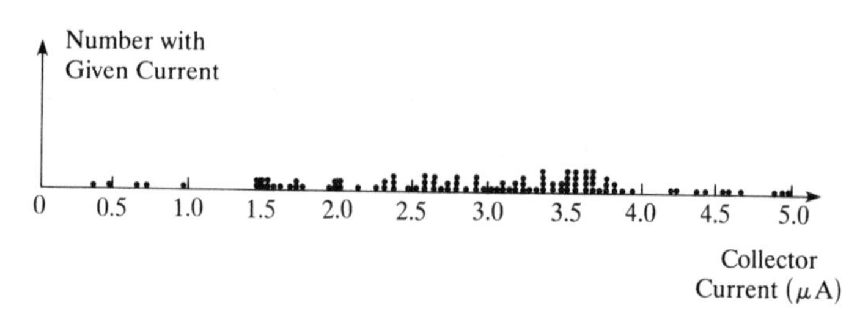

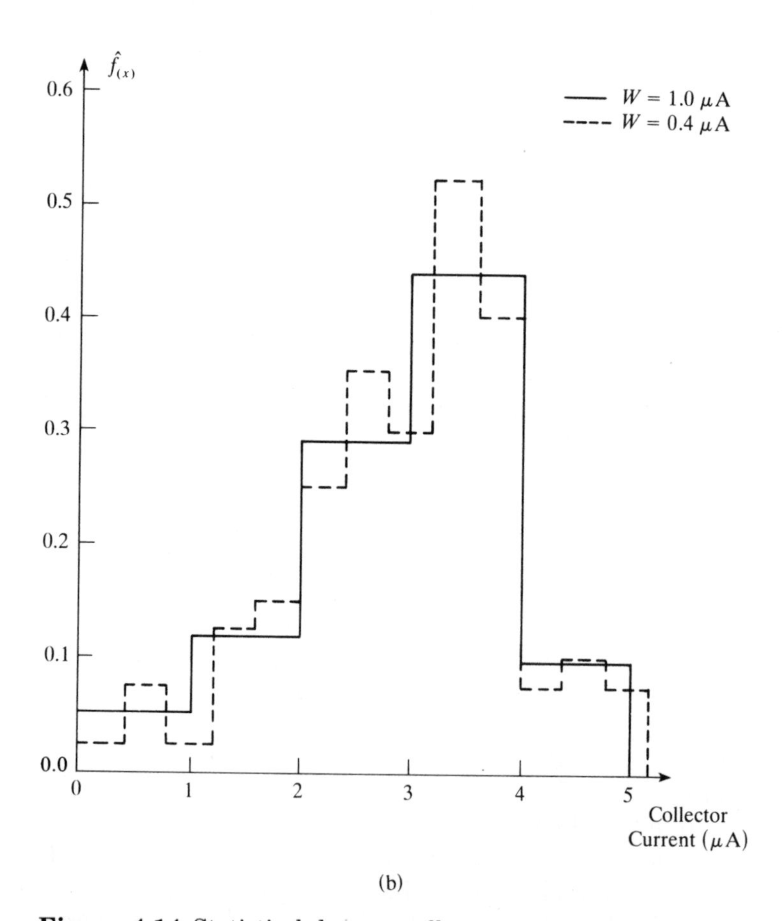

Figure 4.14 Statistical data on collector currents of 100 transistors: (a) histogram with $W = 0.05\ \mu$A;(b) histogram with $W = 1\ \mu$A and estimates of probability density functions with $W = 1.0$ and 0.4 μA. [From Jenkins and Watts (1968), Figure 3.3 and 3.6, with permission.]

computed in the following manner:

1. Initialize N_j to 0, $0 \le j \le k+1$;
2. Sort x_i, $1 \le i \le N$, into classes and increment N_j according to
   a. if $x_i < a$, increment N_0,
   b. if $x_i \ge b$, increment N_{k+1},
   c. for other x_i, calculate $j = \text{INT}((x_i - a)/W) + 1$, and increment N_j.

The INT stands for the integerization operation. The sample probability that a measurement lies within a bin is the relative frequency definition, or

$$\hat{P}\left[d_{j-1} \le x < d_j\right] = \frac{N_j}{N}. \tag{4.53}$$

To approximate the magnitude of the pdf, remember that the pdf is probability per unit value, and divide equation (4.53) by the bin width W, or

$$\hat{f}(x) = \frac{N_j}{NW}, \qquad d_{j-1} \le x < d_j. \tag{4.54}$$

Figure 4.14(b) shows estimates of the histogram and the pdf for the collector current data using two different bin widths. As one can surmise from this figure, if W is too large, then peak magnitudes of $f(x)$ will be underestimated. It has been shown that there is a bias in estimating a pdf [Shanmugan and Breipohl (1988)]. The mean value of the estimator is

$$E\left[\hat{f}(x_{cj})\right] = f(x_{cj}) + f''(x_{cj})\frac{W^2}{24}, \qquad d_{j-1} \le x < d_j, \tag{4.55}$$

where $x_{cj} = (d_{j-1} + d_j)/2$. Thus, some knowledge of the general form of the pdf is important for a good histogram representation.

In signal processing, because most of the variables are continuous, so are the pdf models that one wishes to test as feasible representations. This is handled in a straightforward manner because, for a continuous variable,

$$e_j = NP\left[d_{j-1} \le x < d_j\right]. \tag{4.56}$$

Thus, once one chooses the boundaries for the histogram, one must integrate the hypothesized pdf over the bin boundaries to calculate the bin probabilities.

Example 4.12

In the manufacture of cotton thread, one of the properties being controlled is the tensile strength. Three hundred balls of cotton thread were chosen from a consignment and their breaking tension was measured. The range of tensions was found to be $0.5 \text{ kg} \le x \le 2.3 \text{ kg}$. It is desired to test if a normal distribution is a suitable model for this property. A bin width of 0.14 kg is chosen and results in the creation of 13 bins. A histogram was calculated and the numbers of occurrences are shown in Table 4.1. One practical problem that arises is that the mean and variance of the hypothesized distribution are unknown and must be estimated from the data. These are

$$\hat{m} = 1.41, \qquad s = 0.26.$$

The probabilities are found by standardizing x and using the error

Table 4.1 Histogram of Breaking Tensions (kg)

j	$d_{j-1} - d_j$	N_j	$P[S_j]$
0	0.5–0.64	1	0.0015
1	0.64–0.78	2	0.0065
2	0.78–0.92	9	0.0223
3	0.92–1.06	25	0.0584
4	1.06–1.20	37	0.1205
5	1.20–1.34	53	0.1846
6	1.34–1.48	56	0.2128
7	1.48–1.62	53	0.1846
8	1.62–1.76	25	0.1205
9	1.76–1.90	19	0.0584
10	1.90–2.04	16	0.0223
11	2.04–2.18	3	0.0065
12	2.18–2.32	1	0.0015

function. For the fifth bin,

$$P[S_5] = P[1.20 \le x < 1.34]$$

$$= P\left[-0.81 \le \frac{x - 1.41}{0.26} < -0.27\right] = 0.1846.$$

The probabilities $P[S_j]$ are also listed in the table. The expected number of occurrences are calculated with equation (4.56) and rounding off the result. An empirical rule used is that the number of expected occurrences must at least be 2. This is accommodated by combining bins 0, 1, and 2 and bin 12 with bin 11, resulting in 10 bins. Another practical problem also arises: the use of sample moments affects the degrees of freedom. For each restriction (estimated parameter) in the hypothesized distribution, the degrees of freedom must be reduced by one. Thus $\nu = 10 - 1 - 2 = 7$. For a 5% significance level,

$$\chi^2_{7,0.05} = 14.067.$$

For the breaking strength of the threads,

$$\chi^2 = \sum_{j=2}^{11} \frac{(o_j - e_j)^2}{e_j} = 10 \sum_{j=2}^{11} \frac{\left(\hat{P}[S_j] - P[S_j]\right)^2}{P[S_j]} = 22.07.$$

Thus, the hypothesis is rejected and the thread strengths are not represented by a normal distribution.

Many interesting applications arise when modeling the pdf of time intervals between events is needed. One such application arises when studying the kinetics of eye movement. As one tracks a moving object visually, the position of the eye changes slowly and then it moves quickly back to its initial position. These quick movements are called nystagmus and their times of occurrence are called events. The histogram of time intervals between events (IEI) for one measurement session is shown in Figure 4.15. Two pdf models are proposed, lognormal and gamma, and are fitted by equating the sample moments of the IEI and the models' moments.

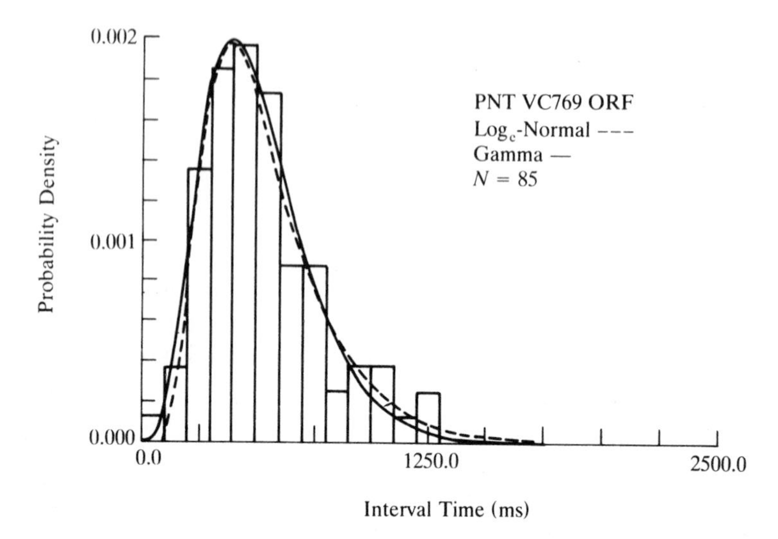

Figure 4.15 An IEI histogram with lognormal and gamma pdf models superimposed. [From Anderson and Correia Figure 7, with permission.]

The resulting models are also plotted in Figure 4.15. Both demonstrate good "fits" by generating χ^2 values with significance regions less than the 0.05 level [Anderson and Correia (1977)].

4.7 General Properties of Estimators

4.7.1 Convergence

As has been studied, an estimator produces estimates that are random variables with a sampling distribution. This distribution cannot always be derived but it is necessary to know if the estimate is close to the true value in some sense. Fortunately, *convergence* can be determined using very important general properties of estimators, the bias and consistency.

The *bias* property describes the relationship between the function being estimated and the mean value of the estimator. For instance, when estimating the mean value of a random variable using equation (4.30), $E[\hat{m}] = m$ and the estimator is said to be *unbiased*. If another estimator for mean value, $\hat{m}_2$, were used, perhaps the result would be $E[\hat{m}_2] \neq m$, and this estimator is said to be *biased*. This is generalized for any estimator, $\hat{g}_N(x)$, being a function of the set of N measurements, $\{x_i\}$. If $E[\hat{g}_N(x)] = g(x)$, then the estimator is unbiased; otherwise the estimator is biased. Equation (4.32) computes the sample variance and is an unbiased estimator because $E[\hat{\sigma}^2] = \sigma^2$. There exists another estimator that is more intuitive because the coefficient of the summation is $1/N$. It is defined as

$$\hat{\sigma}_2^2 = \frac{1}{N} \sum_{i=1}^{N} (x_i - \hat{m})^2. \tag{4.57}$$

However, this estimator is biased because

$$E\left[\hat{\sigma}_2^2\right] = \frac{N-1}{N}\sigma^2 \neq \sigma^2. \tag{4.58}$$

These relationships are derived in many introductory books on probability and statistics and are left for the reader to review. Some references are Hoel (1960) and Larsen and Marx (1986).

The *mean square error* is the mean square difference between the estimator and the true value. It is also the variance $\mathrm{Var}[\cdot]$ of the estimator if the estimator is unbiased. Let $G_N(x)$ equal the sample mean of $g(x)$ using N data points, or

$$G_N(x) = \frac{1}{N} \sum_{i=1}^{N} g(x_i). \tag{4.59}$$

If

$$\lim_{N \to \infty} \mathrm{Var}[G_N(x)] = \lim_{N \to \infty} E\left[(G_N(x) - E[g(x)])^2\right] = 0, \tag{4.60}$$

then the estimator is *consistent*. For the sample mean when the random variable x has a variance σ^2,

$$\mathrm{Var}[\hat{m}] = E\left[(\hat{m}_N - m)^2\right] = \frac{\sigma^2}{N} \tag{4.61}$$

and $\hat{m}$ is a consistent estimator. Essentially, consistency means that the variance of the estimate approaches zero if an infinite amount of data is available. In other words, the estimate will converge to some value if enough data is used. Proving convergence can be an arduous task and its use and implementation are reserved for advanced courses in mathematical statistics and random processes. Some treatments of this topic can be found in Papoulis (1984), Davenport (1970), and Fisz (1963).

4.7.2 Recursion

The general formula for calculating an estimate of the mean of a function of a random variable was given in equation (4.30). If this is implemented directly, it becomes a batch algorithm. In order to reduce truncation errors in machine computation, recursive algorithms can be easily developed for the sample mean. Begin again by rewriting the batch algorithm. Now separate the sum into an $N-1$-point summation and the final value of $g(x_i)$. Then equation (4.30) becomes

$$G_N(x) = \frac{1}{N}\sum_{i=1}^{N-1} g(x_i) + \frac{1}{N}g(x_N). \tag{4.62}$$

Change the summation to become an $N-1$-point sample mean and

$$\begin{aligned} G_N(x) &= \frac{N-1}{N}\frac{1}{N-1}\sum_{i=1}^{N-1} g(x_i) + \frac{1}{N}g(x_N) \\ &= \frac{N-1}{N}G_{N-1}(x) + \frac{1}{N}g(x_N). \end{aligned} \tag{4.63}$$

Thus with an estimate of the mean, one can update the estimate simply in one step with the addition of another sample point by using the weighted sum of equation (4.63). This is the recursive algorithm for calculating a sample mean of $g(x)$.

References

Abramowtiz, M. and I. Stegun (1965). *Handbook of Mathematical Functions*. Dover Publications, New York.

Afifi, A. and S. Azen (1979). *Statistical Analysis, A Computer Oriented Approach*. Academic Press, New York.

Anderson, D. and M. Correia (1977). The detection and analysis of point processes in biological signals. *Proc. IEEE* 65:773–780.

Davenport, W. (1970). *Probability and Random Processes, An Introduction for Applied Scientists and Engineers.* McGraw-Hill Book Co., New York.

de Coulon, F. (1986). *Signal Theory and Processing.* Artech House, Dedham, MA.

Fante, R. (1988). *Signal Analysis.* John Wiley and Sons, New York.

Fisz, M. (1963). *Probability Theory and Mathematical Statistics.* John Wiley and Sons, New York.

Hoel, P. (1960). *Elementary Statistics.* John Wiley and Sons, New York.

Jenkins, G. and D. Watts (1968). *Spectral Analysis and its Applications.* Holden-Day, New York.

Larsen, R. and M. Marx (1986). *An Introduction to Mathematical Statistics.* Prentice-Hall, Englewood Cliffs, NJ.

Larson, H. and B. Shubert (1979). *Probabilistic Models in Engineering Sciences.* John Wiley and Sons, New York.

Maisel, L. (1971). *Probability, Statistics, and Random Processes.* Simon and Schuster, New York.

O'Flynn, M. (1982). *Probabilities, Random Variables, and Random Processes.* Harper and Row Publishers, New York.

Otnes, R. and L. Enochson (1972). *Digital Time Series Analysis.* John Wiley and Sons, New York.

Pandit, S. and S. Wu (1983). *Time Series Analysis and Applications.* John Wiley and Sons, New York.

Papoulis, A. (1984). *Probability, Random Variables, and Stochastic Processes*. McGraw-Hill Book Co., New York.

Peebles, P. (1987). *Probability, Random Variables, and Random Signal Principles*. McGraw-Hill Book Co., New York.

Peerless Engineering Service (1984). *FORTRAN—Scientific Subroutine Library*. John Wiley and Sons, New York.

Shanmugan, K. and A. Breipohl (1988). *Random Signals: Detection, Estimation and Data Analysis*. John Wiley and Sons, New York.

Stark, H. and J. Woods (1986). *Probability, Random Processes, and Estimation Theory for Engineers*. Prentice-Hall, Englewood Cliffs, NJ.

Water Resources Data—Tennessee—Water Year 1981. US Geological Survey Water-Data Report TN-81-1, page 277.

Exercises

4.1 For $f(x) = N(20, 49)$ find the following:
a. $P[x \leq 30]$;
b. $P[x \geq 30]$;
c. $P[15 \leq x \leq 30]$.

4.2 Prove that the coefficient of the exponential density function of equation (4.5) must be $1/b$.

4.3 A Rayleigh pdf is defined over the range $x \geq 0$ as

$$f(x) = \frac{x}{\alpha^2} \exp\left(-\frac{x^2}{2\alpha^2}\right).$$

For $\alpha = 2$, what is $P[4 \leq x \leq 10]$?

4.4 Assume that the radii of the paint droplets in Figure 4.2 can be described by an exponential pdf with $b = 4.1$. What is the probability that the radii are greater than 20 μm (microns)?

4.5 In a certain production line, 1000-Ω (ohm) resistors that have a 10% tolerance are being manufactured. The resistance is described by a normal distribution with $m = 1000\ \Omega$ and $\sigma = 40\ \Omega$. What fraction will be rejected?

4.6 Prove equation (4.12), which relates the variance to the mean and mean square.

4.7 Find the general formulas for the mean and variance of the uniform density function written in equation (4.4).

4.8 What are the mean and variance of the general exponential density function [equation (4.5)]?

4.9 Prove that the variance of the Rayleigh pdf is $\sigma^2 = (2 - \pi/2)\alpha^2$.

4.10 A triangular probability density function is shown in Figure E4.10.

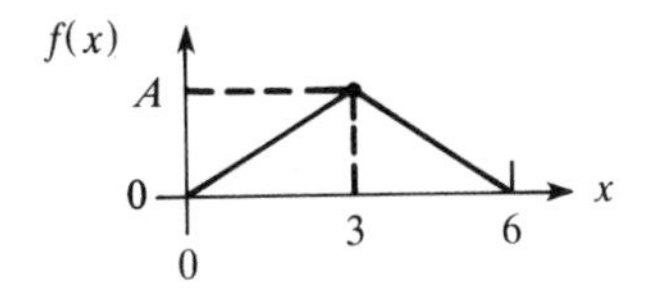

Figure E4.10

a. For it to be a pdf, what is A?
b. Calculate m and σ.
c. What are $P[x \leq 2]$ and $P[x \leq 5]$?

4.11 Derive equation (4.27), the relationship between the covariance and mean product.

4.12 For the bivariate density function in Example 4.7, do the following:
a. Verify that the volume is equal to 1 [equation (4.21b)].
b. Find the conditional mean of variable y.

4.13 For the bivariate density function in Example 4.7, do the following:
a. Derive the marginal density function $f(y)$.
b. Derive the covariance.

4.14 For Example 4.9, verify the critical value of t, the estimate of the mean river flow, and its confidence bounds.

4.15 Prove $-1 \leq \rho \leq 1$. [*Hint*: $E[((x - m_x) + (y - m_y))^2] \leq 0$.]

4.16 For the conditional normal pdf, derive the mean and variance [equations (4.45) and (4.46)].

4.17 Verify the calculations for testing zero correlation and finding the confidence intervals for the correlation coefficient in Example 4.10.

4.18 A test was performed to determine if there is a dependence between the amount of chemical (y, in grams per liter) in solution and its temperature (x, in degrees celsius) for crystallization. The measurements are listed in Table E4.18.

Table E4.18 Chemical Density (y) and Temperature (x)

x	0.3	0.4	1.2	2.3	3.1	4.2	5.3
y	3.2	2.4	4.3	5.4	6.6	7.5	8.7

a. What is the estimate σ_{xy}^{z} of the covariance?
b. Normalize $\hat{\sigma}_{xy}^{z}$ to estimate the sample correlation coefficient, $\hat{\rho}$.
c. Does r test as being zero?
d. What are the 95% confidence limits for ρ?

4.19 Estimate ρ for the water discharge data in Table E2.3. Are they linearly dependent?

4.20 Verify the estimates of the probability density function for the transistor collector current data in Figure 4.14(b).

4.21 Verify the values of χ^2 and $\chi_{\nu,\alpha}^{2}$ for fitting the normal density function to the thread tension data in Example 4.12.

4.22 A neurophysiological experiment to investigate neuronal activity in the visual cortex of a cat's brain is being conducted. The time interval, in seconds, between firings in one neuron is measured and is organized into a histogram shown in Table E4.22.

Table E4.22 Time Interval N_j

Time Interval	N_j
0–15	63
15–30	25
30–45	14
45–60	7
60–75	6
75–90	3
90–105	2

a. What are the sample mean and variance?
b. Fit an exponential pdf to the observations.

c. Test if the exponential distribution is a good model. What is ν the degrees of freedom?

4.23 The accompanying table lists the breaking strength (in kilograms) of plastic rods.

Table E4.23 Breaking Strengths (kg)

6.70	7.04	7.21	7.29	7.35	7.45	7.59	7.70	7.72	7.74
7.84	7.88	7.94	7.99	7.99	8.04	8.10	8.12	8.15	8.17
8.20	8.21	8.24	8.25	8.26	8.28	8.31	8.37	8.38	8.42
8.49	8.51	8.55	8.56	8.58	8.59	8.60	8.66	8.67	8.69
8.70	8.74	8.81	8.84	8.86	8.90	9.01	9.15	9.55	9.80

a. Construct a histogram of these observations with a bin width of 0.5 kg and the lesser boundary of the first bin being 6.25 kg.
b. Scale the histogram to estimate values of a pdf.
c. What types of models might be suitable for these observations?

Computer Exercises

4.24 Repeat Exercise 4.23 using a computer. What are the sample mean, variance, and mean square?

4.25 An electrocardiogram on a normal healthy individual was measured. The signal was sampled with a computer and the time intervals between heartbeats were measured; 180 intervals are listed in Appendix 4.6. It is desired to study the statistical properties of these intervals in order to provide a basis for evaluating the time intervals from an individual with heart disease. Naturally, a computer is needed to perform these tasks.

a. What are $\hat{m}$ and $\hat{\sigma}^2$?
b. Using a suitable bin width, $0.004 \leq W \leq 0.008$, sort these intervals into a histogram.
c. Consider the symmetry, range, and so forth of the histogram. What would be a suitable density function to describe it?
d. Perform a χ^2 test on this proposal. Is it a good model? If not, what might be a better one?

Appendices

Appendix 4.1 Plots and Formulae for Five Probability Density Functions

Beta [Figure A 4.1(a)]

$$f(x) = \begin{cases} Ax^b(1-x)^c, & 0 \le x \le 1, \\ 0 & \text{elsewhere.} \end{cases}$$

It can be shown that

$$\int_0^1 x^b(1-x)^c\,dx = \frac{\Gamma(b+1)\Gamma(c+1)}{\Gamma(b+c+2)}.$$

Hence,

$$A = \frac{\Gamma(b+c+2)}{\Gamma(b+1)\Gamma(c+1)}.$$

Laplace [Figure A 4.1(b)]

$$f(x) = \frac{\alpha}{2}e^{-\alpha|x|}.$$

Cauchy [Figure A 4.1(c)]

$$f(x) = \frac{\alpha/\pi}{\alpha^2 + x^2}.$$

Rayleigh [Figure A 4.1(d)]

$$f(x) = \frac{x}{\alpha^2}e^{-x^2/2\alpha^2}U(x).$$

Maxwell [Figure A 4.1(e)]

$$f(x) = \frac{\sqrt{2}}{\alpha^3\sqrt{\pi}}x^2e^{-x^2/2\alpha^2}U(x).$$

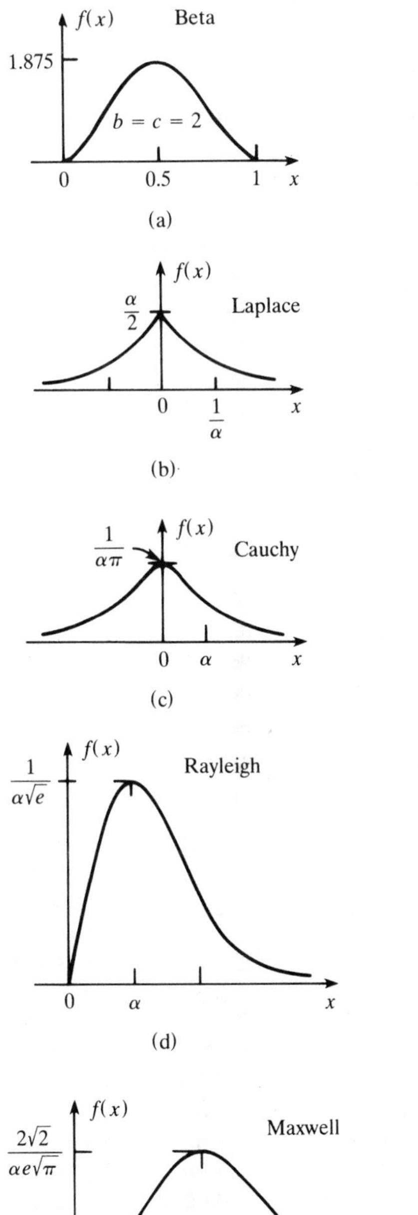

Figure A4.1 Plots of five probability density functions: (a) Beta; (b) Laplace; (c) Cauchy; (d) Rayleigh; (e) Maxwell.

Appendix 4.2 Values of the Error Function

Values of the Error Function, erf(x) **versus** x

$$\text{erf}(x) = \frac{1}{\sqrt{2\pi}} \int_0^x \exp(-\frac{1}{2}t^2)\,dt$$

x	erf(x)	x	erf(x)
0.05	0.01994	2.05	0.47981
0.10	0.03983	2.10	0.48213
0.15	0.05962	2.15	0.48421
0.20	0.07926	2.20	0.48609
0.25	0.09871	2.25	0.48777
0.30	0.11791	2.30	0.48927
0.35	0.13683	2.35	0.49060
0.40	0.15542	2.40	0.49179
0.45	0.17364	2.45	0.49285
0.50	0.19146	2.50	0.49378
0.55	0.20884	2.55	0.49460
0.60	0.22575	2.60	0.49533
0.65	0.24215	2.65	0.49596
0.70	0.25803	2.70	0.49652
0.75	0.27337	2.75	0.49701
0.80	0.28814	2.80	0.49743
0.85	0.30233	2.85	0.49780
0.90	0.31594	2.90	0.49812
0.95	0.32894	2.95	0.49840
1.00	0.34134	3.00	0.49864
1.05	0.35314	3.05	0.49884
1.10	0.36433	3.10	0.49902
1.15	0.37492	3.15	0.49917
1.20	0.38492	3.20	0.49930
1.25	0.39434	3.25	0.49941
1.30	0.40319	3.30	0.49951
1.35	0.41149	3.35	0.49958
1.40	0.41924	3.40	0.49965
1.45	0.42646	3.45	0.49971
1.50	0.43319	3.50	0.49976
1.55	0.43942	3.55	0.49980
1.60	0.44519	3.60	0.49983
1.65	0.45052	3.65	0.49986
1.70	0.45543	3.70	0.49988
1.75	0.45993	3.75	0.49990
1.80	0.46406	3.80	0.49992
1.85	0.46783	3.85	0.49993
1.90	0.47127	3.90	0.49994
1.95	0.47440	3.95	0.49995
2.00	0.47724	4.00	0.49996

Appendix 4.3 Chart of River Flows

River Flow in Duck River (in Cubic Feet per Second), March 1981

Day	Flow	Day	Flow	Day	Flow
1	1590	11	2240	21	1120
2	1500	12	2000	22	1100
3	1390	13	1810	23	1160
4	1490	14	1650	24	1550
5	2110	15	1530	25	2190
6	2900	16	1470	26	2480
7	3610	17	1390	27	2130
8	3550	18	1300	28	1810
9	2990	19	1230	29	1640
10	2550	20	1180	30	2730

Appendix 4.4 Student's *t* Distribution

Probabilities *P* of Exceeding t_c (Table Entries) with ν Degrees of Freedom (d.f.)

			P			
d.f.	0.1	0.05	0.02	0.01	0.002	0.001
1	6.314	12.706	31.821	63.657	318.31	636.62
2	2.920	4.303	6.965	9.925	22.327	31.598
3	2.353	3.182	4.541	5.841	10.214	12.924
4	2.132	2.776	3.747	4.604	7.173	8.610
5	2.015	2.571	3.365	4.032	5.893	6.869
6	1.943	2.447	3.143	3.707	5.208	5.959
7	1.895	2.365	2.998	3.499	4.785	5.408
8	1.860	2.306	2.896	3.355	4.501	5.041
9	1.833	2.262	2.821	3.250	4.297	4.781
10	1.812	2.228	2.764	3.169	4.144	4.587
11	1.796	2.201	2.718	3.106	4.025	4.437
12	1.782	2.179	2.681	3.055	3.930	4.318
13	1.771	2.160	2.650	3.012	3.852	4.221
14	1.761	2.145	2.624	2.977	3.787	4.140
15	1.753	2.131	2.602	2.947	3.733	4.073
16	1.746	2.120	2.583	2.921	3.686	4.015
17	1.740	2.110	2.567	2.898	3.646	3.965
18	1.734	2.101	2.552	2.878	3.610	3.922
19	1.729	2.093	2.539	2.861	3.579	3.883

	P					
d.f.	0.1	0.05	0.02	0.01	0.002	0.001
20	1.725	2.086	2.528	2.845	3.552	3.850
21	1.721	2.080	2.518	2.831	3.527	3.819
22	1.717	2.074	2.508	2.819	3.505	3.792
23	1.714	2.069	2.500	2.807	3.485	3.767
24	1.711	2.064	2.492	2.797	3.467	3.745
25	1.708	2.060	2.485	2.787	3.450	3.725
26	1.706	2.056	2.479	2.779	3.435	3.707
27	1.703	2.052	2.473	2.771	3.421	3.690
28	1.701	1.048	2.467	2.763	3.408	3.674
29	1.699	2.045	2.462	2.756	3.396	3.659
30	1.697	2.042	2.457	2.750	3.385	3.646
40	1.684	2.021	2.423	2.704	3.307	3.551
60	1.671	2.000	2.390	2.660	3.232	3.460
120	1.658	1.980	2.358	2.617	3.160	3.373
∞*	1.645	1.960	2.326	2.576	3.090	3.291

*The last row of the table (∞) gives values of z, the standard normal variable.

Appendix 4.5 Chi-Square Distribution

Probabilities *P* of exceeding χ^2 (Table Entries) with ν Degrees of Freedom* (d.f.)

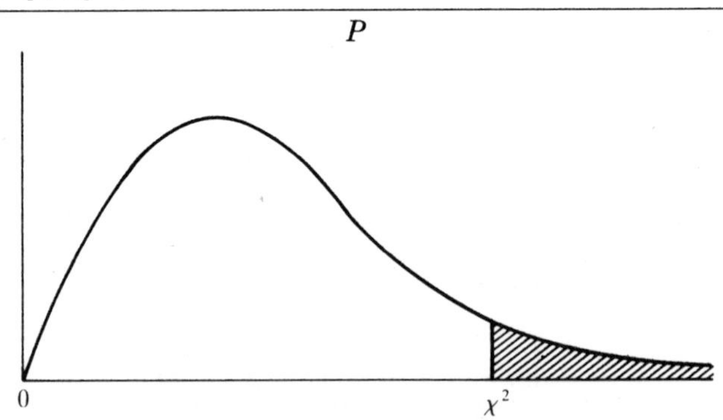

P is the shaded area

d.f.	0.995	0.975	0.050	0.025	0.010	0.005	0.001
1	3.9×10^{-5}	9.8×10^{-4}	3.84	5.02	6.63	7.88	10.83
2	0.010	0.051	5.99	7.38	9.21	10.60	13.81
3	0.071	0.22	7.81	9.35	11.34	12.84	16.27
4	0.21	0.48	9.49	11.14	13.28	14.86	18.47

d.f.	0.995	0.975	0.050	0.025	0.010	0.005	0.001
5	0.41	0.83	11.07	12.83	15.09	16.75	20.52
6	0.68	1.24	12.59	14.45	16.81	18.55	22.46
7	0.99	1.69	14.07	16.01	18.48	20.28	24.32
8	1.34	2.18	15.51	17.53	20.09	21.96	26.13
9	1.73	2.70	16.92	19.02	21.67	23.59	27.88
10	2.16	3.25	18.31	20.48	23.21	25.19	29.59
11	2.60	3.82	19.68	21.92	24.73	26.76	31.26
12	3.07	4.40	21.03	23.34	26.22	28.30	32.91
13	3.57	5.01	22.36	24.74	27.69	29.82	34.53
14	4.07	5.63	23.68	26.12	29.14	31.32	36.12
15	4.60	6.26	25.00	27.49	30.58	32.80	37.70
16	5.14	6.91	26.30	28.85	32.00	34.27	39.25
17	5.70	7.56	27.59	30.19	33.41	35.72	40.79
18	6.26	8.23	28.87	31.53	34.81	37.16	42.31
19	6.84	8.91	30.14	32.85	36.19	38.58	43.82
20	7.43	9.59	31.41	34.17	37.57	40.00	45.32
21	8.03	10.28	32.67	35.48	38.93	41.40	46.80
22	8.64	10.98	33.92	36.78	40.29	42.80	48.27
23	9.26	11.69	35.17	38.08	41.64	44.18	49.73
24	9.89	12.40	36.42	39.36	42.98	45.56	51.18
25	10.52	13.12	37.65	40.65	44.31	46.93	52.62
26	11.16	13.84	38.89	41.92	45.64	48.29	54.05
27	11.81	14.57	40.11	43.19	46.96	49.64	55.48
28	12.46	15.31	41.34	44.46	48.28	50.99	56.89
29	13.12	16.05	42.56	45.72	49.59	52.34	58.30
30	13.79	16.79	43.77	46.98	50.89	53.67	59.70
40	20.71	24.43	55.76	59.34	63.69	66.77	73.40
50	27.99	32.36	67.50	71.42	76.16	79.49	86.66
60	35.53	40.48	79.08	83.30	88.38	91.95	99.61
70	43.28	48.76	90.53	95.02	100.43	104.22	112.32
80	51.17	57.15	101.88	106.63	112.33	116.32	124.84
90	59.20	65.65	113.15	118.14	124.12	128.30	137.21
100	67.33	74.22	124.34	129.56	135.81	140.17	149.44

*For degrees of freedom $\nu > 100$, test $\sqrt{2\chi^2_{(\nu)}}$ as $N(\sqrt{2\nu - 1}, 1)$.

Appendix 4.6 Time Intervals between Heartbeats

Time Intervals between Heartbeats (in Seconds)

I	$T(I)$	I	$T(I)$	I	$T(I)$
1	0.796	44	0.836	87	0.850
2	0.803	45	0.836	88	0.850
3	0.803	46	0.836	89	0.850
4	0.810	47	0.836	90	0.850
5	0.810	48	0.840	91	0.850
6	0.813	49	0.840	92	0.850
7	0.813	50	0.840	93	0.850
8	0.816	51	0.840	94	0.850
9	0.816	52	0.840	95	0.850
10	0.816	53	0.840	96	0.850
11	0.820	54	0.840	97	0.850
12	0.820	55	0.840	98	0.850
13	0.820	56	0.840	99	0.853
14	0.820	57	0.840	100	0.853
15	0.823	58	0.840	101	0.853
16	0.823	59	0.843	102	0.853
17	0.823	60	0.843	103	0.853
18	0.823	61	0.843	104	0.853
19	0.823	62	0.843	105	0.853
20	0.826	63	0.843	106	0.853
21	0.826	64	0.843	107	0.853
22	0.826	65	0.843	108	0.853
23	0.826	66	0.843	109	0.853
24	0.826	67	0.843	110	0.853
25	0.830	68	0.843	111	0.853
26	0.830	69	0.843	112	0.853
27	0.830	70	0.843	113	0.853
28	0.830	71	0.843	114	0.856
29	0.830	72	0.846	115	0.856
30	0.830	73	0.846	116	0.856
31	0.830	74	0.846	117	0.856
32	0.833	75	0.846	118	0.856
33	0.833	76	0.846	119	0.856
34	0.833	77	0.846	120	0.856
35	0.833	78	0.846	121	0.856
36	0.833	79	0.846	122	0.856
37	0.833	80	0.846	123	0.856
38	0.836	81	0.846	124	0.856
39	0.836	82	0.846	125	0.856
40	0.836	83	0.846	126	0.856
41	0.836	84	0.846	127	0.856
42	0.836	85	0.850	128	0.856
43	0.836	86	0.850	129	0.856

Time Intervals between Heartbeats (in Seconds)

I	$T(I)$	I	$T(I)$	I	$T(I)$
130	0.860	147	0.863	164	0.870
131	0.860	148	0.863	165	0.870
132	0.860	149	0.863	166	0.870
133	0.860	150	0.863	167	0.870
134	0.860	151	0.863	168	0.870
135	0.860	152	0.863	169	0.870
136	0.860	153	0.863	170	0.870
137	0.860	154	0.863	171	0.873
138	0.860	155	0.863	172	0.873
139	0.860	156	0.866	173	0.873
140	0.860	157	0.866	174	0.873
141	0.860	158	0.866	175	0.876
142	0.860	159	0.866	176	0.876
143	0.860	160	0.866	177	0.880
144	0.863	161	0.866	178	0.883
145	0.863	162	0.866	179	0.886
146	0.863	163	0.866	180	0.896

Chapter 5 Introduction to Random Processes and Time-domain Description

5.1 Introduction

A *random process* is a random variable with an additional dimension, time. For each measurement or outcome of an experiment there exists a time function instead of a single number. This is true for all signal and time-series measurements. For example, each time there is a large explosion or sudden movement in the earth's tectonic plates, seismic waves are produced that travel considerable distances. These waves are studied and Figure 5.1 shows an example. So, instead of a single data point for each event, there exists a record of movement over time. This situation is described by assigning two independent arguments, t and ζ, to a random process $x(t, \zeta)$. This is depicted in Figure 5.2. The variable ζ_n indicates the time function $x(t)$ for the nth outcome. Each realization $x(t, \zeta_0), \ldots, x(t, \zeta_n)$ is a *sample function* and the set of sample functions is an *ensemble*. The laws of probability and techniques of statistics are applied by describing the behavior of all the processes at a specific time t_0 as a random variable. This is illustrated in Figure 5.2 and $x(t_0, \zeta)$ is a random variable. The behavior of the process at another time, t_1, is also a random

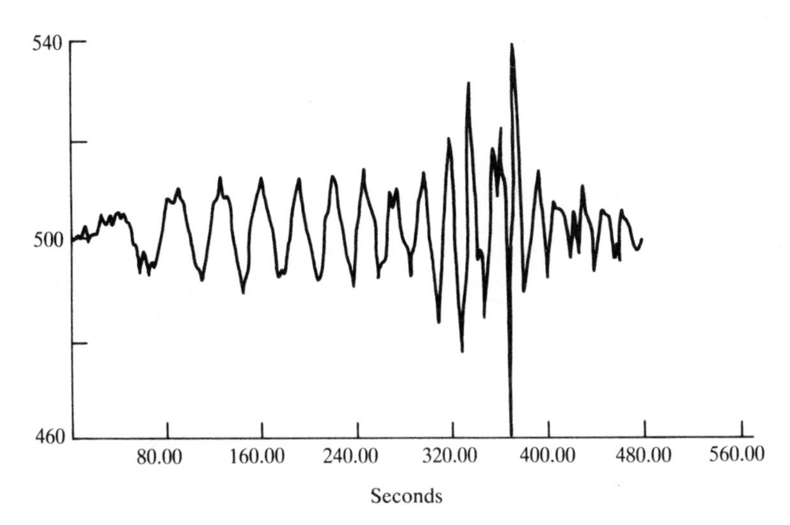

Figure 5.1 Seismic signal (in microns per second) from a nuclear explosion that was recorded at Kongsberg, Norway. [From Burton (1977), Figure 1, with permission.]

variable, $x(t_1, \zeta)$. For simplicity, the argument ζ will be dropped when discussing the ensemble random variable but the time argument will remain explicit; its probability density function is $f_{x(t_1)}(\alpha)$.

The description of the properties of random processes is accomplished by extending the probabilistic description of random variables to include the additional independent variable, time. The presentation of this description will begin with a definition of stationarity and will be followed by a definition of ensemble moment functions for time-domain signals. The estimation of these functions will be considered by using estimators that operate on sample functions. As always, the suitability of the estimators must be evaluated and, most importantly, the validity of using time-domain averages for estimating ensemble averages must be discussed. After this validity is established, several important functions such as the autocorrelation function will be studied. Finally, to develop a better sense of the concept of structure and correlation in a signal, the simulation of signals and some methods to create them will be described. The use of simulation methods will be continued in the next chapter.

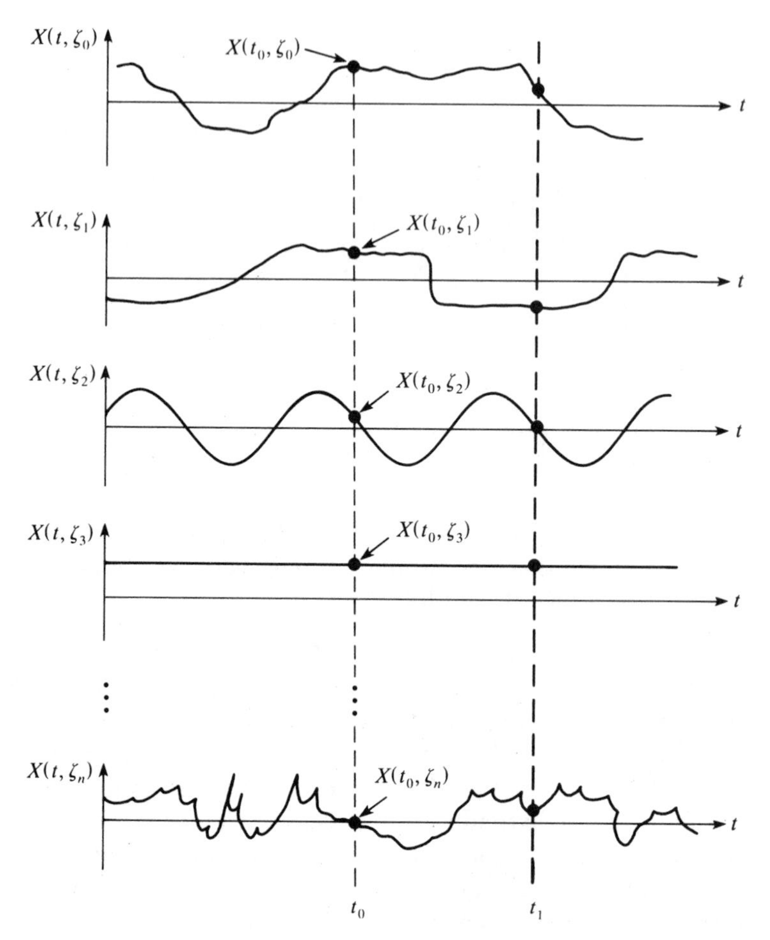

Figure 5.2 Family of sample functions for random process $x(t, \zeta)$.

5.2 Definition of Stationarity

The behavior of any random process is not necessarily the same at different times. For example, consider $x(t)$ at times t_1 and t_2; in general, $f_{x(t_1)}(\alpha) \neq f_{x(t_2)}(\alpha)$. It is highly desirable that this situation be untrue. This leads to the formal definition of *stationarity*: If

$$f_{x(t_1)}(\alpha) = f_{x(t_2)}(\alpha), \tag{5.1}$$

then the process is *first-order stationary*. This is extended to higher orders of joint probability. In the previous chapter, the joint probability between two random variables was defined. This same concept is used to describe the relationship between values of $x(t)$ at two different times, t_1 and t_2. The number of variables being related by a pdf is its order. Thus, a second-order pdf is $f_{x(t_1)x(t_2)}(\alpha_1, \alpha_2)$. Again, it is desirable that the absolute time not be important. Consider the relationship between $x(t_1)$ and $x(t_2)$ and shift the time by τ units. If

$$f_{x(t_1+\tau)x(t_2+\tau)}(\alpha_1, \alpha_2) = f_{x(t_1)x(t_2)}(\alpha_1, \alpha_2), \tag{5.2}$$

then the process is *second-order stationary*. Notice that the second-order proability density function does not depend upon absolute time but only on the time difference between t_1 and t_2. If the pdfs for all orders are equal with a time shift, that is,

$$f_{x(t_1+\tau)\ldots x(t_n+\tau)\ldots}(\alpha_1, \ldots, \alpha_n, \ldots) = f_{x(t_1)\ldots x(t_n)\ldots}(\alpha_1, \ldots, \alpha_n, \ldots), \tag{5.3}$$

then the process is *strictly stationary*, that is, stationary in all orders. Rarely is it necessary or feasible to demonstrate or have strict stationarity. In fact, it is usually difficult even to ascertain second-order stationarity. It is usually sufficient to have only stationarity of several moments for most applications. A very useful form of stationarity is when the mean and covariance do not change with time. This is called *wide-sense stationarity* or *weak stationarity*. It is defined when

$$E[x(t_1)] = E[x(t_2)] \tag{5.4}$$

and

$$\text{Cov}[x(t_1), x(t_2)] = \text{Cov}[x(t_1), x(t_1+\tau)]. \tag{5.5}$$

Equation (5.5) defines the autocovariance function for a stationary process, and τ is the time difference between values of the process $x(t)$. Thus, a wide-sense stationary process is stationary in the mean and covariance. Many types of stationarity can be defined by

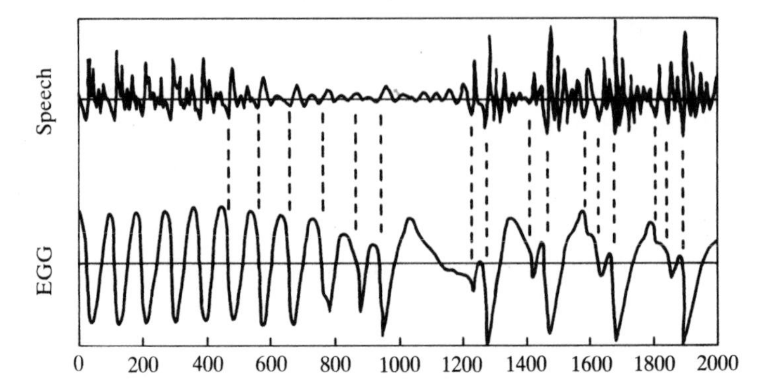

Figure 5.3 The speech and EGG waveforms during the utterance of the phrase "should we chase those cowboys?". A more positive EGG magnitude indicates a greater glottal opening. [From Childers and Labar (1984, IEEE), Figure 6, with permission.]

they are not relevant to the material in this book. If one is interested in this topic, consult textbooks on stochastic processes, such as Papoulis (1984).

An omnipresent example of a nonstationary signal is speech. Figure 5.3 shows the sound intensity from the utterance of the phrase "should we chase those cowboys?" and the corresponding electroglottograph (EGG). The EGG is a measure of the amount of opening in the glottis. Notice how the characteristics of both waveforms, particularly the spread of amplitude values in the speech, change with time. Most signals are stationary over longer periods of time. Figure 5.4 shows an electroencephalogram (EEG) that is stationary over longer time intervals; the stationary periods are separated by the vertical lines.

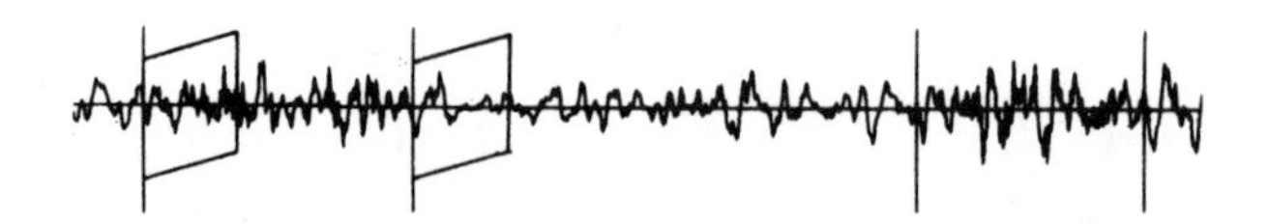

Figure 5.4 EEG signal. Vertical lines indicate periods within which the EEG is stationary. [From Bodenstein and Praetorius (1977, IEEE), Figure 3, with permission.]

5.3 Definition of Moment Functions

5.3.1 General Definitions

The general definition of an ensemble moment is an extension of the definition of a moment of a random variable with the addition of a time argument. This is

$$E[g(x(t_1))] = \int_{-\infty}^{\infty} g(\alpha) f_{x(t_1)}(\alpha)\, d\alpha. \tag{5.6}$$

Several particular ensemble moments are extremely important and must be explicitly defined and studied. The moments now become functions of time. The *ensemble average* is

$$m_x(t_1) = E[x(t_1)] = \int_{-\infty}^{\infty} \alpha f_{x(t_1)}(\alpha)\, d\alpha. \tag{5.7}$$

If the process is first-order stationary, then equation (5.4) is true and the time argument can be dropped, that is, $m_x(t_1) = m_x$. As can be easily seen, all first-order moments, such as the variance and mean square, would be independent of time. The converse is not necessarily true. Stationarity of the mean does not imply first-order stationarity. The *ensemble covariance* in equation (5.5) is

$$\operatorname{Cov}[x(t_1), x(t_2)] = \int_{-\infty}^{\infty}\int_{-\infty}^{\infty} (\alpha_1 - m_x(t_1))(\alpha_2 - m_x(t_2)) \times f_{x(t_1)x(t_2)}(\alpha_1, \alpha_2)\, d\alpha_1\, d\alpha_2. \tag{5.8}$$

The notation in equation (5.8) is quite cumbersome. The simplified notation is

$$\gamma_x(t_1, t_2) = \operatorname{Cov}[x(t_1), x(t_2)], \tag{5.9}$$

and $\gamma_x(t_1, t_2)$ is more commonly known as the *autocovariance function* (ACVF). Just as with random variables, with random processes there is a relationship between the covariance and the mean product. It is

$$\gamma_x(t_1, t_2) = E[x(t_1)x(t_2)] - E[x(t_1)]E[x(t_2)] \tag{5.10}$$

and its proof is left as an exercise. The mean product is called the *autocorrelation function* (ACF) and is symbolized by

$$\varphi_x(t_1, t_2) = E[x(t_1)x(t_2)]. \tag{5.11}$$

When $t_2 = t_1$, $\gamma_x(t_1, t_1) = \sigma_x^2(t_1)$, the variance as a function of time.

When the process is second-order stationary, all the second-order moments are functions of the time difference $\tau = t_2 - t_1$. This can be proved through the general definition of a moment with the times expressed explicitly. Then,

$$E[g(x(t_1), x(t_2))]$$

$$= \int_{-\infty}^{\infty}\int_{-\infty}^{\infty} g(x(t_1), x(t_2)) f(x(t_1), x(t_2))\, dx(t_1)\, dx(t_2). \tag{5.12}$$

Let $t_2 = t_1 + \tau$ and reexamine equation (5.12). The integrand is now a function of the time parameters t_1 and τ. Because, by definition of stationarity, the moment cannot be a function of t_1, it is only a function of the time parameter τ.

5.3.2 Moments of Stationary Processes

Several moments similar to the covariance are extremely important in signal processing for stationary situations. A process that is stationary in autocovariance and autocorrelation has

$$\gamma_x(t_1, t_2) = \gamma_x(t_2 - t_1) = \gamma_x(\tau),$$

$$\sigma_x^2(t_1) = \sigma_x^2, \tag{5.13}$$

and

$$\varphi_x(t_1, t_2) = \varphi_x(t_2 - t_1) = \varphi_x(\tau).$$

The autocovariance function indicates any *linear dependence* between values of the random process $x(t)$ occurring at different times or, when $x(t)$ is stationary, between values separated by τ time units. This dependence is common and occurs in the sampled signal plotted in Figure 5.5. Successive points are correlated and this can be easily ascertained by studying the scatter plots in Figure 5.6. Notice that, for points separated by one time unit, the

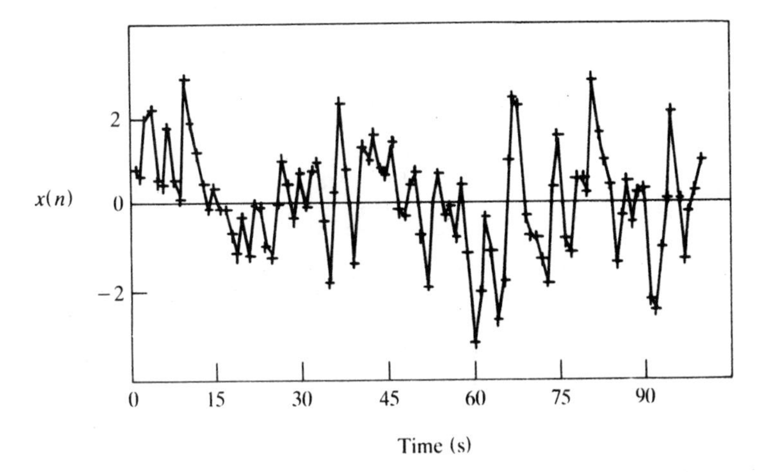

Figure 5.5 A time series containing 100 points. [Adapted from Fuller (1976), Figure 2.1.1, with permission.]

scatter plot can be described with a regression line indicating a positive correlation. For points separated by two time units there is no correlation because Figure 5.6(b) shows a circular distribution of points, which is indicative of zero correlation. The ACVF can show this dependence relationship succinctly. Just as in probability theory, this function needs to be normalized to the range of the correlation coefficient as shown in the previous chapter. The *normalized autocovariance function* (NACF) is defined as

$$\xi_x(\tau) = \frac{\varphi_x(\tau) - m_x^2}{\sigma_x^2} = \frac{\gamma_x(\tau)}{\sigma_x^2}. \tag{5.14}$$

Figure 5.7 shows an example of a stationary random process and its NACF. The signal is an electromyogram (EMG) measured from the biceps muscle while the elbow is not moving and the muscle's force is constant. Notice how the signal dependence is a function of the time separation, or time lag, τ. There seems to be dependence between EMG values that are separated by less than 4 ms.

The three types of correlation function are interrelated by equation (5.14) and have some important properties. These will be described for the wide-sense stationary situation because we will be dealing with signals having at least this condition of stationarity.

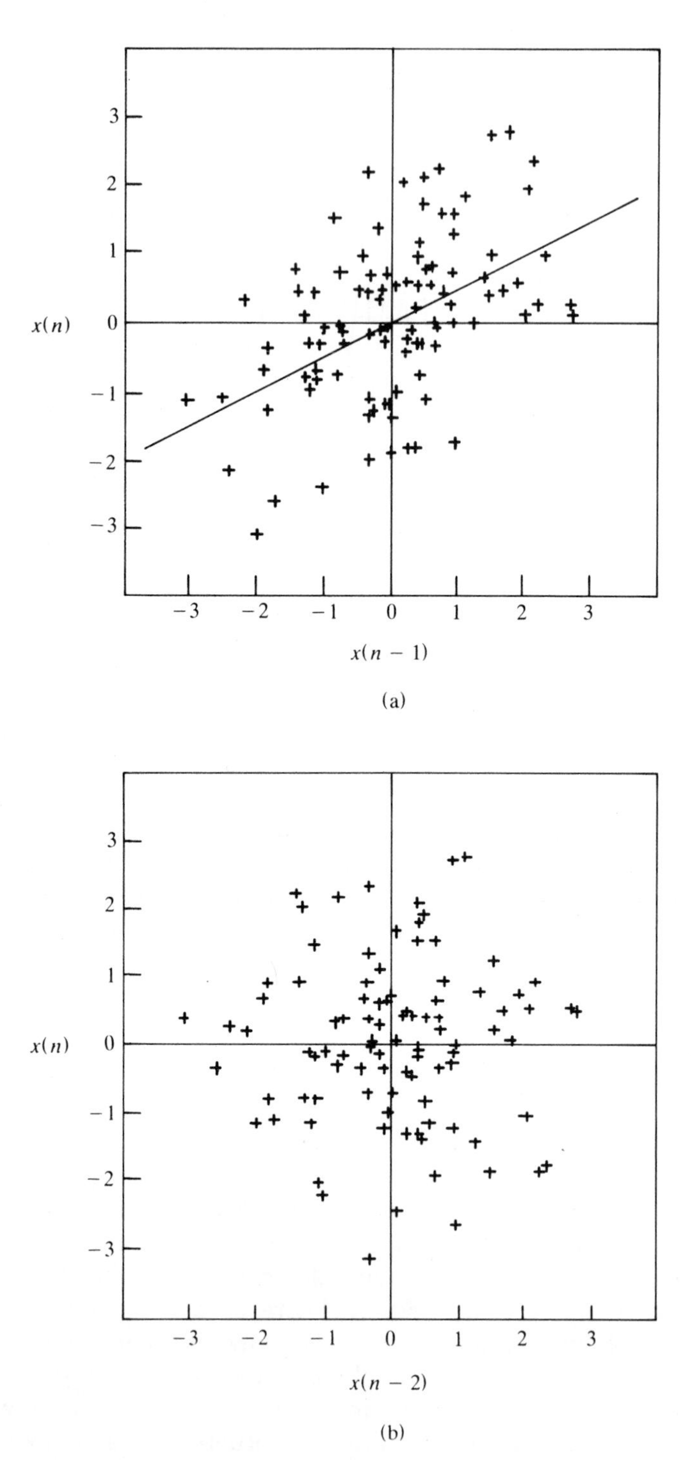

Figure 5.6 Scatter plots for the time series $x(n)$ in Figure 5.5: (a) $x(n)$ versus $x(n-1)$; (b) $x(n)$ versus $x(n-2)$. [Adapted from Fuller (1976), Figures 2.2.2 and 2.1.3, with permission.]

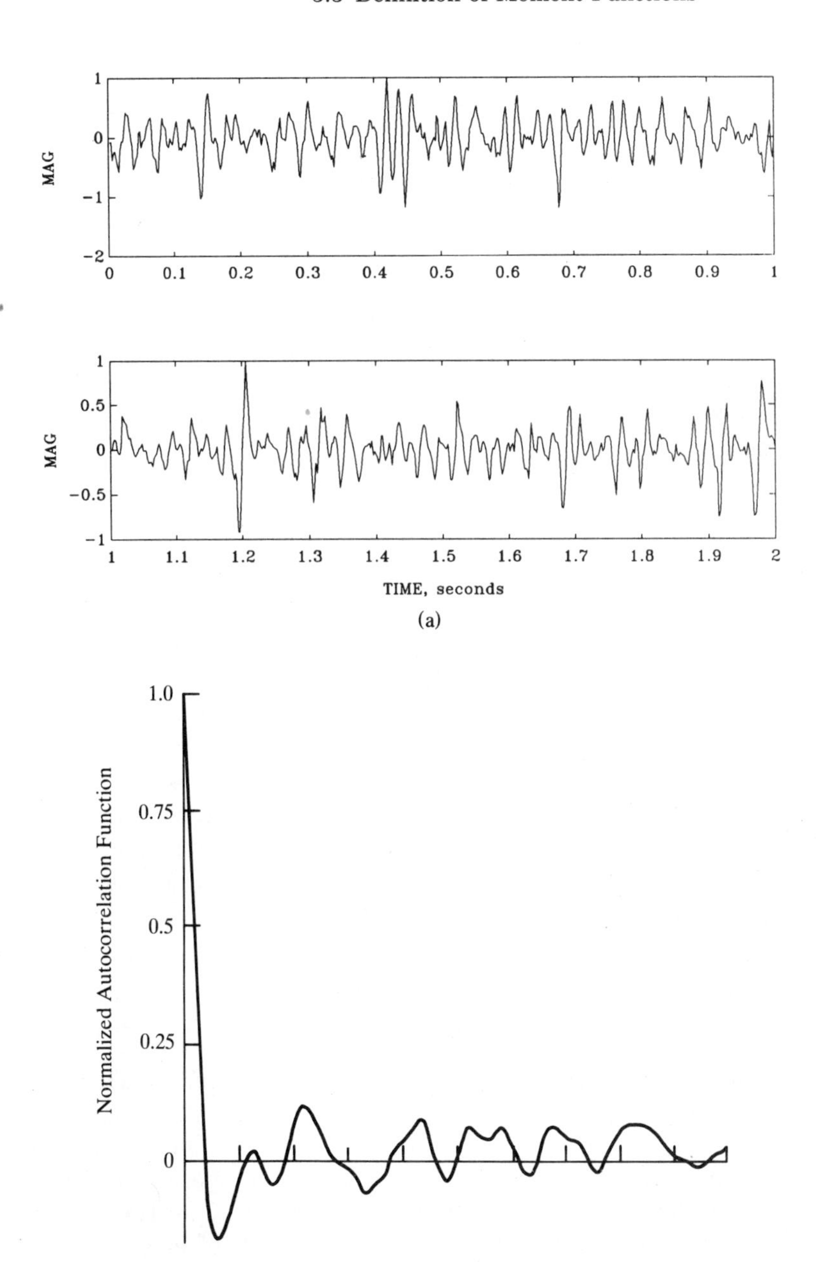

Figure 5.7 (a) A surface EMG during an isometric, constant-force contraction and (b) its estimated NACF. [Adapted from Kwatney, Thomas and Kwatney (1970, IEEE), Figure 5, with permission.]

These properties, in terms of the autocovariance, are as follows:

1. $\gamma_x(0) = \sigma_x^2$;
2. $\gamma_x(\tau) = \gamma_x(-\tau)$, even function;
3. If $\gamma_x(\tau) = \gamma_x(\tau + P)$ for all τ, then $x(t)$ is periodic with period P;
4. $|\gamma_x(\tau)| \leq \gamma_x(0)$.

A process with no linear dependence among values is a *white noise* (WN) process. It usually has a zero mean and $\gamma_x(\tau) = 0$ for $\tau \neq 0$. More will be stated about correlation functions and their uses in subsequent sections of this chapter.

5.4 Time Averages and Ergodicity

In signal analysis, just as in statistics, the goal is to extract information about a random process by estimating the moments and characteristics of its probability description. However, it is much more difficult to measure or often unfeasible to develop a large ensemble of sample functions. It is thus desirable to ascertain properties and parameters of a random process from a single sample function. This means using discrete time averages as estimators for ensemble averages. For instance, for estimating the mean of a signal, the time average is

$$\hat{m} = \lim_{N \to \infty} \frac{1}{2N + 1} \sum_{n=-N}^{N} x(n), \tag{5.15}$$

where $2N + 1$ is the number of sample points. Two questions immediately arise: first, is equation (5.15) convergent to some value? and second, if so, does $\hat{m} = E[x]$? One aspect becomes obvious; the process must be stationary in the mean because only one value of $\hat{m}$ is produced. So these two questions simplify to determining if $\hat{m} = E[x]$, which is the topic of *ergodicity*. Much of the proof of the ergodic theorems requires knowledge of convergence theorems that are beyond the goals of this book. However, some relevant results will be summarized. Good and readable proofs can be found in Papoulis (1984), Davenport (1970), and Gray and Davisson (1986).

A random process is said to satisfy an ergodic theorem if the time average converges to some value. This value is not necessarily the ensemble average. However, for our purposes, this discussion will focus on convergence of time averages to ensemble averages in processes that are at least wide-sense stationary. Essentially, we will consider the bias and consistency of time averages. For the time domain, the sample mean is

$$\hat{m}_N = \frac{1}{N} \sum_{n=0}^{N-1} x(n), \tag{5.16}$$

where $x(n)$ is the sampled sequence of the continuous process $x(t)$. The mean value of the sample mean is

$$\begin{aligned} E[\hat{m}_N] &= E\left[\frac{1}{N} \sum_{n=0}^{N-1} x(n)\right] \\ &= \frac{1}{N} \sum_{n=0}^{N-1} E[x(n)] \\ &= \frac{1}{N} \sum_{n=0}^{N-1} m = m \end{aligned} \tag{5.17}$$

and the estimate is unbiased. For convergence it must be shown that equation (5.16) is a consistent estimator. Its variance is

$$\begin{aligned} \sigma_{\hat{m}}^2 &= E\left[(\hat{m}_N - m)^2\right] \\ &= E\left[\left(\left(\frac{1}{N} \sum_{n=0}^{N-1} x(n)\right) - m\right)^2\right] \\ &= E\left[\left(\frac{1}{N} \sum_{n=0}^{N-1} (x(n) - m)\right)^2\right]. \end{aligned} \tag{5.18}$$

Using dummy variables for the summing indices which indicate the time sample, equation (5.18) becomes

$$\sigma_{\hat{m}}^2 = \frac{1}{N^2} E\left[\sum_{i=0}^{N-1} (x(i) - m) \sum_{j=0}^{N-1} (x(j) - m)\right]$$

$$= \frac{1}{N^2} E\left[\sum_{i=0}^{N-1} \sum_{j=0}^{N-1} (x(i) - m)(x(j) - m)\right]$$

$$= \frac{1}{N^2} \left(\sum_{i=0}^{N-1} \sum_{j=0}^{N-1} E[(x(i) - m)(x(j) - m)]\right)$$

$$= \frac{1}{N^2} \sum_{i=0}^{N-1} \sum_{j=0}^{N-1} \gamma_x(iT, jT). \tag{5.19}$$

The function $\gamma_x(iT, jT)$ is the discrete-time representation of the autocovariance function, equation (5.10). Equation (5.19) can be simplified into a useful form using two properties of the ACVF of a stationary process. First, the time difference is $kT = jT - iT$ and k represents the *lag interval*, the number of sampling intervals within the time difference; thus, $\gamma_x(iT, jT)$ becomes $\gamma_x(k)$ because the sampling interval is understood. Second, $\gamma_x(k) = \gamma_x(-k)$. The variance of the time series estimator for the mean becomes

$$\sigma_{\hat{m}}^2 = \frac{1}{N^2} \sum_{i=0}^{N-1} \sum_{j=0}^{N-1} \gamma_x(i - j)$$

$$= \frac{\sigma_x^2}{N} + \frac{2}{N} \sum_{k=1}^{N-1} \left(1 - \frac{k}{N}\right) \gamma_x(k). \tag{5.20}$$

For the mean value, the convergence in the limit as $N \to \infty$ depends on the ACVF. The first term on the right-hand side of equation (5.20) approaches zero if the variance of the process is finite. The second term approaches zero if $\gamma_x(k) = 0$ for $k > M$ and M is finite. In other words, the values of the random process that are far apart in time must be uncorrelated. Both of these conditions are satisfied in almost all engineering applications. Formalizing this situation, it is defined that a weakly stationary process is *ergodic in the mean*, that is,

$$\lim_{N\to\infty} \hat{m}_N = = E[x(t)] \tag{5.21}$$

if and only if (iff)

$$\gamma_x(0) < \infty \quad \text{and} \quad \lim_{k\to\infty} \gamma_x(k) \to 0. \tag{5.22}$$

Note that because the estimator is unbiased, the variance of the estimate is also the mean squared error. Thus the time average in equation (5.16) is also a consistent estimator and can be used to estimate the mean of a random process under the conditions just started. All of the concepts discussed in Section 4.4 for establishing confidence intervals and so forth can be applied.

Example 5.1

A simple example of ergodicity can be demonstrated using the ensemble of constant processes with random amplitudes such as sample function $x(t, \zeta_3)$ in Figure 5.2. Assume that the ensemble consists of six sample functions with constant amplitudes. The amplitudes are integers between 1 and 6 inclusive and are chosen by the toss of an honest die. The process is strictly stationary because nothing changes with time and the ensemble moments are

$$m_x(t) = 3.5, \qquad \sigma_x^2(t) = 2.92, \qquad \gamma_x(\tau) = 2.92.$$

However, a time-average estimate of the mean using any sample function will yield integers and does not converge to $m_x(t)$, regardless of the number of samples used. This is because the second condition of equation (5.22) is not satisfied.

It is also necessary to estimate other functions or moments of a random process from a sample function, besides the mean value. There are also ergodic theorems for them. Again, fortunately for most applications, the conditions naturally exist. In this section, only one other function will be mentioned. The probability distribution function of the amplitudes of a random process can be estimated from the histogram of amplitude values of a sample function as in Section 4.6 iff

$$\text{the process is stationary,} \tag{5.23a}$$

$$F_{x(t)x(t+\tau)}(\alpha_1, \alpha_2) = F_{x(t)}(\alpha_1) F_{x(t+\tau)}(\alpha_2) \quad \text{as } \tau \to \infty. \tag{5.23b}$$

Condition (5.23b) means that the process values must be independent for large time separations [Papoulis (1984)].

5.5 Estimating Correlation Functions

5.5.1 Estimator Definition

The three types of correlation function are used for a wide range of applications, including investigating the structure of a signal, estimating delay times of reflected signals for ranging, and system modeling. Before discussing the estimation of these functions, consider the use of the normalized autocovariance function for studying the behavior of wind speeds at a site used to generate electric power through harnessing the wind's energy. Figures 5.8(a) and 5.8(b) show the hourly wind speeds and the estimated NACFs. The wind speeds do vary much over several hours and the NACF shows that the correlation coefficient is at least 0.6 between speeds occurring at time intervals of 10 hours or less. This is a vastly different structure from the average wind speeds considered on a daily basis. The daily average wind speed and its estimated NACF are plotted in Figures 5.8(c) and 5.8(d). The signal has many more fluctuations and the NACF shows that only wind speeds of consecutive days have an appreciable correlation.

In order to obtain accurate estimates of the correlation functions so that good interpretations can be made from them, the statistical properties of their estimators must be understood. Because the properties of the three types of correlation function are similar, only the properties of the sample autocovariance function will be studied in detail. Use of these estimators will be emphasized in subsequent sections. Three time-domain correlation functions

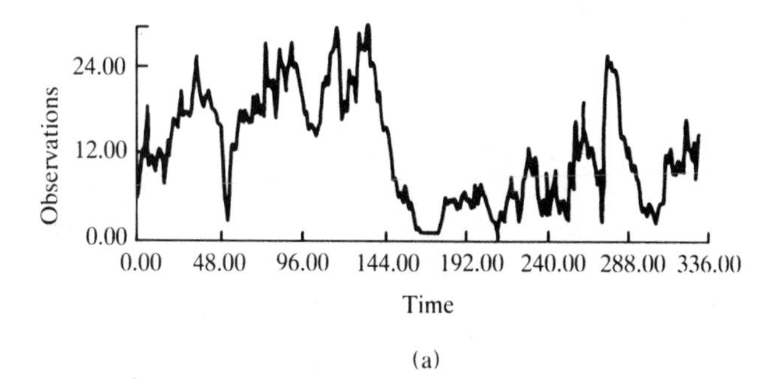

(a)

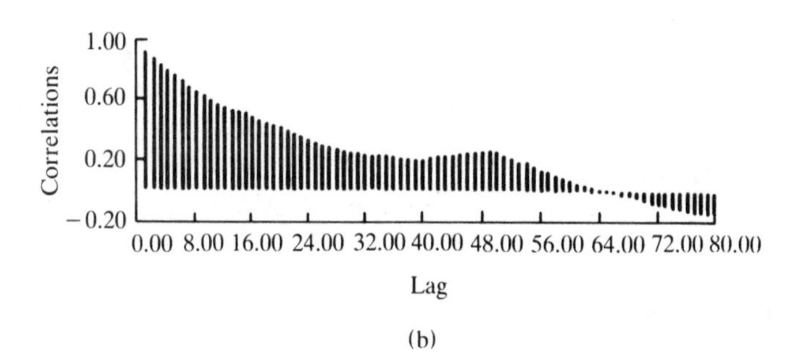

(b)

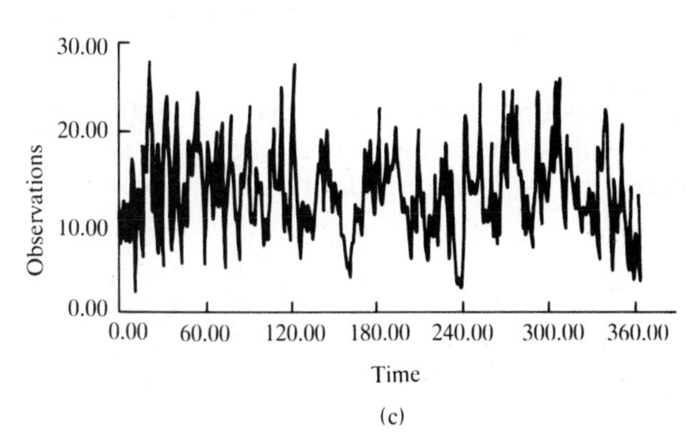

(c)

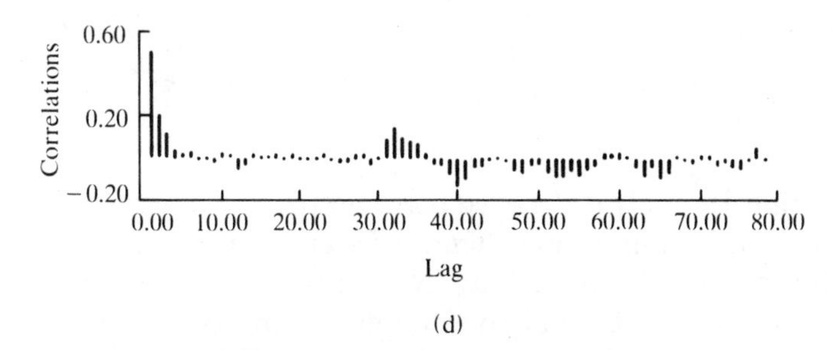

(d)

Figure 5.8 Wind speeds (in knots) at Belmullet, Ireland, during 1970: (a) hourly wind speeds during December; (b) the correlation function of (a); (c) daily average wind speeds during 1970; (d) the autocorrelation function of (c). [From Raferty, Haslett, and McColl (1982), Figures 4–7, with permission.]

are defined that are analogous to the ensemble correlation functions. These are the *time autocorrelation function* $R_x(k)$, the *time autocovariance function* $C_x(k)$, and the *time normalized autocovariance function* $\rho_x(k)$. The subscripts are not used when there is no confusion about the random variable being considered. These functions are defined mathematically as

$$R(k) = \lim_{N \to \infty} \frac{1}{2N+1} \sum_{n=-N}^{N} x(n)x(n+k), \tag{5.24}$$

$$C(k) = R(k) - m^2, \tag{5.25}$$

and

$$\rho(k) = \frac{C(k)}{C(0)}. \tag{5.26}$$

For the present assume that the mean is zero. Adjustments for a nonzero mean will be mentioned later. Two time-average estimators are commonly used and are finite sum versions of equation (5.24). These are

$$\hat{C}(k) = \frac{1}{N} \sum_{n=0}^{N-k-1} x(n)x(n+k) \tag{5.27}$$

and

$$\hat{C}'(k) = \frac{1}{N-k} \sum_{n=0}^{N-k-1} x(n)x(n+k). \tag{5.28}$$

Notice that the only difference between these two estimators is the divisor of the summation. This difference affects their sampling distribution greatly for a finite N. Naturally, it is desired that the estimators be ergodic and then unbiased and consistent. The bias properties are relatively easy to evaluate. Ergodicity and consistency are more difficult to evaluate but are derived within the same procedure.

5.5.2 Estimator Bias

The mean estimator in equation (5.27) when $k \geq 0$ is

$$E[\hat{C}(k)] = \frac{1}{N} E\left[\sum_{n=0}^{N-k-1} x(n)x(n+k)\right]$$

$$= \frac{1}{N} \sum_{n=0}^{N-k-1} E[x(n)x(n+k)]$$

$$= \frac{1}{N} \sum_{n=0}^{N-k-1} \gamma(k) = \left(1 - \frac{k}{N}\right)\gamma(k). \qquad (5.29)$$

The estimator is biased but for an infinite sample is unbiased, that is, $\hat{C}(k)$ is an *asymptotically unbiased* estimator of the autocovariance function. For the estimator of equation (5.28) it can be shown that it is an unbiased estimator, or that

$$E[\hat{C}'(k)] = \gamma(k). \qquad (5.30)$$

5.5.3 Consistency and Ergodicity

The variance of these time-average estimators is examined in order to determine their ergodicity and convergence for finite time samples. The solution is quite complex. The biased estimator is examined in detail and elaborations of these results are made to study the sampling properties of the unbiased estimator. By definition, the variance of the biased covariance estimator for $k \geq 0$ is

$$\mathrm{Var}[\hat{C}(k)] = E[\hat{C}^2(k)] - E^2[\hat{C}(k)]$$

$$= E\left[\frac{1}{N^2} \sum_{i=0}^{N-k-1} \sum_{j=0}^{N-k-1} x(i)x(i+k)x(j)x(j+k)\right]$$

$$- \left(1 - \frac{k}{N}\right)^2 \gamma^2(k)$$

$$= \frac{1}{N^2} \sum_{i=0}^{N-k-1} \sum_{j=0}^{N-k-1} E[x(i)x(i+k)x(j)x(j+k)]$$

$$- \left(1 - \frac{k}{N}\right)^2 \gamma^2(k). \qquad (5.31)$$

The crucial term in the evaluation is the fourth-order moment within the summation. For processes that are fourth-order stationary and Gaussian, this moment can be expressed as a sum of autocovariance functions and is

$$E[x(i)x(i+k)x(j)x(j+k)]$$
$$= \gamma^2(k) + \gamma^2(i-j) + \gamma(i-j+k)\gamma(i-j-k). \quad (5.32)$$

For other, non-Gaussian processes of practical interest, equation (5.32) is a good approximation [Bendat and Piersol (1986)]. The variance expression is simplified by inserting equation (5.32) into equation (5.31) and letting $r = i - j$. The double summation reduces to a single summation and the variance expression becomes

$$\mathrm{Var}[\hat{C}(k)]$$
$$= \frac{1}{N} \sum_{r=-N+k+1}^{N-k-1} \left(1 - \frac{|r| + k}{N}\right)\left(\gamma^2(r) + \gamma(r+k)\gamma(r-k)\right).$$
$$(5.33)$$

This solution is derived in Anderson (1971), Priestley (1981), and in Appendix 5.3. An alternate expression for equation (5.33) is

$$\mathrm{Var}[\hat{C}(k)] \approx \frac{1}{N} \sum_{r=-\infty}^{\infty} \left(\gamma^2(r) + \gamma(r+k)\gamma(r-k)\right), \quad (5.34)$$

for all values of k and large values of N. It can easily be seen that

$$\lim_{N\to\infty} \mathrm{Var}[\hat{C}(k)] \to 0$$

iff

$$\lim_{k\to\infty} \gamma(k) \to 0 \quad \text{and} \quad |\gamma(0)| \le M, \quad \text{where } M \text{ is finite.}$$

Thus the random process is ergodic in autocovariance for $\hat{C}(k)$, and $\hat{C}(k)$ is also a consistent estimator. If the unbiased estimator is used, the term $(N - |k|)$ is substituted for N in the variance expression in equation (5.34). Thus the process is ergodic and consistent for this estimator as well and both are valid estimators for the ensemble autocovariance of a zero-mean, fourth-order stationary process.

5.5.4 Sampling Properties

We can now proceed to examine the properties of the finite time estimators. For the remainder of this text we will study only estimators of functions that satisfy ergodic properties. Thus, the symbols that have been used for time correlation functions will be used because they are commonly used in the engineering literature. The estimator $\hat{C}'(k)$ is unbiased and seems to be the better estimator. The other estimator, $\hat{C}(k)$, has the bias term

$$b(k) = \hat{C}(k) - C(k) = -\frac{|k|}{N}C(k). \tag{5.35}$$

The magnitude of $b(k)$ is affected by the ratio $|k|/N$ and the value of $C(k)$. To keep the bias small, care must be taken to keep the ratio small when values of $C(k)$ are appreciable. Fortunately, for the processes concerned $C(k) \to 0$ as $k \to \infty$ and the bias will be small for large lag. So the number of samples must be large for small lags.

For a finite N, the variance for $\hat{C}(k)$ is equation (5.34) and for $\hat{C}'(k)$ is

$$\operatorname{Var}\left[\hat{C}'(k)\right] \approx \frac{1}{N - |k|} \sum_{r=-\infty}^{\infty} \left(\gamma^2(r) + \gamma(r+k)\gamma(r-k)\right). \tag{5.36}$$

Examination of this expression shows that the variance increases greatly as $|k|$ approaches N. In fact, it is greater than the variance of the biased estimator by a factor of $N/(N - |k|)$. This is the reason that so much is given to the biased estimator. It is the *estimator of choice* and is implemented in most software algorithms. In order to compromise with the bias produced for large lag values, the typical *rule-of-thumb* is to limit the maximum lag such that $|k| \leq 0.1N$. If the mean value must also be estimated so that the estimator is

$$\hat{C}(k) = \frac{1}{N} \sum_{n=0}^{N-k-1} (x(n) - \hat{m}_N)(x(n+k) - \hat{m}_N), \tag{5.37}$$

the bias term is changed approximately by the amount $-\sigma^2(1 - |k|/N)$.

The NACF is estimated by

$$\hat{\rho}(k) = \frac{\hat{C}(k)}{\hat{C}(0)}. \tag{5.38}$$

The biased estimator must be used here to insure that

$$|\hat{\rho}(k)| \leq 1 \quad \text{for all } k. \tag{5.39}$$

Equation (5.38) is a biased estimator with the same bias form as that for the autocovariance, that is,

$$E[\hat{\rho}(k)] = \left(1 - \frac{|k|}{N}\right)\rho(k). \tag{5.40}$$

The variance expression is much more complicated [Priestley (1981)]. It is

$$\text{Var}[\hat{\rho}(k)] \approx \frac{1}{N}\sum_{r=-\infty}^{\infty}\left(\rho^2(r) + \rho(r+k)\rho(r-k) + 2\rho^2(k)\rho^2(r) - 4\rho(k)\rho(r)\rho(r-k)\right). \tag{5.41}$$

5.5.5 Asymptotic Distributions

Fortunately, in time series analysis we are almost always concerned with sample functions that have a large number of samples. Therefore, the central limit theorem becomes appropriate because all of the estimators require summing many terms. The sampling distributions for the mean, autocovariance, and normalized autocovariance functions become Gaussian asymptotically with large N and for certain boundary conditions on the random process [Priestley, (1981)]. For our purposes these conditions are satisfied if the process is weakly stationary and the estimator is ergodic, as discussed in the preceding sections. The major difference between these asymptotic distributions and the exact statistical sampling distributions in Chapter 4 arise from the size of the sample. All of the information is now available to perform hypothesis testing on desired values of the correlation functions. In practice, $\hat{C}(k)$ is used and it is assumed that the sample function is large enough that the extra bias term is negligible.

Another factor arises with correlation functions. These are functions of lag time and it is desired to test all of the values of the estimate simultaneously. This is a complex multivariate problem that is studied in more advanced courses. The multivariate attribute arises because $\hat{\rho}(k)$ and $\hat{\rho}(k+1)$ are themselves correlated. This attribute disappears for one very useful process, the white noise process. Thus, in the case of white noise, we can test whether a sample function has a structure by knowing only the first-order sampling distribution. From equations (5.40) and (5.41) the mean and variance of the sampling distribution of the correlation function of a white noise process are

$$E[\hat{\rho}(k)] = 0, \qquad \mathrm{Var}[\rho(k)] = \frac{1}{N} \quad \text{for } k \neq 0. \tag{5.42}$$

Example 5.2

In an industrial process involving distillation, it is desired to know if there is any correlation between batches. Figure 5.9(a) shows the normalized yields over time. Notice that high values follow low values. This implies that there may be a negative correlation between successive batches. The correlation function is estimated using equation (5.38) and is plotted in Figure 5.9(b). The magnitude of $\hat{\rho}(1)$ is negative and the magnitudes of $\hat{\rho}(k)$ alternate in sign, which is consistent with our observation of the trends. To test for no structure in the process, the white-noise test is applied. The sampling distribution for $\rho(k)$ of the null hypothesis is Gaussian with the mean and variance given in equation (5.42). Because there are 70 series points in the time series, the variance of $\hat{\rho}(k)$ is 0.0143. The 95% confidence limits for any correlation are

$$|\hat{\rho}(k) - \rho(k)| \leq 1.96/\sqrt{N} = 0.234.$$

For zero correlation this becomes

$$-0.234 \leq \hat{\rho}(k) \leq 0.234.$$

These limits are indicated in Figure 5.9(b) by the horizontal dashed lines. The correlations at lags 1 and 2 are nonzero although low in magnitude. This indicates that the initial impressions are correct; successive yields are weakly negatively correlated and the yields of every second batch are weakly positively correlated. To obtain the confidence limits for the actual value of correlation, the same

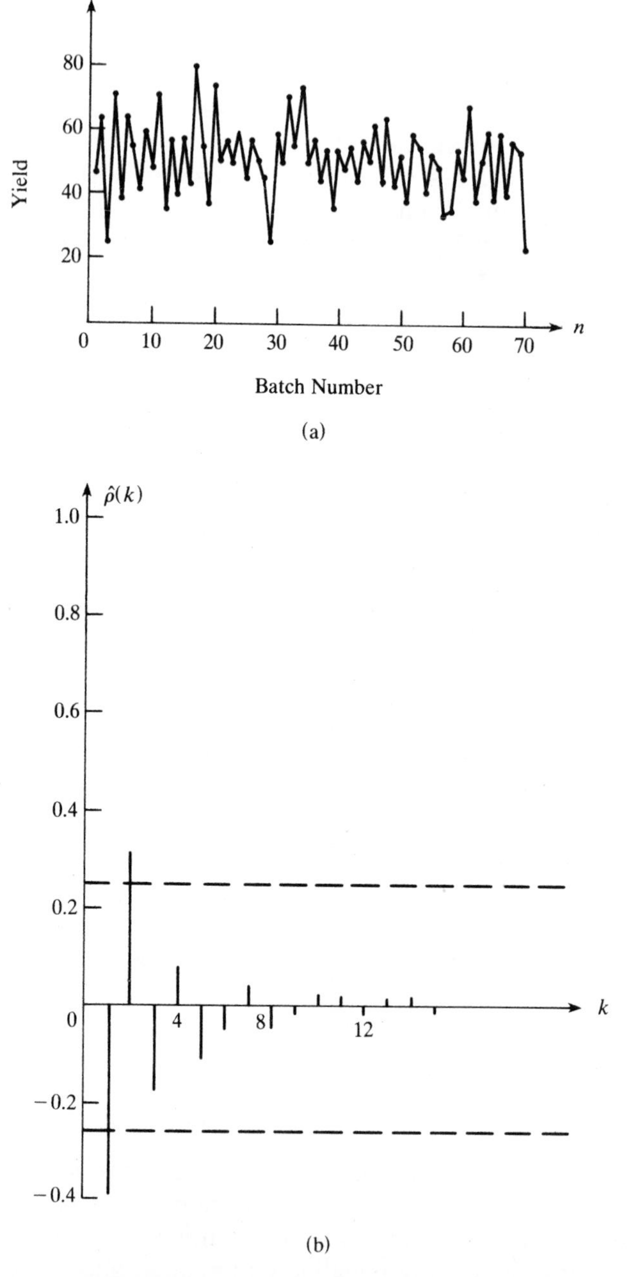

Figure 5.9 (a) The yield of a distillation process over time and (b) its sample NACF. The dashed lines indicate the zero-correlation confidence limits. [Adapted from Jenkins and Watts (1968), Figures 5.2 and 5.6, with permission.]

boundary relationship prevails and

$$\hat{\rho}(k) - 0.234 \le \rho(k) \le \hat{\rho}(k) + 0.234.$$

For the second lag, $\hat{\rho}(2) = 0.3$, and

$$0.069 \le \rho(2) \le 0.534.$$

5.6 Random Numbers and Random Process Simulation

Random processes with known characteristics are often needed to test signal processing procedures or to simulate signals with known properties. This section will introduce methods to generate random processes with different first-order probability characteristics and to produce a correlational structure.

5.6.1 Random Number Generation

Random number generators can be used to generate random processes by simply assuming that successive numbers are also successive time samples. Almost all higher-level languages, such as FORTRAN and C, have random number function calls in the system library. The most used algorithm is the *linear congruential generator*. It has the recurrence relationship

$$I(n+1) = aI(n) + c\ [\text{modulo } m], \tag{5.43}$$

where $I(n)$, a, and c are integers, b is the computer's word length, and $m = 2^b$. Therefore, the integers range in value from 0 to $m - 1$ inclusive. The sequence is initiated with a seed number $I(0)$ and the recursion relationship is used to generate subsequent numbers. All system functions return a floating-point number, $y(n) = I(n)/m$; that is, $y(n)$ has a magnitude range $0 \le y(n) < 1$ and is uniformly distributed. Thus for these types of algorithms, $m_y = 0.5$ and $\sigma_y^2 = 0.0833$. The goal is to produce a series of random numbers that are independent of one another. Proper choice of the seed number, multiplier, and increment in equation (5.43) insure this. For a good sequence, choose $I(0)$ to be an odd integer, $c = 0$, and $a = 8 \cdot \text{INT} \pm 3$, where INT is some integer.

The generator will produce 2^{b-2} numbers before repetition begins. This type of number generator is called *pseudorandom* because of this repetition and because the same sequence of numbers will be produced every time the same seed number is used. The time series produced are usually very good if one is only concerned with several sample functions. (It is usually advisable to check the uncorrelatedness of any generated sequence when using a generator for the first time.) If it is necessary to generate a large ensemble of functions, an additional shuffling manipulation must be employed to insure complete representation of the sample space. A subroutine for producing this shuffling can be found in Press et al. (1986). Consult Press et al. (1986) and Schwartz and Shaw (1975) for more details on random number generation.

Example 5.3

Figure 5.10(a) shows a 100-point random sequence generated using the congruence algorithm with a multiplicative factor a of 317 and an offset c of 0. The sample moments are reasonable and are $\hat{m} = 0.453$ and $s^2 = 0.0928$. The estimate of the NACF is plotted in Figure 5.10(b) and is indicative of a white noise process because $|\hat{\rho}(k)| \leq 0.2$.

Example 5.4

A second uniform uncorrelated random sequence is generated with the same algorithm used in the previous example but is initiated with a different seed number. The random sequence and estimate of the NACF are shown in Figure 5.11. The sample moments and are $\hat{m} = 0.5197$ and $s^2 = 0.08127$. Compare these results with those of the previous example. Notice that they differ in specific numbers, as expected, but that the statistical properties are the same. Something else occurs that at first may seem contradictory. Perform the white-noise test on the sample NACF. The magnitude of $\hat{\rho}(8)$ exceeds the threshold magnitude of 0.2. Should

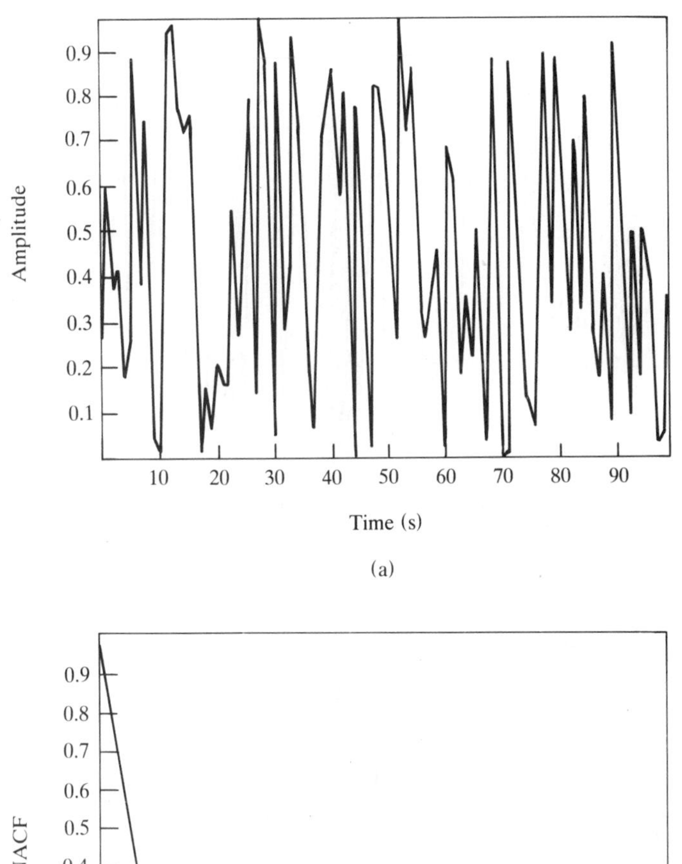

Figure 5.10 (a) A sample function of a uniformly distributed white noise process and (b) the estimate of its NACF.

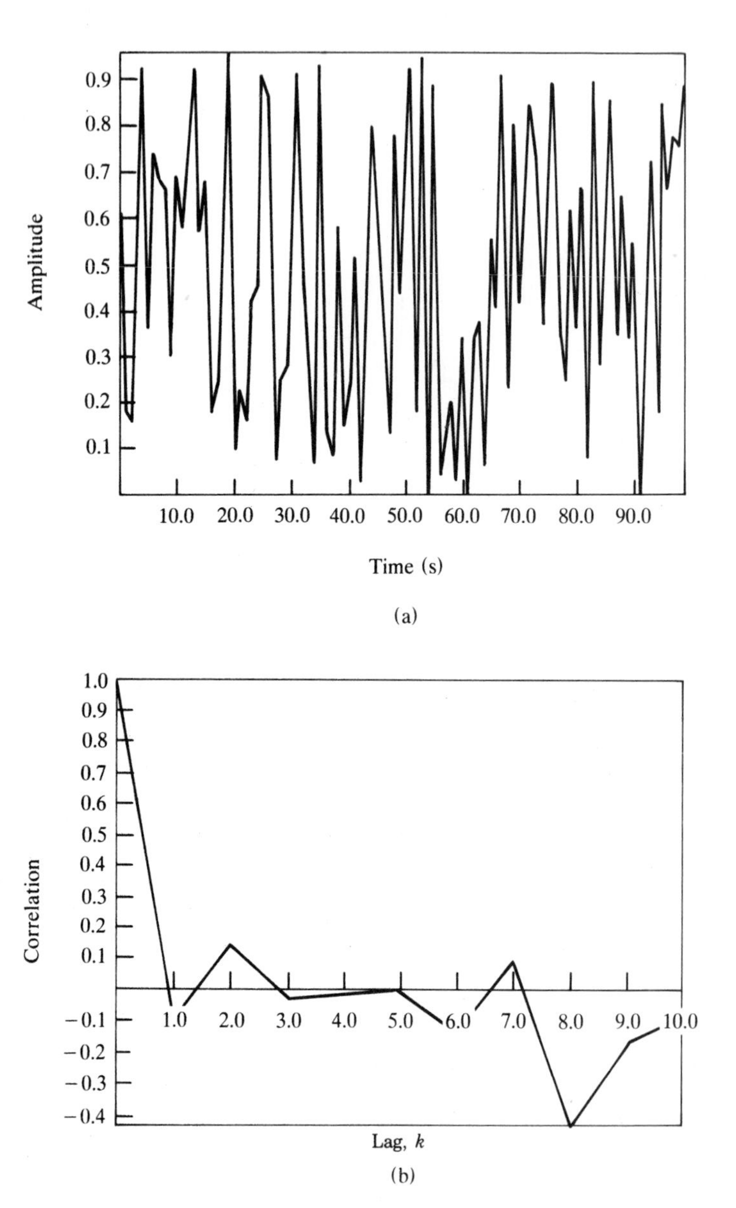

Figure 5.11 (a) A sample function of a uniformly distributed white noise process and (b) the estimate of its NACF.

it then be concluded that this second sequence is indeed correlated? No, it should not be. Remember that a significance level of 5% means that the testing statistic will not adhere to the test inequality for 5% of the tests when the hypothesis is true. Thus, for a white noise process, it should be expected that 1 out of 20 values of $\hat{\rho}(k)$ will test as nonzero. The usual indicator of this is when $\hat{\rho}(1)$ is close to zero and another value of $\hat{\rho}(k)$, $k \geq 1$, is not. As has been demonstrated in the previous examples, when an actual signal is not white noise, the NACFs for low lags tend to have the greater values.

5.6.2 Change of Mean and Variance

To be more useful, one would like to control the mean and variance of the random process. For this purpose a linear transformation is very useful. Let $x(n)$ represent the signal produced by the random number generator and $y(n)$ represent the signal resulting from a transformation, that is,

$$y(n) = a + bx(n). \tag{5.44}$$

The mean and variance of $y(n)$ are easily derived as

$$E[y(n)] = E[a + bx(n)] = a + bE[x(n)] = a + bm_x, \tag{5.45}$$

$$\begin{aligned}
\sigma_y^2 &= E[y^2(n)] - E^2[y(n)] \\
&= E\left[(a + bx(n))^2\right] - (a + bm_x)^2 \\
&= E\left[a^2 + 2abx(n) + b^2x^2(n)\right] - a^2 - 2abm_x - b^2m_x^2 \\
&= b^2\sigma_x^2.
\end{aligned} \tag{5.46}$$

Notice that this transformation is independent of the density functions of the random processes $x(n)$ and $y(n)$.

5.6.3 Density Shaping

As was defined in a previous section, the term "white noise" describes a process with an uncorrelated time structure. Thus, a random number generator produces a white noise process with a uniform pdf. However, the amplitude distribution of a white noise process can be any pdf. Several techniques can be used to transform a uniform process into another process. The most simple approximation is to use the central limit theorem. Simulation tests have shown that summing 12 uniformly distributed numbers will produce a set of numbers with a Gaussian distribution. The mean and variance must be determined to characterize fully the process. For now, neglecting the time indicator and concentrating on the summing process,

$$y = \sum_{i=1}^{N} x_i,$$

$$E[y] = m_y = E\left[\sum_{i=1}^{N} x_i\right] \tag{5.47}$$

$$= \sum_{i=1}^{N} E[x_i] = Nm_x,$$

for a stationary process. Notice that the mean of a sum is the sum of the means for any process. The variance of y is

$$\mathrm{Var}[y] = E[y^2] - E^2[y]$$

$$= E\left[\left(\sum_{i=1}^{N} x_i\right)^2\right] - \left(\sum_{i=1}^{N} E[x_i]\right)^2$$

$$= E\left[\sum_{i=1}^{N} \sum_{j=1}^{N} x_i x_j - E[x_i]E[x_j]\right]$$

$$= \sum_{i=1}^{N} \sum_{j=1}^{N} \mathrm{Cov}[x_i x_j]. \tag{5.48}$$

For a stationary, uncorrelated process this becomes

$$\mathrm{Var}[y] = \sum_{i=1}^{N} \sigma_x^2 = N\sigma_x^2. \tag{5.49}$$

This derivation is very similar to the one in Section 5.4 concerning ergodicity of the mean.

Example 5.5

A Gaussian white noise process with zero mean is approximated by summing six uniformly distributed points and using a linear transformation. The process in Example 5.3 with 600 points is used. For $N = 6$ a random process is produced with

$$m_y = 6 \cdot \tfrac{1}{2} = 3; \qquad \sigma_y^2 = 6 \cdot \tfrac{1}{12} = \tfrac{1}{2}.$$

The linear transformation is simply $z = y - 3$. A 100-point sample function of the process $z(n)$ is shown in Figure 5.12(a). Its sample mean and variance are 0.0687 and 0.47083, respectively. These statistics match very well to the desired process. The sample NACF is plotted in Figure 5.12(b). It is indicative of a white noise process.

In order to produce random processes with other pdfs, single-valued probability transformations are utilized. This topic is covered in detail in textbooks on introductory probability theory and statistics, such as Hoel (1971) and Brownlee (1960). It is assumed that the initial process $f(x)$ is independent and has the [0, 1] uniform pdf. The transformation is

$$\int_0^x f_x(\alpha)\, d\alpha = \int_{-\infty}^{y} f_y(\beta)\, d\beta, \qquad 0 \le x \le 1, \tag{5.50}$$

where x is the number in the uniform process and y is the number of the new process. The solution is

$$x = F_y(y) \quad \text{or} \quad y = F_y^{-1}(x), \tag{5.51}$$

where $F_y^{-1}(x)$ indicates the inverse solution of the desired probability distribution function.

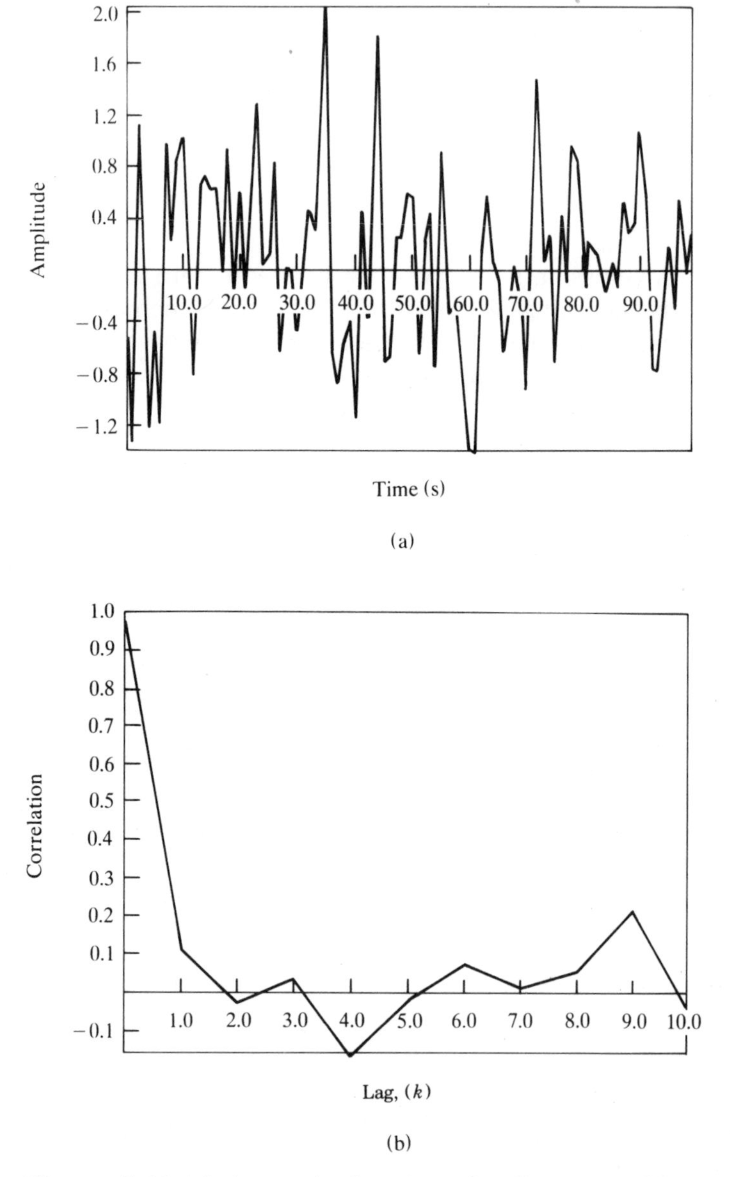

Figure 5.12 (a) A sample function of a Gaussian white noise process generated by summing points in the process in Figure 5.11 and (b) the estimate of its NACF.

Example 5.6

Find the transformation to create a random variable with a Rayleigh distribution from a uniformly distributed random variable.

$$x = \int_0^y \frac{\beta}{a^2} \exp\left(\frac{-\beta^2}{2a^2}\right) d\beta = 1 - \exp\left(\frac{-y^2}{2a^2}\right)$$

or

$$y = \left(2a^2 \ln\left(\frac{1}{1-x}\right)\right)^{0.5}.$$

Figure 5.13 shows the result of such a transformation, with $a = 2.0$, on the uniform process of Example 5.3. The sample mean and variance are 2.605 and 1.673, respectively. These sample moments

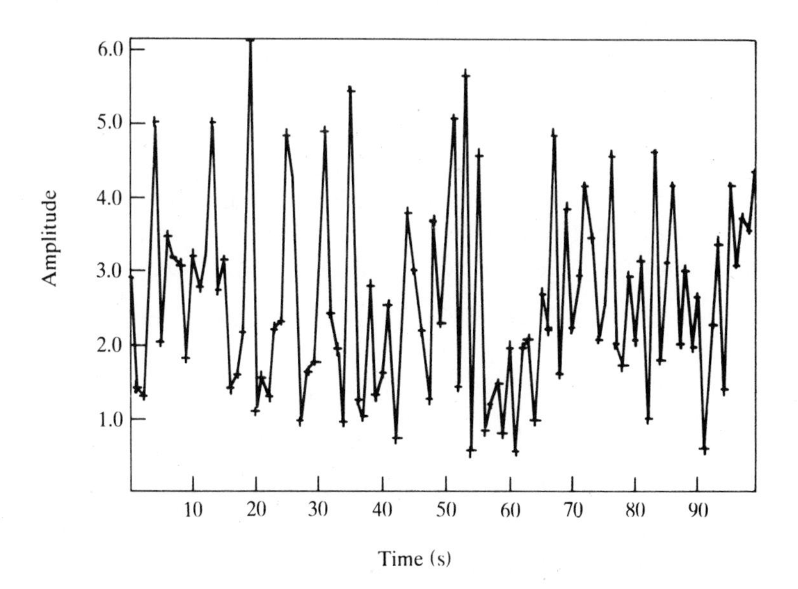

Figure 5.13 A sample function of a Rayleigh process generated by transforming the uniform process in Figure 5.11.

correspond well to what is expected theoretically, that is, $m_y = 2.5$ and $\sigma_y^2 = 1.72$.

There will be situations in which either the inverse transformation [equation (5.51)] cannot be solved or there is no mathematical form for the desired pdf. In these situations a random variable with the desired distribution can be created by using a numerical technique called the rejection method [Press et al. (1986)].

5.6.4 Correlation and Structure

5.6.4.1 General Moving Average

Many signals that are encountered are processes which are uncorrelated and have characteristics similar to the white noise examples. However, most signals measured have a correlation between successive points. Refer again to Figures 5.7–5.9. For purposes of simulation, it is necessary to take a white noise process with zero mean, $x(n)$, and induce a correlation within it. This can be done using algorithms involving moving averages and autoregression. More will be stated about these in subsequent chapters. Presently, the basic concepts will be used to create a structure in a random process. A *moving average* (MA) *process* $y(n)$ requires averaging successive points of $x(n)$ using a sliding window. This is different from the procedure for creating a Gaussian white noise process as in Example 5.5. In the latter procedure, successive points of $y(n)$ were not a function of common points of $x(n)$. Remember that, for a white noise process, $R_x(k) = \sigma_x^2\delta(k)$. A three-point process is created by the relationship

$$y(n) = x(n) + x(n-1) + x(n-2). \qquad (5.52)$$

Notice that this is simply a sliding, or moving, sum. For time instant 3,

$$y(3) = x(3) + x(2) + x(1). \qquad (5.53)$$

The most general form for a $(p+1)$-point moving average also has weighting coefficients and is

$$y(n) = a_0x(n) + a_1x(n-1) + \cdots + a_px(n-p). \quad (5.54)$$

The parameter p is the *order* of the moving average. Because this is a type of system, $x(n)$ and $y(n)$ are usually called the input and output processes, respectively. The mean and variance of this structure for weighting coefficients equalling one have already been derived in equations (5.47) and (5.49). For generality, the weighting coefficients need to be included. The mean and variance are

$$m_y = m_x \sum_{i=0}^{p} a_i, \tag{5.55}$$

$$\sigma_y^2 = \sigma_x^2 \sum_{i=0}^{p} a_i^2. \tag{5.56}$$

The derivation is left as an exercise for the reader. The output process has an autocovariance function that depends upon the order of the MA process and on the values of the coefficients.

5.6.4.2 First-Order MA

The autocorrelation function for a two-point MA process will be derived in detail, to illustrate the concepts necessary for understanding the procedure. Remember $m_y = m_x = 0$.

$$R_y(0) = \sigma_y^2 = \sigma_x^2 \sum_{i=0}^{1} a_i^2, \tag{5.57}$$

$$\begin{aligned} R_y(1) &= E[y(n)y(n-1)] \\ &= E[(a_0x(n) + a_1x(n-1))(a_0x(n-1) + a_1x(n-2))] \\ &= E[a_0x(n)a_0x(n-1) + a_1x(n-1)a_0x(n-1) \\ &\quad + a_0x(n)a_1x(n-2) + a_1x(n-1)a_1x(n-2)]. \end{aligned}$$

Because the mean of a sum is a sum of the means, the preceding expectation is a sum of autocorrelations, or

$$R_y(1) = a_0^2R_x(1) + a_1a_0R_x(0) + a_1a_0R_x(2) + a_1^2R_x(1). \tag{5.58}$$

Because $x(n)$ is a white noise process, $R_x(k) = 0$ for $k \neq 0$ and equation (5.58) simplifies to

$$R_y(1) = a_1 a_0 R_x(0) = a_1 a_0 \sigma_x^2. \tag{5.59}$$

Thus, the MA procedure has induced a correlation between successive points. For the second lag,

$$\begin{aligned} R_y(2) &= E[y(n)y(n-2)] \\ &= E[(a_0 x(n) + a_1 x(n-1))(a_0 x(n-2) + a_1 x(n-3))] \\ &= E[a_0 x(n) a_0 x(n-2) + a_1 x(n-1) a_0 x(n-2) \\ &\qquad + a_0 x(n) a_1 x(n-3) + a_1 x(n-1) a_1 x(n-3)] \end{aligned}$$

and

$$R_y(2) = a_0^2 R_x(2) + a_1 a_0 R_x(1) + a_1 a_0 R_x(3) + a_1^2 R_x(2). \tag{5.60}$$

Again, because $x(n)$ is white noise, $R_y(2) = 0$. All of the other autocorrelation function values have the same form as equation (5.60), thus,

$$R_y(k) = 0 \quad \text{for} \quad |k| \geq 2. \tag{5.61}$$

Naturally, $R_y(-1) = R_y(1)$. The NACF is $\rho_y(k) = R_y(k)/\sigma_y^2$ and for the two-point MA process is, in general,

$$\begin{aligned} \rho_y(0) &= 1, \\ \rho_y(\pm 1) &= \frac{a_0 a_1}{a_0^2 + a_1^2}, \\ \rho_y(k) &= 0 \quad \text{for } |k| \geq 2. \end{aligned} \tag{5.62}$$

Example 5.7

Form a two-point average process that is the mean of a white noise process; that is, $a_0 = a_1 = 0.5$. Thus,

$$y(n) = 0.5x(n) + 0.5x(n-1)$$

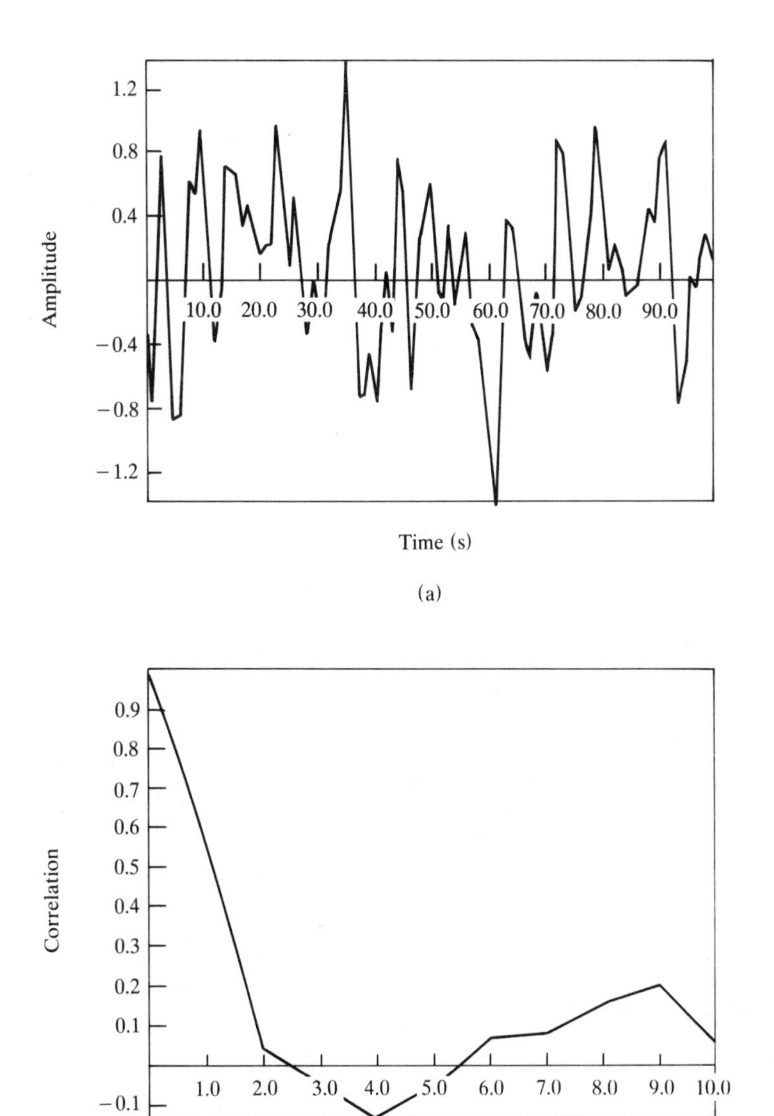

Figure 5.14 (a) A sample function of a correlated Gaussian process and (b) the estimate of its NACF.

and $\sigma_y^2 = 0.5\sigma_x^2$, $\rho_y(1) = 0.5$. The process in Example 5.5 was used as the input process and the output process $y(n)$ is shown in Figure 5.14(a). The sample variance is 0.263 and is close to what is expected. The estimate of the NACF is plotted in Figure 5.14(b). $\hat{\rho}_y(1) = 0.55$ and tests to be nonzero and is well within the 95% confidence level for the desired correlation.

Example 5.8

The effect of a finite sample on the estimation of the NACF can be demonstrated through simulating a signal and calculating the parameters of the sampling distribution. The correlated random process generated in the previous example is used. Naturally, from the equations, $\hat{\rho}(0) = 1$ always. From equations (5.40) and (5.41) the sampling mean and variance can be approximated. The sampling mean for lag $= \pm 1$ is

$$E[\hat{\rho}(\pm 1)] = \left(1 - \frac{1}{N}\right)\rho(\pm 1) = 0.5\left(1 - \frac{1}{100}\right) = 0.495.$$

For practical purposes this is almost unbiased. The variance expression for $k = 1$ is

$$\mathrm{Var}[\hat{\rho}(1)] \approx \frac{1}{N}\sum_{r=-\infty}^{\infty}(\rho^2(r) + \rho(r+1)\rho(r-1) + 2\rho^2(1)\rho^2(r) - 4\rho(1)\rho(r)\rho(r-1))$$

and easily reduces to

$$\mathrm{Var}[\hat{\rho}(1)] \approx \frac{1}{N}\sum_{r=-1}^{1}(\rho^2(r) + \rho(r+1)\rho(r-1) + 2\rho^2(1)\rho^2(r) - 4\rho(1)\rho(r)\rho(r-1)).$$

On a term-by-term basis this summation becomes

(term 1) $0.25 + 1 + 0.25 = 1.5,$

(term 2) $0 + 0.25 + 0 = 0.25,$

(term 3) $2 \cdot 0.25 \cdot (\text{term } 1) = 0.75,$

(term 4) $-4 \cdot 0.5 \cdot (0 + 0.5 + 0.5) = -2.0,$

and Var $[\hat{\rho}(1)] \approx \frac{1}{100}0.5 = 0.005$. The sampling mean and variance for $\hat{\rho}(-1)$ is the same. For $|k| \geq 2$,

$$E[\hat{\rho}(k)] = \left(1 - \frac{|k|}{N}\right)\rho(k) = 0$$

and the estimates are unbiased. The variance for all the lags greater than 1 is the same because only 1 term in the summation is nonzero and is

$$\text{Var}[\hat{\rho}(k)] \approx \frac{1}{N} \sum_{r=-\infty}^{\infty} \rho^2(r) = \tfrac{1}{100}1.5 = 0.015, \qquad |k| \geq 2.$$

Compared to a white noise process, the variance of the estimates of the correlation function of the first-order MA process has two differences:

1. At lags with no correlation the variances are larger;
2. At a lag of ± 1 the variance is smaller.

5.6.4.3 Second-Order MA

The general formula for the autocorrelation function of a pth-order MA process can be derived in a direct manner. Study again the equations in which the ACF is derived for the first-order process. Notice that only nonzero terms arise from those terms containing $R_x(0)$. Consider a second-order process,

$$y(n) = a_0 x(n) + a_1 x(n-1) + a_2 x(n-2). \qquad (5.63)$$

Its value of the ACF at lag 1 is derived from

$$\begin{aligned}R_y(1) &= E[y(n)y(n-1)]\\ &= E\big[(a_0x(n) + a_1x(n-1) + a_2x(n-2))\\ &\qquad \times(a_0x(n-1) + a_1x(n-2) + a_2x(n-3))\big].\end{aligned}$$

As can be seen from the previous equation, the terms producing nonzero expectations are

$$a_1x(n-1)\cdot a_0x(n-1) \quad \text{and} \quad a_2x(n-2)\cdot a_1x(n-2),$$

therefore,

$$R_y(1) = (a_1a_0 + a_2a_1)\sigma_x^2. \tag{5.64}$$

Notice that there are two terms and the indices in the subscripts of the products of the coefficients differ by a value of 1. For lag 2 the autocorrelation value is

$$R_y(2) = a_2a_0\sigma_x^2. \tag{5.65}$$

Notice that there is one term and the indices of the subscript in the coefficient differ by a value of 2.

5.6.4.4 Overview

The ACF for the general pth-order MA process is

$$R_y(k) = \begin{cases}\sigma_x^2\left(\sum_{i=0}^{p-k} a_ia_{i+k}\right), & |k| \le p,\\ 0, & |k| > p.\end{cases} \tag{5.66}$$

Its proof is left as an exercise. Does equation (5.66) produce equations (5.64) and (5.65)?

There are many types of structure in a time series. The material just presented is actually the simplest manner in which to create a correlation in a time series. More complex correlation functions can be created by using higher-order moving averages or another process, called *autoregression*. In this process an output value depends on its previous values. One form is

$$y(n) = a_1y(n-1) + b_0x(n). \tag{5.67}$$

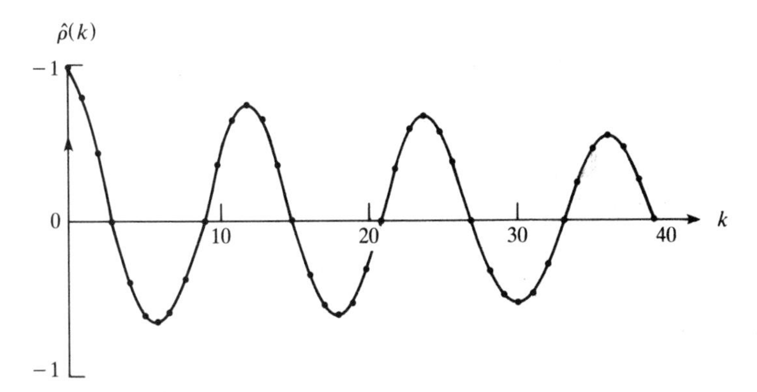

Figure 5.15 The NACF of monthly measurements of air temperature in a city. The measurements are shown in Figure 1.1. [Adapted from Chatfield (1975), Figure 2.4a, with permission.]

These processes will be studied in detail in a subsequent chapter. Naturally occurring processes usually have a more complex structure than the ones represented by first- and second-order MA processes. Review again several examples. The time series of hourly wind speeds in Figure 5.8 has a correlation function with nonzero values at many more lags than in the last example. A different structure is the distillation time series in Figure 5.9, which alternates in magnitude; it has an alternating NACF to reflect this behavior. $\rho(1)$ is negative, which indicates that successive points will have opposite sign. A signal with an oscillation or a periodicity will have a periodic correlation function. An example is the air temperature record plotted in Figure 1.1 and its NACF which is shown in Figure 5.15.

5.7 Assessing Stationarity of Signals

Because the techniques for investigating the properties of signals require them to be at least weakly stationary, there must be methods to assess signal stationarity. The simplest way to assess the stationarity of a sampled signal is to consider the physics and environment that produced it. Again, the speech waveform is an obvious situation. Simple observation of the waveform in Figure 5.3 shows that different words and utterances are composed of different signals called *formants*. If one needs to investigate the properties of a formant, the use of short time series is necessary. In distinction, many situations are similar to the one reflected in the EEG record in Figure 5.4. Here the signal is stationary for periods greater than 1 s. What would cause the signal's characteristics to change? Because the EEG is a measurement of brain activity, anything that could affect the activity could cause the EEG to change: It is well known that the state of alertness or drowsiness or the type of mental task causes changes. Atmospheric noise that affects electromagnetic communication is nonstationary and depends upon the time of day and cloud conditions. Another situation is the intensity of vibration of an airplane's wing during changes in airspeed.

What if the state of stationarity cannot be assessed by considering physical and environmental conditions? Then one can resort to statistical tests. Fortunately, in many real-world measurements, if the more complex properties (such as the ACF or pdf) of a signal change, so do some simple properties (such as the mean or variance). Thus, testing whether there is a change in these moments over time is an adequate method for assessing wide-sense stationarity. If the signal also has a Gaussian pdf, testing these moments is a test of complete stationarity. The most direct procedure is to divide a signal's record into multiple segments and to compare the means and variances of each segment among each other. Standard statistical tests are available for testing the equality of sample means and variances. Two tests are summarized in Appendix 5.4. Care must be taken to insure that the segments are long enough to reflect the signal properties. In particular, if a signal has oscillating components, several cycles of the oscillation must be included in each segment. Additional discussions of stationarity testing using these tests and nonparametric tests can be found in Bendat and Piersol (1986), Otnes and Enochson (1972), and Shanmugan and Breipohl (1988).

Again, consider the speech signal in Figure 5.3. Divide it into three time segments S_i of 600 ms such that

$$\begin{aligned} S_1 &\Rightarrow 0 \le t \le 600 \text{ ms}, \\ S_2 &\Rightarrow 600 \le t \le 1200 \text{ ms}, \\ S_3 &\Rightarrow 1200 \le t \le 1800 \text{ ms}. \end{aligned} \tag{5.68}$$

Qualitatively compare the mean and variance among the three

segments. The means of the segments are slightly positive and probably are equal. However, the variances are different. One can assess this by considering the range of amplitudes. The variance of S_2 is much less than those for S_1 and S_3 and most likely the variance of S_1 is less than that of S_3. Thus the signal is nonstationary.

Example 5.9

The stationarity of the distillation signal used in Example 5.2 and tabulated in Appendix 5.1 is being assessed using the Student's t and F tests to compare the means and variances, respectively. (Refer to Appendix 5.4.) The signal is divided into two segments, such that

$$S_1 \Rightarrow 1 \leq n \leq 35,$$

$$S_2 \Rightarrow 36 \leq n \leq 70,$$

where $N_1 = N_2 = 35$. The sample means and variances for each segment are $\hat{m}_1 = 53.17$, $\hat{\sigma}_1^2 = 176.78$, $m_2 = 49.11$, and $\hat{\sigma}_2^2 = 96.17$.

The T statistic for testing the equality of means has a t distribution with degrees of freedom $\nu = N_1 + N_2 - 2$,

$$T = \frac{\hat{m}_1 - \hat{m}_2}{\left(\left(\frac{1}{N_1} + \frac{1}{N_2}\right)\left(\frac{N_1\hat{\sigma}_1^2 + N_2\hat{\sigma}_2^2}{N_1 + N_2 - 2}\right)\right)^{1/2}}$$

$$= \frac{53.17 - 49.11}{\left(\frac{2}{35}\frac{35(176.78 + 96.17)}{68}\right)^{1/2}} = 1.43.$$

The significance region is $|T| \geq t_{68;\,0.025} = 2$. The signal tests as stationary in the mean.

The F statistic for testing the equality of variances uses the unbiased estimates of variances. The number of degrees of freedom

$\nu_1 = N_1 - 1$ and $\nu_2 = N_2 - 1 = 34$. The F statistic is

$$F = \frac{(N_1/\nu_1)\hat{\sigma}_1^2}{(N_2/\nu_2)\hat{\sigma}_2^2} = 1.84.$$

The 95% confidence internal for a two-sided test is

$$F_{34,\,34;\,0.975} \le F \le F_{34,\,34;\,0.025} \quad \text{or} \quad 0.508 \le F \le 1.97.$$

Because F falls within this interval, the variances are equal at the 95% confidence level and the signal is stationary in variance.

The tests show that the distillation signal is stationary in the mean and variance at a confidence level of 95%. One of the working assumptions is that any nonstationarity in the covariance functions would be reflected in the nonstationarity of the variance. Thus, it is concluded that the signal is wide-sense stationary.

References

Anderson, T. (1971). *The Statistical Analysis of Time Series*. John Wiley and Sons, New York.

Bendat, J. and A. Piersol (1986). *Random Data, Analysis and Measurement Procedures*. John Wiley and Sons, New York.

Bodenstein, G. and H. Praetorius (1977). Feature extraction from the electroencephalogram by adaptive segmentation. *Proc. IEEE* 65:642–652.

Brownlee, K. (1960). *Statistical Theory and Methodology in Science and Engineering*. John Wiley and Sons, New York.

Burton, P. (1977). Analysis of a dispersed and transient signal. In *Applications of Time Series Analysis*. Proceedings of conference held at Southampton University, September, 1977 (sponsored by Institute of Sound and Vibration Research).

Chatfield, C. (1975). *The Analysis of Time Series: Theory and Practice*. John Wiley and Sons, New York.

Childers, D. and J. Labar (1984). Electroglottography for laryngeal function assessment and speech analysis. *IEEE Trans. Biomed. Engrg.* 31:807–817.

Davenport, W. (1970). *Probability and Random Processes*. McGraw-Hill Book Co., New York.

Fuller, W. (1976). *Introduction to Statistical Time Series*. John Wiley and Sons, New York.

Gray, R. and L. Davisson (1986). *Random Processes, A Mathematical Approach for Engineers*. Prentice-Hall, Englewood Cliffs, NJ.

Hoel, P. (1971). *Introduction to Mathematical Statistics*. John Wiley and Sons, New York.

Jenkins, G. and D. Watts (1968). *Spectral Analysis and Its Applications*. Holden-Day, San Francisco.

Komo, J. (1987). *Random Signal Analysis in Engineering Systems*. Academic Press, Harcourt Brace Jovanovich, Publishers, Orlando, FL.

Kwatney, E., D. Thomas, and H. Kwatney (1970). An application of signal processing techniques to the study of myoelectric signals. *IEEE Trans. Biomed. Engrg.* 17:303–312.

Otnes, R. and L. Enochson (1972). *Digital Time Series Analysis*. John Wiley and Sons, New York.

Papoulis, A. (1984). *Probability, Random Variables and Stochastic Processes*. McGraw-Hill Book Co., New York.

Press, W., B. Flannery, S. Teukolsky, and W. Vetterling (1986). *Numerical Recipes—The Art of Scientific Computing*. Cambridge University Press, Cambridge, UK.

Priestley, M. (1981). *Spectral Analysis and Time Series*, Vol. 1: *Univariate Series*. Academic Press, London.

Raferty, A., J. Haslett, and E. McColl (1982). Wind power: A space-time approach. In *Time Series Analysis: Theory and Practice*, Vol. 2 (O. Anderson, ed.) 191–200. North-Holland Publishing Co., Amsterdam.

Schwartz, M. and L. Shaw (1975). *Signal Processing: Discrete Spectral Analysis, Detection, and Estimation*. McGraw-Hill Book Co., New York.

Shanmugan, K. and A. Breipohl (1988). *Random Signals: Detection, Estimation and Data Analysis*. John Wiley and Sons, New York.

Exercises

5.1 Prove $\gamma(t_1, t_2) = E[x(t_1)x(t_2)] - E[x(t_1)]E[x(t_2)]$ from equation (5.8).

5.2 One of the steps for deriving the ergodic conditions for the time-average sample mean is equation (5.20).
a. Prove the equality for $N = 4$.
b. Prove the equality for any N.

5.3 Let some signal $x(n)$ be ergodic in the mean. Suppose it could have a Gaussian pdf. Use the pdf of a bivariate Gaussian random variable that is in Section 4.3 and let $x \Rightarrow x(n)$ and $y \Rightarrow x(n + k)$. Show that condition (5.23b) is true and, thus, the estimate of the pdf from the histogram would be valid.

5.4 In Example 5.1 let $x(t, \zeta)$ be the random process described by the probabilities

$$P[x(t, \zeta_i)] = \tfrac{1}{6}, \qquad 1 \le i \le 6,$$

$$P\left[x(t, \zeta_i), x(t + \tau, \zeta_j)\right] = \begin{cases} \tfrac{1}{6}, & i = j, \\ 0, & i \ne j. \end{cases}$$

Show that $m(t) = 3.5$, $\sigma^2(t) = 2.92$, and $\gamma(\tau) = 2.92$.

5.5 Prove that, for positive and negative values of lag, the mean of the estimator $\hat{R}(k)$ is $(1 - |k|/N)\varphi(k)$. (Refer to Section 5.5.2.)

5.6 Prove that the estimator $\hat{R}(k)$ of the autocorrelation function is unbiased.

5.7 Study the derivation of the single-summation expression for $\text{Var}[\hat{C}(k)]$ [equation (5.33)] in Appendix 5.3. Combine all of the equations as stated and show that equation (5.33) is true.

5.8 For the short-duration time series in Table E5.8 do the following:

a. Plot $x(n)$ versus $x(n + 1)$ and guess the approximate value of $\hat{\rho}(1)$.

b. Calculate the following sample parameters: mean, variance, $\hat{R}(1)$, and $\hat{\rho}(1)$.

c. Does $\hat{\rho}(1)$ reflect that $\rho(1) = 0$?

Table E5.8 Time Series

n	$x(n)$
0	0.102
1	0.737
2	0.324
3	0.348
4	0.869
5	0.279
6	0.361
7	0.107
8	0.260
9	0.962

5.9 Duplicate the results in Example 5.2. The time series is listed in Appendix 5.1.

5.10 Generate 100 points of a white noise process using any random number generator.

a. Estimate the autocovariance function to a maximum lag of 50, using both biased and unbiased estimators.

b. Plot the results and qualitatively compare them.

5.11 For the random process in Example 5.4, use the Gaussian sampling distribution and a two-sided test to determine the following:

a. What are the 95% and 99% confidence limits for $\rho(8)$?

b. Does either level suggest a zero correlation?
c. What is the maximum significance level for the values of lag 8 to be uncorrelated?

5.12 For the random process generated in Example 5.5, show from the graph of the sample NACF that the process is uncorrelated.

5.13 Verify the derivation of the mean and variance of the linear transformation in Section 5.6.2.

5.14 Prove that, in general, for a stationary signal, the variance of a sum in equation (5.48) reduces to

$$\text{Var}[y] = N\sigma^2 + 2N\sum_{k=1}^{N-1}\left(1 - \frac{k}{N}\right)C(k).$$

5.15 It is desired to produce a Gaussian white noise process $z(n)$ with a mean of 2 and a variance of 0.5 by summing 12 uniformly distributed numbers from a random number generator. This is similar to the purpose of Example 5.5.
a. What are the mean and variance of the summing process $y(n)$?
b. What linear transformation is needed so that $m_z = 2$ and $\sigma_z^2 = 0.5$?

5.16 Find the transformation necessary to create random variables with the following pdfs from a uniformly distributed random variable:
a. an exponential pdf;
b. a Maxwell pdf with $a = 2$.

5.17 For the Rayleigh random process produced in Example 5.6, test that the sample moments are consistent with what is desired by the transformation.

5.18 Generate and plot independent random time series containing 100 points and having the following characteristics:
a. uniform pdf, $m = 0.5$, $\sigma = 1$;
b. Maxwell pdf, $a = 2$.
Calculate the sample mean and variance. Are they consistent with what was desired? Statistically test these sample moments.

5.19 For the three-point moving average process of equation (5.52), write the algorithms for $y(4)$, $y(5)$, and $y(6)$. How many points of the process $x(n)$ do they have in common with $y(3)$?

5.20 Derive the general forms for the mean and variance of the moving average process [equations (5.55) and (5.56)].

5.21 Derive in detail the value of $R(2)$, equation (5.65), for the second-order MA process.

5.22 Find the autocorrelation functions for the following moving average processes:
a. $y(n) = x(n) + x(n-1) + x(n-2)$;
b. $y(n) = x(n) + 0.5x(n-1) - 0.3x(n-2)$.
How do they differ?

5.23 Prove that the autocorrelation function for the pth-order MA process is

$$R_y(k) = \begin{cases} \sigma_x^2\left(\sum_{i=0}^{p-k} a_i a_{i+k}\right), & |k| \le p, \\ 0, & |k| > p. \end{cases}$$

5.24 Verify the results of the two tests made on the estimate of $\rho(1)$ in Example 5.7.

Computer Exercises

5.25 An alternative method for generating pseudo-Gaussian random numbers uses the linear multiplicative congruential generator with the Box–Muller algorithm [Komo (1987)]. The method is as follows: (a) Generate two uniform random numbers, $u(1)$ and $u(2)$, as conventionally defined; (b) make the angular and magnitude transformations,

$$\text{ANG} = 2\pi u(1),$$

$$R = \sqrt{-2\ln(u(2))};$$

(c) two independent pseudo-Gaussian numbers are

$$x(1) = R\cos(\text{ANG}) \text{ and } x(2) = R\sin(\text{ANG}).$$

Generate 500 random numbers with this method. What are the mean and variance? Plot the histogram. Does it appear to be Gaussian?

5.26 The time series of sunspot numbers listed in Appendix 5.2 is going to be tested for stationarity of the mean.
a. Divide the sample function into 7 segments of 25 points. Calculate the sample mean for each segment.
b. Test to see if the sample means of the segments are equal to one another. Use the Student's t distribution with a 95% confidence region.
c. Is the signal stationary in the mean?

5.27 Perform a correlation analysis of the sunspot numbers in Appendix 5.2.
a. Estimate the NACF for lags from 0 to 20 and plot it with appropriate labels.
b. Describe it qualitatively and state what it may indicate.
c. Do any correlation values test as being nonzero? If so, what are their lag times?

Appendices

Appendix 5.1 Yields from a Distillation Process

1–18	19–36	37–54	55–70
47	37	45	53
64	74	54	49
23	51	36	34
71	58	54	35
38	50	48	54
65	60	55	45
55	44	45	68
41	57	57	38
59	50	50	50
48	45	62	60
71	25	44	39
35	59	64	59
56	50	43	40
40	71	52	57
58	56	38	54
44	74	60	23
80	50	55	
55	58	41	

Appendix 5.2 Wolfer's Sunspot Numbers

Year	Wolfer's Number	Year	Wolfer's Number	Year	Wolfer's Number	Year	Wolfer's Number
1761	85.9	1805	42.2	1849	95.9	1893	84.9
1762	61.2	1806	28.1	1850	66.5	1894	78.0
1763	45.1	1807	10.1	1851	64.5	1895	64.0
1764	36.4	1808	8.1	1852	54.2	1896	41.8
1765	20.9	1809	2.5	1853	39.0	1897	26.2
1766	11.4	1810	0.0	1854	20.6	1898	26.7
1767	37.8	1811	1.4	1855	6.7	1899	12.1
1768	69.8	1812	5.0	1856	4.3	1900	9.5
1769	106.1	1813	12.2	1857	22.8	1901	2.7
1770	100.8	1814	13.9	1858	54.8	1902	5.0
1771	81.6	1815	35.4	1859	93.8	1903	24.4
1772	66.5	1816	45.8	1860	95.7	1904	42.0
1773	34.8	1817	41.1	1861	77.2	1905	63.5
1774	30.6	1818	30.4	1862	59.1	1906	53.8
1775	7.0	1819	23.9	1863	44.0	1907	62.0
1776	19.8	1820	15.7	1864	47.0	1908	48.5
1767	37.8	1811	1.4	1855	6.7	1899	12.1
1768	69.8	1812	5.0	1856	4.3	1900	9.5
1769	106.1	1813	12.2	1857	22.8	1901	2.7
1770	100.8	1814	13.9	1858	54.8	1902	5.0
1771	81.6	1815	35.4	1859	93.8	1903	24.4
1772	66.5	1816	45.8	1860	95.7	1904	42.0
1773	34.8	1817	41.1	1861	77.2	1905	63.5
1774	30.6	1818	30.4	1862	59.1	1906	53.8
1775	7.0	1819	23.9	1863	44.0	1907	62.0
1776	19.8	1820	15.7	1864	47.0	1908	48.5
1777	92.5	1821	6.6	1865	30.5	1909	43.9
1778	154.4	1822	4.0	1866	16.3	1910	18.6
1779	125.9	1823	1.8	1867	7.3	1911	5.7
1780	84.8	1824	8.5	1868	37.3	1912	3.6
1781	68.1	1825	16.6	1869	73.9	1913	1.4
1782	38.5	1826	36.3	1870	139.1	1914	9.6
1783	22.8	1827	49.7	1871	111.2	1915	47.4
1784	10.2	1828	62.5	1872	101.7	1916	57.1
1785	24.1	1829	67.0	1873	66.3	1917	103.9
1786	82.9	1830	71.0	1874	44.7	1918	80.6
1787	132.0	1831	47.8	1875	17.1	1919	63.6
1788	130.9	1832	27.5	1876	11.3	1920	37.6
1789	118.1	1833	8.5	1877	12.3	1921	26.1
1790	89.9	1834	13.2	1878	3.4	1922	14.2
1791	66.6	1835	56.9	1879	6.0	1923	5.8
1792	60.0	1836	121.5	1880	32.3	1924	16.7

Appendix 5.3 Variance of Autocovariance Estimate

By definition, the variance of the biased covariance estimator for $k \geq 0$ is

$$\operatorname{Var}[\hat{C}(k)] = E[\hat{C}^2(k)] - E^2[\hat{C}(k)]$$

$$= E\left[\frac{1}{N^2}\sum_{i=0}^{N-k-1}\sum_{j=0}^{N-k-1} x(i)x(i+k)x(j)x(j+k)\right]$$

$$-\left(1-\frac{k}{N}\right)^2 \gamma^2(k)$$

$$= \frac{1}{N^2}\sum_{i=0}^{N-k-1}\sum_{j=0}^{N-k-1} E[x(i)x(i+k)x(j)x(j+k)]$$

$$-\left(1-\frac{k}{N}\right)^2 \gamma^2(k). \tag{A5.1}$$

Using the expression in equation (5.32) for the fourth-order moments, equation (A5.1) becomes

$$\operatorname{Var}[\hat{C}(k)] = \frac{1}{N^2}\sum_{i=0}^{N-k-1}\sum_{j=0}^{N-k-1} (\gamma^2(k) + \gamma^2(i-j)$$

$$+\gamma(i-j+k)\gamma(i-j-k))$$

$$-\left(1-\frac{k}{N}\right)^2 \gamma^2(k). \tag{A5.2}$$

Equation (A5.2) will be resolved into a single-summation form by simplifying each term separately.

Term 1: Because $\gamma^2(k)$ is not a function of the summing indices,

$$\sum_{i=0}^{N-k-1}\sum_{j=0}^{N-k-1} \gamma^2(k) = (N-k)^2\gamma^2(k). \tag{A5.3}$$

Term 2: The second term,

$$\sum_{i=0}^{N-k-1} \sum_{j=0}^{N-k-1} \gamma^2(i-j), \tag{A5.4}$$

has $(i-j)$ in all of the arguments. All of the values of $\gamma^2(i-j)$ that are summed can be represented in a matrix as

$$\begin{bmatrix} \gamma^2(0) & \gamma^2(1) & \cdots & \gamma^2(p-1) \\ \gamma^2(-1) & \gamma^2(0) & \cdots & \gamma^2(p-2) \\ \vdots & \vdots & \cdots & \vdots \\ \gamma^2(-p+1) & \gamma^2(-p+2) & \cdots & \gamma^2(0) \end{bmatrix}, \tag{A5.5}$$

with $p = N - k$. All of the elements along a diagonal are equal and the equivalent summation can be made by summing over the diagonals. The value of the lag is represented by the substitution $r = i - j$. Now r ranges from $-p + 1$ to $p - 1$ and the number of elements along each diagonal is $N - |r| + k$. Thus,

$$\sum_{i=0}^{N-k-1} \sum_{j=0}^{N-k-1} \gamma^2(i-j) = \sum_{r=-N+k+1}^{N-k-1} \gamma^2(r)(N-|r|+k). \tag{A5.6}$$

Term 3: The resolution for the third term

$$\sum_{i=0}^{N-k-1} \sum_{j=0}^{N-k-1} \gamma(i-j+k)\gamma(i-j-k) \tag{A5.7}$$

follows the same procedure as that for term 2, with the matrix element now being $\gamma(r+k)\gamma(r-k)$. Its summation is

$$\sum_{r=-N+k+1}^{N-k-1} \gamma(r+k)\gamma(r-k)(N-|r|+k). \tag{A5.8}$$

By combining equations (A5.8), (A5.6), and (A5.3) with equation

(A5.2), the result is

$$\operatorname{Var}\left[\hat{C}(k)\right] = \frac{1}{N} \sum_{r=-N+k+1}^{N-k-1} \left(1 - \frac{|r| + k}{N}\right)$$
$$\times\left(\gamma^2(r) + \gamma(r+k)\gamma(r-k)\right). \quad \text{(A5.9)}$$

Appendix 5.4 Parametric Stationarity Tests

A5.4.1 Equality of Two Means

The hypothesis being tested is whether the means in two independent sets of measurements are equal. Because this test is used in conjunction with the test for equality of variances, the variances of the two sets of measurements will be assumed to be equal. The test statistic is a t statistic that uses a normalized value of the difference of the sample means of the two populations. Let the sample means and variances for populations 1 and 2 be represented by the symbols $\hat{m}_1$, $\hat{\sigma}_1^2$, $\hat{m}_2$, and $\hat{\sigma}_2^2$. The test statistic is

$$T = \frac{\hat{m}_1 - \hat{m}_2}{\left(\left(\frac{1}{N_1} + \frac{1}{N_2}\right)\left(\frac{N_1\hat{\sigma}_1^2 + N_2\hat{\sigma}_2^2}{N_1 + N_2 - 2}\right)\right)^{1/2}}. \quad (5.10)$$

T has a Student's t distribution with $\nu = N_1 + N_2 - 2$ degrees of freedom [Anderson (1971)]. The $(1 - \alpha)$ confidence interval is

$$t_{\nu,1-\alpha/2} \leq T \leq t_{\nu,\alpha/2}. \quad \text{(A5.11)}$$

The table for the t test is in Appendix 4.4.

A5.4.2 Equality of Variances

The second parametric test for assessing the stationarity of signals requires testing the equality of variances of two sets of measurements. It is assumed that the true mean values, m_1 and m_2, are unknown. The F distribution is appropriate for this task [Anderson (1971)]. The test statistic F is the ratio of the two unbiased estimates of the variance,

$$F = \frac{(N_1/\nu_1)\hat{\sigma}_1^2}{(N_2/\nu_2)\hat{\sigma}_2^2}. \qquad \text{(A5.12)}$$

F has two degrees of freedom, $\nu_1 = N_1 - 1$ and $\nu_2 = N_2 - 1$. The $(1 - \alpha)$ confidence interval for a two-sided test is

$$F_{\nu_1, \nu_2; 1-\alpha/2} \leq F \leq F_{\nu_1, \nu_2; \alpha/2}. \qquad \text{(A5.13)}$$

The tables for the F distribution are quite voluminous and the values are usually only tabulated for the upper confidence limit. Because F is a ratio of positive numbers, the lower limit can be determined from the tables because

$$F_{\nu_1, \nu_2; 1-\alpha/2} = \frac{1}{F_{\nu_1, \nu_2; \alpha/2}}. \qquad \text{(A5.14)}$$

For tables of values of the F distribution, please consult either the references Bendat and Piersol (1986), Shanmugan and Breipohl (1988) and Jenkins and Watts (1968) or any book of statistical tables.

Chapter 6 Random Signals, Linear Systems, and Power Spectra

6.1 Introduction

This chapter is devoted to some basic principles and definitions that are essential for understanding additional approaches for analyzing random signals. Direct Fourier transformation is no longer sufficient for performing frequency analysis. Other concepts must be used and are based on the definition of signal power. Systems concepts are presented in the context of random signals. They are important because systems can modify the properties of any random input signal and because discrete-time systems can be models for signals. In fact, this latter capability is the basis for the modern approaches to signal analysis. Comprehensive treatments of discrete-time systems can be found in books such as Jong (1982), Zeimer, Tranter, and Fannin (1987), and DeFatta, Lucas, and Hodgkiss (1988). Please consult them for additional information.

6.2 Power Spectra

Harmonic analysis for deterministic waveforms and signals requires only windowing and Fourier transformations of the desired time series. However, implementing *spectral analysis*, frequency analysis for random signals, requires understanding the probabilistic properties of the spectral estimates. Equation (6.1) shows the Fourier transform of a discrete-time random signal $x(n)$ with a sampling interval of T units.

$$X(f) = T \sum_{n=-\infty}^{\infty} x(n) e^{-j2\pi f n T}. \tag{6.1}$$

What is implied, and not shown directly, is that $X(f)$ is also a random variable. So equation (6.1) alone will not suffice as a mechanism for performing the frequency analysis. The major problem is that the Fourier transform of a wide-sense stationary random signal does not exist. Let us examine this. For $X(f)$ to exist the energy must be finite, or

$$\text{energy} = T \sum_{n=-\infty}^{\infty} x(n)^2 < \infty. \tag{6.2}$$

Because $x(n)$ is at least wide-sense stationary, the energy is infinite for every sample function [Priestley (1981)]. In fact, the average energy is also infinite, that is,

$$E[\text{energy}] = T \sum_{n=-\infty}^{\infty} E\left[x(n)^2\right] = \infty. \tag{6.3}$$

However, equation (6.3) suggests that if average power is considered, it would be a finite quantity on which to base a definition of a frequency transformation. The average power is defined as

$$\begin{aligned} E[\text{power}] &= \lim_{N\to\infty} \sum_{n=-N}^{N} \frac{TE\left[x(n)^2\right]}{(2N+1)T} \\ &= E\left[x(n)^2\right] < \infty. \end{aligned} \tag{6.4}$$

The methodology for incorporating a frequency variable into equation (6.4) requires an additional definition. Define a signal $x_P(n)$

that equals a finite duration portion of $x(n)$, that is,

$$x_P(n) = \begin{cases} x(n), & |n| \leq N, \\ 0, & |n| > N, \end{cases} \tag{6.5}$$

such that $E[x(n)^2] < \infty$. Frequency is introduced by using *Parseval's theorem*,

$$T \sum_{n=-\infty}^{\infty} x_P(n)^2 = \int_{-1/2T}^{1/2T} X_P(f) X_P^*(f)\, df. \tag{6.6}$$

Because the $x_P(n)$ sequence is finite, the summation limits can be changed to $-N$ and N and equation (6.6) can be inserted directly into equation (6.4). The order of mathematical operations is changed as shown. Now

$$E[\text{power}] = \lim_{N\to\infty} E\left(\sum_{n=-N}^{N} \frac{T x_P(n)^2}{(2N+1)T} \right)$$

$$= \lim_{N\to\infty} E\left(\frac{\int_{-1/2T}^{1/2T} X_P(f) X_P^*(f)\, df}{(2N+1)T} \right)$$

$$= \int_{-1/2T}^{1/2T} \lim_{N\to\infty} E\left(\frac{X_P(f) X_P^*(f)}{(2N+1)T} \right) df. \tag{6.7}$$

The integrand defines the power distributed over frequency and is the *power spectral density* (PSD) function $S(f)$,

$$S(f) = \lim_{N\to\infty} E\left(\frac{X_P(f) X_P^*(f)}{(2N+1)T} \right) df. \tag{6.8}$$

The PSD is the Fourier transform of the autocorrelation function. The proof of this relationship is called the *Wiener–Khintchin theorem* for continuous-time processes and *Wold's theorem* for discrete-time processes. Their proofs and contributions are summarized in detail by Priestley (1981) and Koopmans (1974). Because only wide-sense stationary signals are being considered, as stated in

Section 5.5, $R(k) = \varphi(k)$ and $C(k) = \gamma(k)$. The transform pair formed using the DTFT is

$$S(f) = T \sum_{k=-\infty}^{\infty} R(k) e^{-j2\pi fkT}, \tag{6.9}$$

$$R(k) = \int_{-1/2T}^{1/2T} S(f) e^{j2\pi fkT}\, df. \tag{6.10}$$

This relationship is logical because $R(0) = E[x(n)^2]$ and $R(k)$ contains information about frequency content. One of the properties of $R(k)$, stated in Chapter 5, is that if $R(k)$ is periodic, then $x(n)$ is a periodic random process. Because the ACF is a moment function, hence deterministic, and asymptotically approaches zero for large lag values, it is Fourier-transformable.

An alternative name for $S(f)$ is the *variance spectral density* function. This definition is best understood if one considers the inverse transform in equation (6.10) with $k = 0$ and the mean value of the process being zero. Under those conditions,

$$R(0) = \int_{-1/2T}^{1/2T} S(f)\, df = \sigma^2. \tag{6.11}$$

The PSD function has three important properties:

a. it is a real function, $S(f) = S^*(f)$;
b. it is an even function, $S(f) = S(-f)$;
c. it is nonnegative, $S(f) \geq 0$.

The proofs of properties a and b are left as exercises for the reader. Notice that, because of property a, all phase information is lost.

6.3 System Definition Review

6.3.1 Basic Definitions

Linear time-invariant discrete-time systems are studied in courses that treat filters, systems, and deterministic signals. These systems have an important role in the framework of random

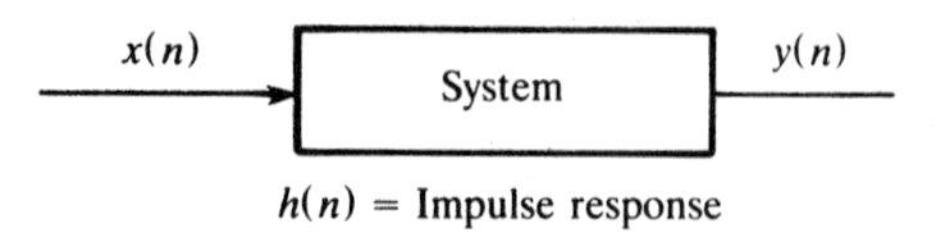

Figure 6.1 System block diagram; $x(n)$ is the input signal, $y(n)$ is the output signal, and $h(n)$ is the impulse response.

signals as well because random signals are subject to filtering and are inputs to measurement systems, and so forth. The general concept is the same as with deterministic signals. Figure 6.1 shows the conventional block diagram with $x(n)$ as the *input signal*, $y(n)$ as the *output signal*, and $h(n)$ as the *unit impulse response*. The fundamental relationship between all of these is given by the *convolution sum* and is

$$y(n) = \sum_{i=-\infty}^{\infty} h(n-i)x(i) = \sum_{i=-\infty}^{\infty} x(n-i)h(i). \quad (6.12)$$

For the system to be *stable* it is necessary that

$$\sum_{n=-\infty}^{\infty} |h(n)| < \infty. \quad (6.13)$$

A system is *causal* if

$$h(n) = 0, \qquad n < 0. \quad (6.14)$$

The complementary relationship in the frequency domain among these signals and the impulse response is also well known from the *convolution theorem* and is

$$Y(f) = H(f)X(f), \qquad -\frac{1}{2T} \leq f \leq \frac{1}{2T}, \quad (6.15)$$

where $Y(f)$, $H(f)$, and $X(f)$ are the DTFT of $y(n)$, $h(n)$, and $x(n)$, respectively. The function $H(f)$ describes the frequency domain properties of the system and is called the *transfer function*, or *frequency response*. As with most Fourier transforms it is a complex function and can be written in polar form,

$$H(f) = |H(f)|e^{j\phi(f)}, \quad (6.16)$$

where $|H(f)|$ and $\phi(f)$ are the *magnitude* and *phase responses*, respectively.

The impulse response has two general forms. One is when the convolution sum has finite limits, that is,

$$y(n) = \sum_{i=-s}^{q} x(n-i)h(i), \tag{6.17}$$

where s and q are finite. This structure represents a *nonrecursive* system and has the same form as the moving-average signal model. If $s \leq 0$, the system is causal. When a nonrecursive system is noncausal, usually the impulse response is an even function, that is, $s = q$ and $h(i) = h(-i)$. If either s or q is infinite, the system can take more parsimonious representation. Most often in this latter situation the systems are causal and therefore $s = 0$ and $q = \infty$. The structure becomes

$$y(n) = b(0)x(n) - \sum_{i=1}^{p} a(i)y(n-i). \tag{6.18}$$

This is a *recursive* structure and is identical mathematically to the autoregressive signal model. A system can have recursive and nonrecursive components simultaneously and takes the structure

$$y(n) = \sum_{l=0}^{q} b(l)x(n-l) - \sum_{i=1}^{p} a(i)y(n-i). \tag{6.19}$$

The relationship among the system coefficients, $a(i)$ and $b(l)$, and the components of the impulse response are complicated except for the nonrecursive system.

The general frequency response is obtained from the DTFT of equation (6.19). After rewriting the equation as

$$\sum_{i=0}^{p} a(i)y(n-i) = \sum_{l=0}^{q} b(l)x(n-l)$$

with $a(0) = 1$ and taking the DTFT, it becomes

$$\sum_{i=0}^{p} a(i)Y(f)e^{-j2\pi fiT} = \sum_{l=0}^{q} b(l)X(f)e^{-j2\pi flT},$$

and, finally,

$$H(f) = \frac{Y(f)}{X(f)} = \frac{\Sigma_{l=0}^{q} b(l) e^{-j2\pi f l T}}{\Sigma_{i=0}^{p} a(i) e^{-j2\pi f i T}}. \tag{6.20}$$

The general signal model represented by equation (6.19) and whose transfer function is represented by equation (6.20) is the *autoregressive moving average* (ARMA) model of order (p, q). Sometimes this is written as ARMA (p, q). Two very useful models are simplifications of the ARMA model. When $a(0) = 1$ and $a(i) = 0$ for $i \geq 1$, the moving average model of order q is produced. This was used in Chapter 5. Because the denominator portion of its transfer function is 1, the MA model MA(q) is represented by an *all-zero* system. When $b(0) = 1$ and $b(l) = 0$ for $l \geq 1$, the model is the *autoregressive* (AR) model of order p, also written as AR(p). Because the numerator portion of the transfer function is 1, the AR model is represented by an *all-pole* system.

Because the input and output signals are random, these deterministic relationships per se do not describe all the relationships between $x(n)$ and $y(n)$. They must be used as a framework to find other relationships between the probabilistic moments and nonrandom functions that describe the signals. The purpose of this chapter is to define some of the more important relationships and to define some functional forms for the system. For the presentations in this book, it shall be assumed that *the signals are at least wide-sense stationary* and that *all systems generating signals are causal, stable, and minimum-phase* [Oppenheim and Willsky (1983)].

6.3.2 Relationships Between Input and Output

The mean value of the output is found directly from equation (6.12) and for causal systems is

$$E[y(n)] = E\left[\sum_{j=0}^{\infty} x(n-j) h(j)\right]$$

$$= \sum_{j=0}^{\infty} E[x(n-j)] h(j).$$

Because $x(n)$ is stationary, let $m_x = E[x(n-j)]$ and

$$E[y(n)] = m_y = m_x \sum_{j=0}^{\infty} h(j). \tag{6.21}$$

Thus, the mean values have a simple relationship and if $m_x = 0$ and $\Sigma_{j=0}^{\infty}|h(j)| < \infty$, then $m_y = 0$. If the summation is considered from the DTFT perspective, then

$$\sum_{j=0}^{\infty} h(j) = \sum_{j=0}^{\infty} h(j) e^{-j2\pi f n T}\Big|_{f=0} = \frac{H(0)}{T} \tag{6.22}$$

and $m_y = m_x H(0)/T$.

The simplest relationship between correlation functions is derived also from the convolution relationship. Now multiply both sides of equation (6.12) by $x(n-k)$ and take the expectation, or

$$\begin{aligned} R_{xy}(k) &= E[x(n-k)y(n)] \\ &= E\left[\sum_{i=-\infty}^{\infty} h(i)x(n-k)x(n-i)\right] \\ &= \sum_{i=-\infty}^{\infty} h(i)E[x(n-k)x(n-i)]. \end{aligned} \tag{6.23}$$

The term $R_{xy}(k)$ in equation (6.23) is the mean of a cross product and defines the *cross correlation function* (CCF) between signals $y(n)$ and $x(n)$ for a time difference (kT). The variable k defines the number of time units that the signal $y(n)$ is delayed, or lagged, with respect to $x(n)$. Hence kT is also called the *lag time*. The more conventional definition is

$$R_{xy}(k) = E[x(n)y(n+k)]. \tag{6.24}$$

The ordering of the terms within the expectation brackets is very important. The expectation within the summation of equation (6.23) is an autocorrelation function. Because only time differences are important, $(n-i)-(n-k) = (k-i)$ and

$$R_x(k-i) = E[x(h-k)x(n-i)]. \tag{6.25}$$

Equation (6.23) is rewritten as

$$R_{xy}(k) = \sum_{i=-\infty}^{\infty} R_x(k-i)h(i) = R_x(k) * h(k). \quad (6.26)$$

That is, the CCF between the output and input signals is equal to the convolution of the ACF of the input signal with the system's impulse response. The Fourier transform of equation (6.26) is simple and produces several new functions. Now

$$\mathrm{DTFT}\left[R_{xy}(k)\right] = \mathrm{DTFT}[R_x(k)]\mathrm{H}(f) \quad (6.27)$$

and

$$\mathrm{DTFT}[R_x(k)] = S_x(f) \quad (6.28)$$

is the PSD of the signal $x(n)$, and

$$\mathrm{DTFT}\left[R_{xy}(k)\right] = S_{xy}(f) \quad (6.29)$$

is the *cross power spectral density function* (CPSD) between signals $y(n)$ and $x(n)$. A detailed explanation of the meaning of these functions is contained in subsequent sections in this chapter.

A more complicated but also a much more useful relationship is that between the PSD of input and output signals. The basic definition is

$$\begin{aligned}
R_y(k) &= E[y(n)y(n+k)] \\
&= E\left[\left(\sum_{i=-\infty}^{\infty} x(n-i)h(i)\right)\left(\sum_{l=-\infty}^{\infty} x(n+k-l)h(l)\right)\right] \\
&= \sum_{i=-\infty}^{\infty}\sum_{l=-\infty}^{\infty} h(i)h(l)E[x(n-i)x(n+k-l)] \\
&= \sum_{i=-\infty}^{\infty}\sum_{l=-\infty}^{\infty} h(i)h(l)R_x(k+i-l). \quad (6.30)
\end{aligned}$$

Reordering the terms in equation (6.30) and using equation (6.26) yields

$$R_y(k) = \sum_{i=-\infty}^{\infty} h(i) \sum_{l=-\infty}^{\infty} h(l) R_x(\{k+i\} - l), \qquad (6.31)$$

$$R_y(k) = \sum_{i=-\infty}^{\infty} h(i) R_{xy}(k+i). \qquad (6.32)$$

Use the substitution $u = -i$ in equation (6.32) and it becomes

$$R_y(k) = \sum_{u=-\infty}^{\infty} h(-u) R_{xy}(k-u). \qquad (6.33)$$

The relationships between the PSDs are found by taking the DTFT of equation (6.33)

$$S_y(f) = H^*(f) S_{xy}(f) \qquad (6.34)$$

and substituting equation (6.30)

$$S_y(f) = H^*(f) H(f) S_x(f) = |H(f)|^2 S_x(f). \qquad (6.35)$$

The term $|H(f)|^2$ is the *power transfer function*.

6.4 Systems and Signal Structure

In the previous chapter it was seen that a moving average of a sequence of uncorrelated numbers can produce a correlation in the resulting sequence. The concept of using a discrete-time system to induce structure in a sequence will be formalized in this section. Essentially, this is accomplished by using white noise as the input to the system and the structured sequence is the output of the system.

6.4.1 Moving Average Process

The general form for a moving average process of order q is

$$y(n) = \sum_{l=0}^{q} b(l) x(n-l). \qquad (6.36)$$

The mean, variance, and correlation structure of the output process $y(n)$ can be derived in general from this system and the properties of the input process. The mean value of the output is

$$E[y(n)] = m_y = E\left(\sum_{l=0}^{q} b(l)x(n-l) \right) = \sum_{l=0}^{q} b(l)E[x(n-l)]$$

$$= E[x(n)] \sum_{l=0}^{q} b(l) = m_x \sum_{l=0}^{q} b(l). \tag{6.37}$$

If $m_x = 0$, then $m_y = 0$. The derivation of the ACF and variance of $y(n)$ is more complex. Fortunately, the general form has already been developed in Section 6.3.2. Compare the form of the MA process, equation (6.36), to the general definition of a system's output, equation (6.12). It can be seen that $b(l) = h(l)$. Thus, from equation (6.30),

$$R_y(k) = E[y(n)y(n+k)]$$

$$= \sum_{i=0}^{q} \sum_{l=0}^{q} b(i)b(l)R_x(k+i-l). \tag{6.38}$$

The variance becomes

$$\sigma_y^2 = R_y(0) - m_y^2$$

$$= \sum_{i=0}^{q} \sum_{l=0}^{q} b(i)b(l)R_x(i-l) - \left(m_x \sum_{l=0}^{q} b(l) \right)^2. \tag{6.39}$$

For signaling modeling, the input process $x(n)$ is often a zero-mean white noise process, $R_x(k) = \sigma_x^2 \delta(k)$. For this situation, let $u = k + i$ in equation (6.38) and the ACF and variance simplify to

$$R_y(k) = \sigma_x^2 \sum_{l=k}^{q} b(l)b(l-k) = \sigma_x^2 \sum_{u=0}^{q-k} b(u)b(u+k), \tag{6.40}$$

$$\sigma_y^2 = \sigma_x^2 \sum_{l=0}^{q} b(l)^2. \tag{6.41}$$

One can ascertain that a qth-order MA process has an autocorrelation function with magnitudes of zero for lags greater than q and lesser than $-q$. This is left as an exercise. Examples and Exercises are provided in Chapter 5.

6.4.2 Structure With Autoregressive Systems

The AR model with stationary inputs can also be used to develop structure in an uncorrelated sequence and produce a sequence with stationary signal-like properties. The derivation of the correlational structure of the output process is more complex than those produced through MA processes. Thus, a first-order process will be studied in great detail. The essential concepts will be covered and extension to higher-order AR processes is straightforward but the manipulations are more complicated as the order is increased. For a first-order model,

$$y(n) = -a(1)y(n-1) + b(0)x(n). \tag{6.42}$$

The mean of the output signal is

$$\begin{aligned} E[y(n)] = m_y &= E[-a(1)y(n-1) + b(0)x(n)] \\ &= -a(1)E[y(n-1)] + b(0)E[x(n)] \\ &= -a(1)m_y + b(0)m_x. \end{aligned} \tag{6.43}$$

Solving for m_y yields

$$m_y = \frac{b(0)}{1+a(1)} m_x. \tag{6.44}$$

This is a more complex relationship between the means of the input and output than equation (6.37). Again m_x, and thus m_y, is usually zero. The variance is

$$\begin{aligned} \sigma_y^2 &= E\left[y(n)^2\right] = E\left[(-a(1)y(n-1) + b(0)x(n))^2\right] \\ &= E\left[a(1)^2 y(n-1)^2 - 2a(1)b(0)y(n-1)x(n) + b(0)^2 x(n)^2\right] \\ &= a(1)^2 E\left[y(n\text{-}1)^2\right] - 2a(1)b(0)E[y(n-1)x(n)] \\ &\quad + b(0)^2 E\left[x(n)^2\right]. \end{aligned} \tag{6.45}$$

The middle term of equation (6.45), $E[y(n-1)x(n)]$, represents the cross correlation between the present input and previous output values, $R_{yx}(1)$. Because this is a causal system, there can be no correlation and $R_{yx}(1) = m_y m_x$. For $m_x = 0$, equation (6.45) becomes

$$\sigma_y^2 = \frac{b(0)^2}{1 - a(1)^2}\sigma_x^2. \qquad (6.46)$$

The autocorrelation function is, by definition,

$$\begin{aligned} R_y(k) &= E[y(n)y(n+k)] \\ &= E[(-a(1)y(n-1) + b(0)x(n)) \\ &\qquad \times(-a(1)y(n-1+k) + b(0)x(n+k))]. \qquad (6.47) \end{aligned}$$

Several lag values will be considered. For $k = 1$,

$$\begin{aligned} R_y(1) = E\big[&a(1)^2 y(n-1)y(n) - a(1)b(0)y(n-1)x(n+1) \\ &-a(1)b(0)x(n)y(n) + b(0)^2 x(n)x(n+1)\big]. \qquad (6.48) \end{aligned}$$

Notice that, when the expectation is distributed, the following hold:

a. The first and fourth terms will result in autocorrelation function values of signals $y(n)$ and $x(n)$, respectively.
b. The second and third terms will result in cross correlation function values between $x(n)$ and $y(n)$.

Writing this explicitly yields

$$\begin{aligned} R_y(1) = a(1)^2 R_y(1) &- a(1)b(0)R_{yx}(2) \\ &- a(1)b(0)R_{xy}(0) + b(0)^2 R_x(1). \qquad (6.49) \end{aligned}$$

As we seek to simplify this expression, it is apparent that $R_x(1) = 0$.

The values of the cross correlations depend upon the lag. Consider the third term,

$$R_{xy}(0) = E[x(n)y(n)]$$

$$= -a(1)E[x(n)y(n-1)] + b(0)E\left[x(n)^2\right]. \quad (6.50)$$

The first term on the right-hand side of equation (6.50) contains $R_{xy}(-1)$. For this model it is known from a previous paragraph that $R_{xy}(-1) = R_{yx}(1) = m_y m_x = 0$. Strictly speaking, one could carry this one more step and obtain

$$E[x(n)y(n-1)] = -a(1)E[x(n)y(n-2)]$$

$$+ b(0)E[x(n)x(n-1)]$$

$$= -a(1)E[x(n)y(n-2)]. \quad (6.51)$$

Continuing this procedure, it terminates with an expression

$$E[x(n)y(0)] = y(0)m_x = 0. \quad (6.52)$$

Thus $R_{xy}(0) = b(0)\sigma_x^2$. The same reasoning shows that the value of the second term of equation (6.49) is zero. Demonstrating this is left as an exercise. Therefore,

$$R_y(1) = a(1)^2 R_y(1) - a(1)b(0)^2\sigma_x^2$$

and

$$R_y(1) = -\frac{a(1)b(0)^2\sigma_x^2}{1 - a(1)^2}. \quad (6.53)$$

Referring to the expression for the variance of $y(n)$, then

$$R_y(1) = -a(1)\sigma_y^2$$

and

$$\rho_y(1) = -a(1). \quad (6.54)$$

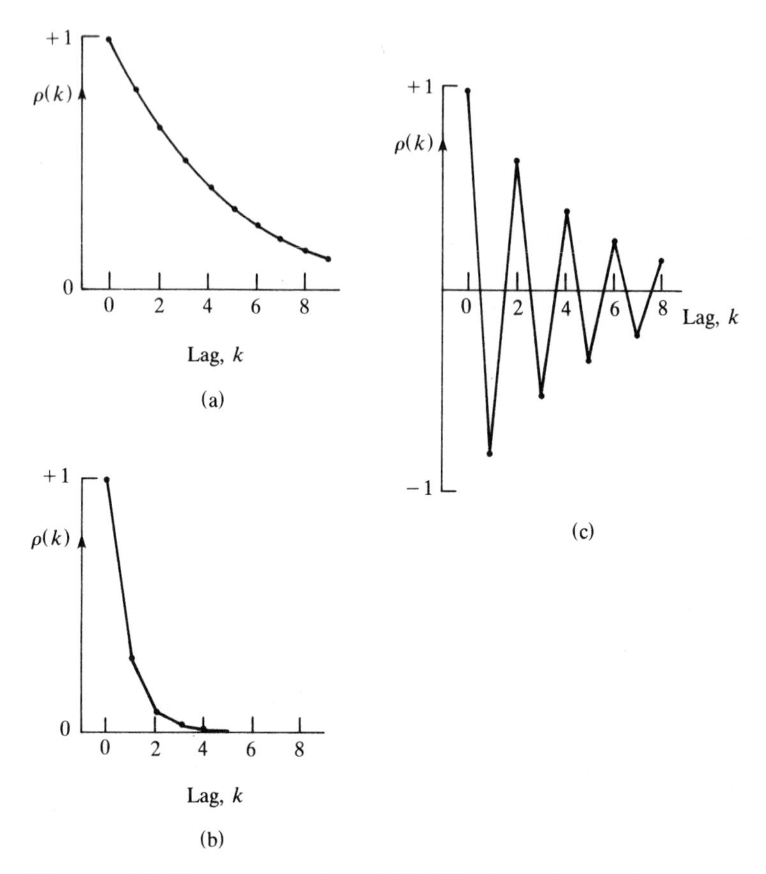

Figure 6.2 Three examples of NACFs for a first-order AR process: (a) $a(1) = -0.8$; (b) $a(1) = -0.3$; (c) $a(1) = 0.8$. [Adapted from Chatfield (1975), Figure 3.1, with permission.]

This procedure can be continued to provide a closed-form solution for the ACF,

$$R_y(k) = \sigma_y^2(-a(1))^{|k|}. \tag{6.55}$$

Plots of several NACFs for different values of $a(1)$ are shown in Figure 6.2. Notice that, in contrast to the correlation functions for an MA process, these NACFs have nonzero values at all lags and approach zero only asymptotically. The ACF in equation (6.55) can also be derived through a frequency-domain approach, using equation (6.35) and the IDFT of $S_y(f)$.

Example 6.1

Because the ACF reflects structure in a signal, interpreting these should indicate some qualitative properties of the signal Compare Figures 6.3(a) and 6.3(b). One ACF has a monotonically decreasing trend in magnitude, whereas the other has values that oscillate positively and negatively. The first ACF indicates that signal points that are one, two, and so forth time units apart are positively correlated to one another. Thus, it should be expected that the signal should have some short-term trends over time. A

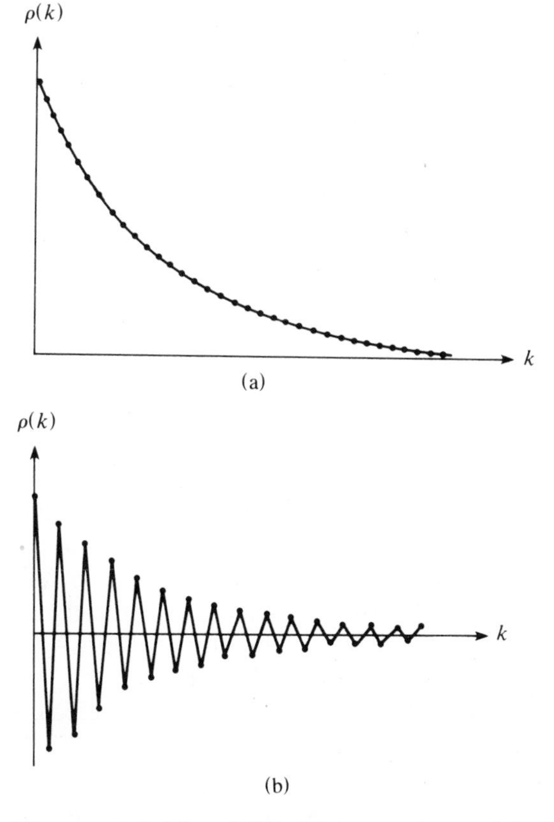

Figure 6.3 The ACFs from a first-order AR process with (a) $a(1) = -0.9$ and (b) $a(1) = 0.9$; (c) and (d) realizations of these processes. [Adapted from Jenkins and Watts (1968), Figures 6.4 and 6.5, with permission.]

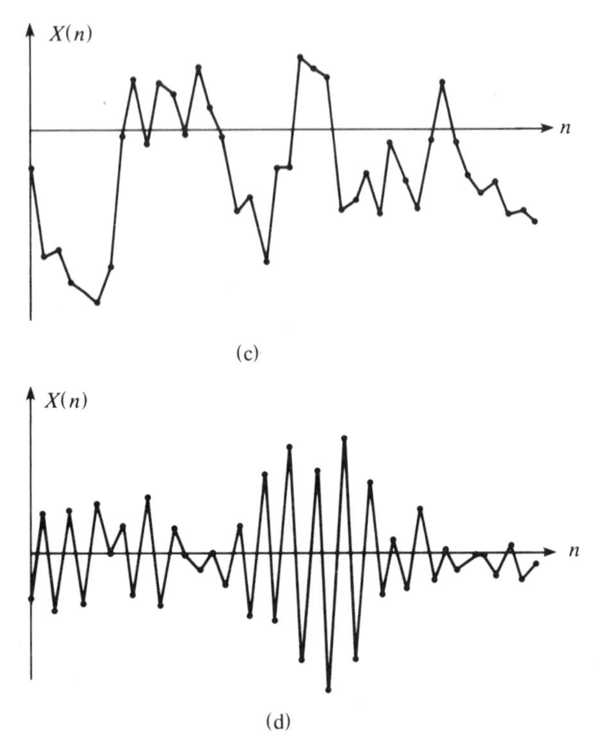

Figure 6.3 (*Continued*)

signal containing 100 points was simulated and is plotted in Figure 6.3(c). Examining it reveals that groupings of four to five consecutive points are either positive or negative. The other ACF indicates that signal points that are odd time units apart are negatively correlated. Thus, any positive signal values should be followed by negative ones and *vice versa*. In Figure 6.3(d) is plotted a simulated time series from such a system. Again, its qualitative properties correspond to what is expected from the ACF.

6.4.3 Higher-Order AR Systems

The description of the ACF for higher-order AR models is much more complex and the derivation of closed-form solutions is the subject of the study of difference equations [Pandit and Wu

(1983)]. However, a recursive solution for $R_y(k)$ can be derived in a simple manner. Begin with the equation for a pth-order AR model,

$$y(n) + a(1)y(n-1) + a(2)y(n-2)$$
$$+ \cdots + a(p)y(n-p) = x(n). \qquad (6.56)$$

The usual convention of letting $a(0) = b(0) = 1$ has been adopted. The ACF of $y(n)$ is obtained by simply premultiplying equation (6.56) by $y(n-k)$ and taking the expectations, that is,

$$R_y(k) = E[y(n-k)y(n)]$$
$$= E[-a(1)y(n-k)y(n-1)$$
$$- \cdots - a(p)y(n-k)y(n-p) + y(n-k)x(n)]$$
$$= -a(1)R_y(k-1) - a(2)R_y(k-2)$$
$$- \cdots - a(p)R_y(k-p) + E[y(n-k)x(n)]. \qquad (6.57)$$

It is known from the previous paragraph that, for $m_x = 0$ and $k > 0$, $R_{yx}(k) = 0$ and equation (6.57) becomes

$$R_y(k) + a(1)R_y(k-1) + a(2)R_y(k-2)$$
$$+ \cdots + a(p)R_y(k-p) = 0. \qquad (6.58)$$

This is the recursive expression for the ACF of process $y(n)$. For the situation when $k = 0$, equation (6.57) becomes

$$R_y(0) + a(1)R_y(1) + a(2)R_y(2) + \cdots + a(p)R_y(p) = \sigma_x^2. \qquad (6.59)$$

The proof of the latter expression is left as an exercise. Equations (6.59) and (6.58) will also be very important in the study of signal modeling.

Example 6.2

Consider the second-order AR process,

$$y(n) - 1.0y(n-1) + 0.5y(n-2) = x(n).$$

By inspection the recursive equations for the ACF and NACF are

$$R_y(k) - 1.0R_y(k-1) + 0.5R_y(k-2) = 0$$

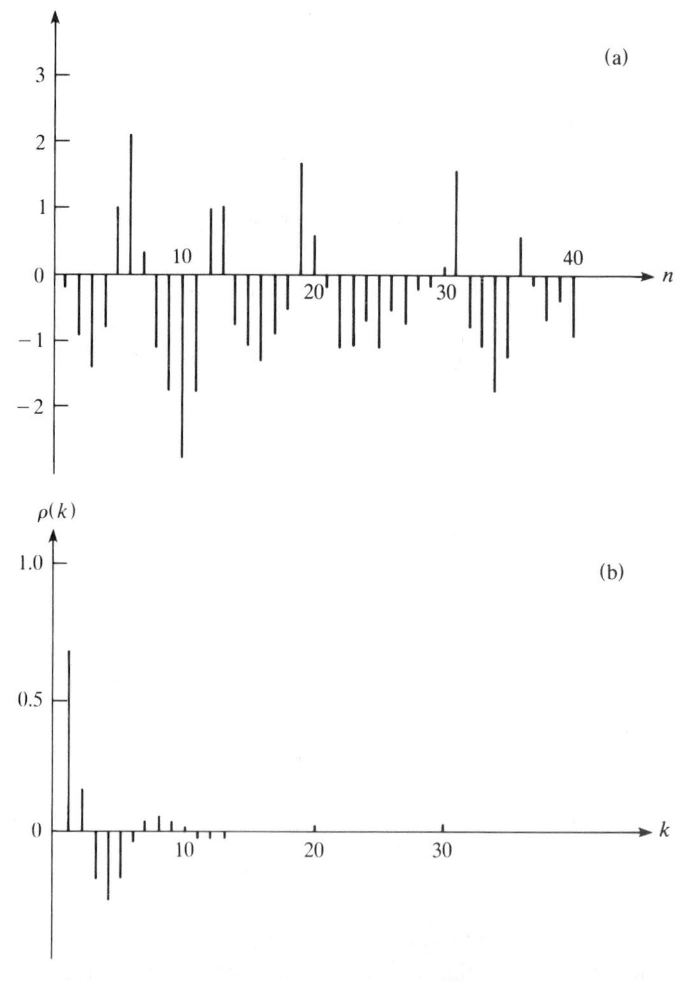

Figure 6.4 For a second-order AR process in Example 6.2, (a) a sample function, $N = 40$, and (b) the NACF are plotted. [Adapted from Jenkins and Watts (1968), Figure 5.9, with permission.]

and

$$\rho_y(k) - 1.0\rho_y(k-1) + 0.5\rho_y(k-2) = 0.$$

Simply using the latter equation, one can generate the NACF. However, one must have the initial conditions because to begin,

$$\rho_y(2) = 1.0\rho_y(1) - 0.5\rho_y(0).$$

The subscript of $\rho(k)$ is dropped for now. Because $\rho(0) = 1$, $\rho(1)$ must be ascertained. Let $k = 1$ and

$$\rho(1) - 1.0\rho(0) + 0.5\rho(-1) = 0.$$

Solving for $\rho(1)$ yields $\rho(1) = 0.667$ and the NACF can be found. For this signal a sample function and its NACF are plotted in Figure 6.4.

Example 6.3

The power transfer function for the AR process in Example 6.2 is to be derived. The transfer function is

$$H(f) = \frac{1}{1 - e^{-j2\pi fT} + 0.5e^{-j4\pi fT}}.$$

Therefore,

$$|H(f)|^2 = \frac{1}{1 - e^{-j2\pi fT} + 0.5e^{-j4\pi fT}} \frac{1}{1 - e^{j2\pi fT} + 0.5e^{j4\pi fT}}$$

$$= \frac{1}{2.25 - 3\cos(2\pi fT) + \cos(4\pi fT)}.$$

This is plotted in Figure 6.5 for $T = 1$.

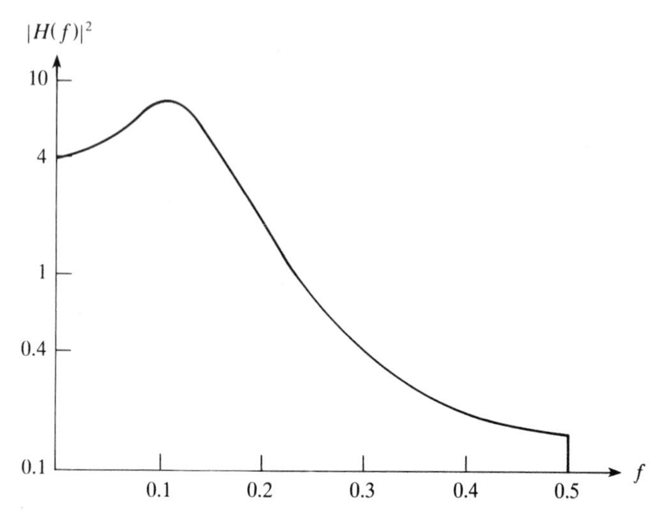

Figure 6.5 Power transfer function for a second-order AR system with $a(1) = -1.0$, $a(2) = 0.5$, and $T = 1$ (Example 6.3).

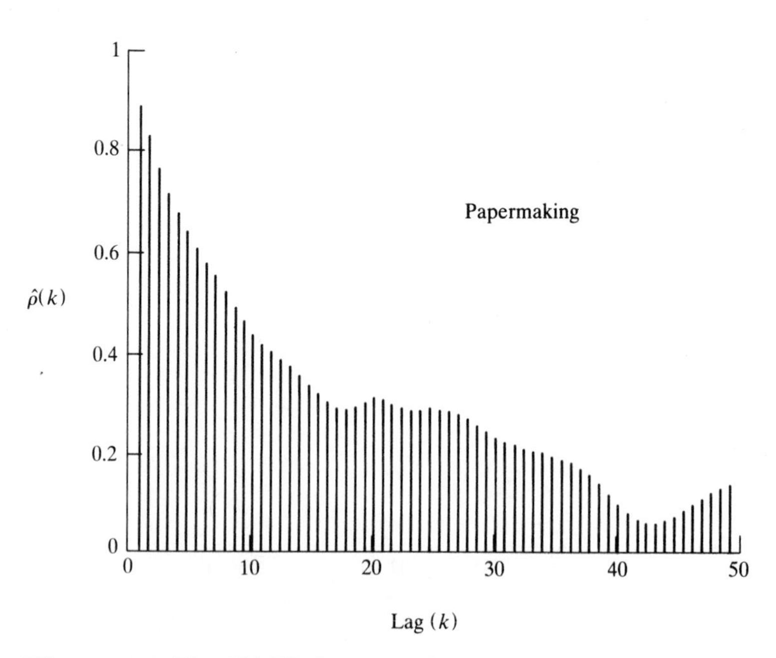

Figure 6.6 The NACF for an industrial papermaking process represented by an ARMA (2, 1) process, $a(1) = -1.76$, $a(2) = 0.76$, and $b(1) = -0.94$. [Adapted from Pandit and Wu (1983), Figure 3.15, with permission.]

As might be expected, AR models with different parameters generate signals with different characteristics. An industrial paper-making process can be represented by an ARMA(2, 1) model,

$$y(n) - 1.76y(n-1) + 0.76y(n-2) = x(n) - 0.94x(n-1). \tag{6.60}$$

Its NACF is plotted in Figure 6.6. It indicates some long-duration positive correlations in the output time series. Other processes can be very complex. Models for speech generation can have orders of 14 or higher. The NACFs of several speech sounds are plotted in Figure 6.7. They are definitely different from each other and have

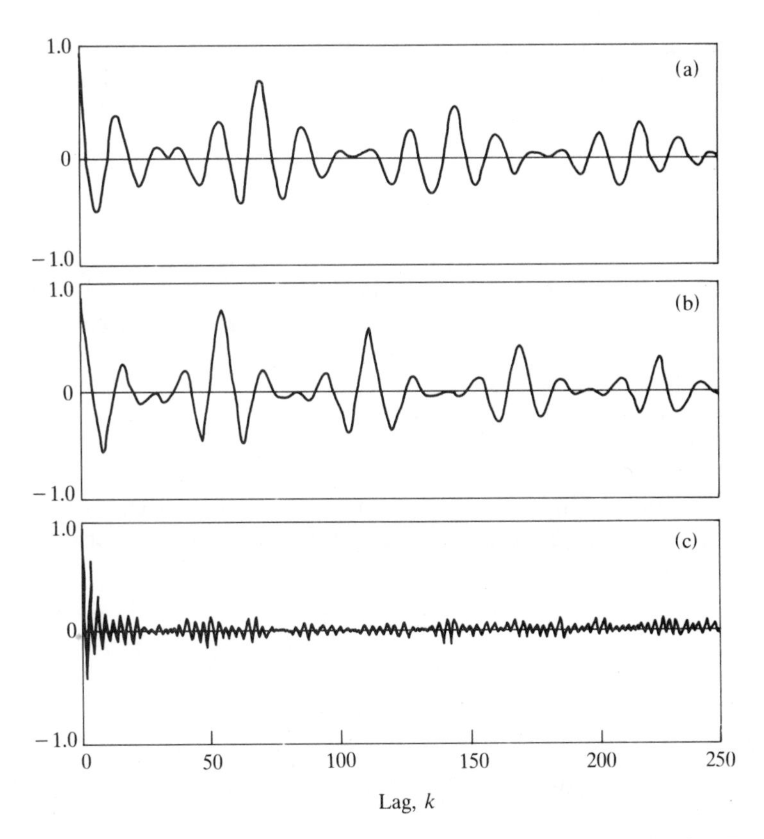

Figure 6.7 The NACFs for three different speech signals, T = 0.1 ms. [From Rabiner and Schafer (1978), Figure 4.24, with permission.]

different model parameters. Their interpretation is not easy as with the first- and second-order models. However, it can be appreciated that the autocorrelation function represents some of the time-domain characteristics of a signal in a compact manner.

6.5 Time-Series Models for Spectral Density

Although one of the main goals of signal analysis is to learn techniques to estimate the PSD of a measured time series, it is instructive to derive PSDs of several models of theoretical signals and to study them. In this manner one can develop a practical understanding of the concepts involved. Also, the evaluation of many estimation techniques is based on the estimation of the PSD from sample functions generated from model processes. At this point, consider the white noise process and several MA processes.

Example 6.4

A zero-mean white noise process is defined by the ACF $R(k) = \sigma^2\delta(k)$. Its PSD is found using the DTFT in Section 6.2 and is

$$S(f) = T \sum_{k=-\infty}^{\infty} R(k)e^{-j2\pi fkT} = T \sum_{k=-\infty}^{\infty} \sigma^2\delta(k)e^{-j2\pi fkT} = \sigma^2 T.$$

This is plotted in Figure 6.8(a). The spectral density is constant for all frequencies. This only occurs for white noise. Notice that the area under $S(f)$ equals σ^2.

Example 6.5

A second-order MA Process was studied in Chapter 5 in Example 5.7:

$$y(n) = 0.5x(n) + 0.5x(n-1),$$

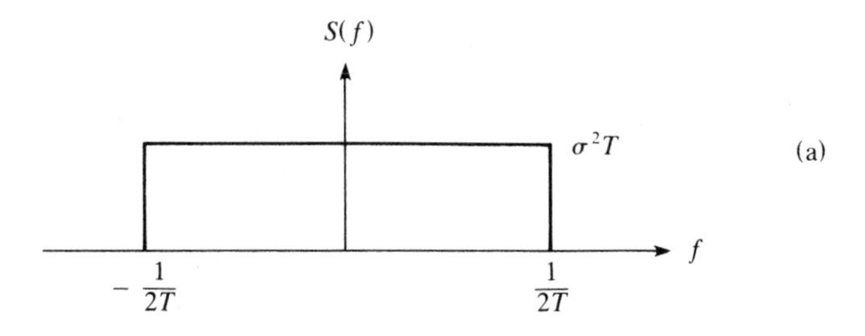

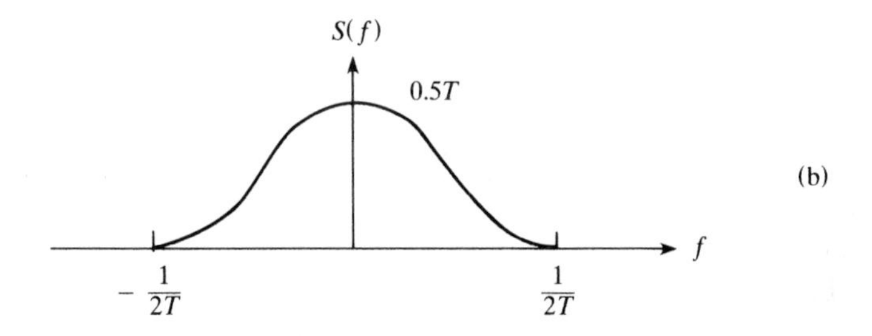

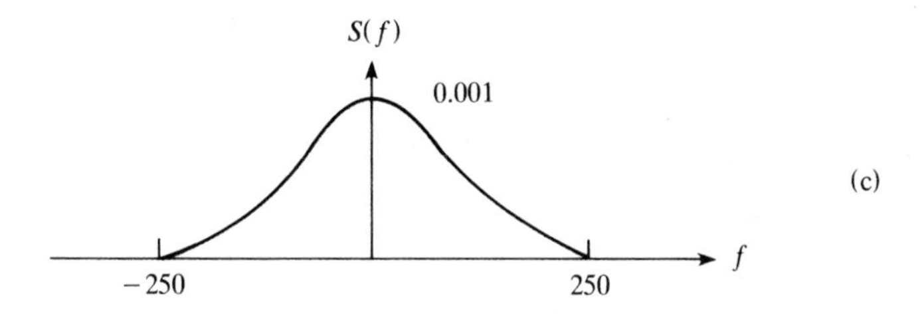

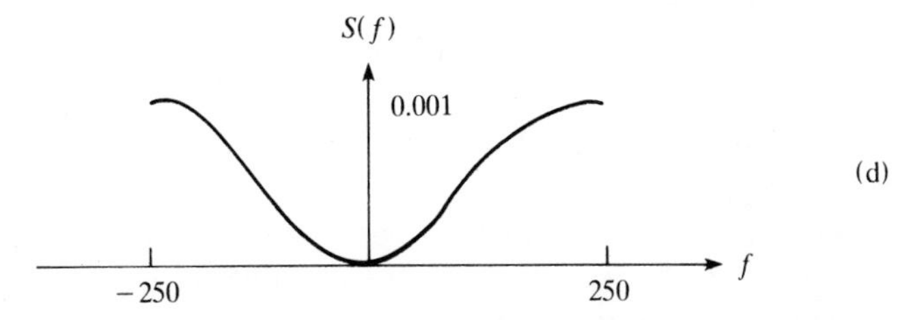

Figure 6.8 The PSDs of several moving average processes: (a) white noise; (b) first-order, low-frequency in parametric form; (c) first-order, low-frequency with sampling interval = 2 ms; (d) first-order, high-frequency with sampling interval = 2 ms.

with autocorrelation function values

$$R(0) = 0.25, \qquad R(1) = 0.125, \qquad R(k) = 0 \quad \text{for } |k| \geq 2.$$

The PSD is

$$\begin{aligned} S(f) &= T(0.125e^{j2\pi fT} + 0.25 + 0.125e^{-j2\pi fT}) \\ &= T(0.25 + 0.25\cos(2\pi fT)) \end{aligned}$$

and is plotted in Figure 6.8(b). Notice that most of the power is concentrated in the lower frequency range. This process is, in general, a low-pass process.

The spectrum in Example 6.5 was left in parametric form to illustrate two observations: First, in most of the theoretical presentations of the PSD, $T = 1$. Second, the shape of the PSD is independent of the sampling frequency. Figure 6.8(c) shows the spectrum of Figure 6.8(b) with $T = 2$ ms. Using equation (6.11), the variance is

$$\begin{aligned} \sigma^2 &= \int_{-1/2T}^{1/2T} S(f)\,df \\ &= \int_{-1/2T}^{1/2T} T(0.25 + 0.25\cos(2\pi fT))\,df \\ &= 2T\left(0.25f + 0.25\frac{\sin(2\pi fT)}{2\pi T}\right)\Bigg|_0^{1/2T} = 0.25. \end{aligned} \tag{6.61}$$

In both cases, the area under the PSD functions is 0.25. The major difference is the range of the frequency axis.

The complement of the low-frequency process is the high-pass process, where the majority of the variance is concentrated in the higher frequency range. This is easily formed from the first-order MA process of Example 6.5 by changing the sign of the coefficient of the term with lag 1, that is,

$$y(n) = 0.5x(n) - 0.5x(n-1). \tag{6.62}$$

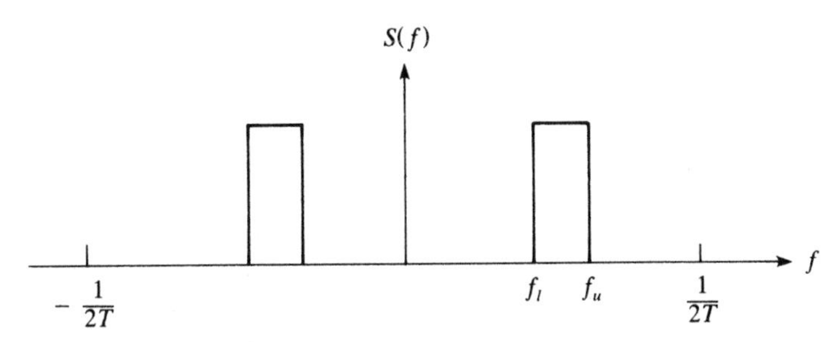

Figure 6.9 General schematic of the PSD of a band-pass process.

Its PSD is shown in Figure 6.8(d). The derivation is left as an exercise.

Another type of process that is frequently encountered is the *band-pass* process. It contains power in the middle range of frequencies. A hypothetical PSD for such a process is shown in Figure 6.9. The *lower and upper frequency bounds* are designated in f_l and f_u, respectively. Second- and higher-order MA processes can have this general type of PSD. In fact, both MA and AR processes of high order can produce spectra with multiple bands of frequencies. Figure 6.10 shows the PSD of the output of an AR (20) system used to simulate speech in a person with muscle

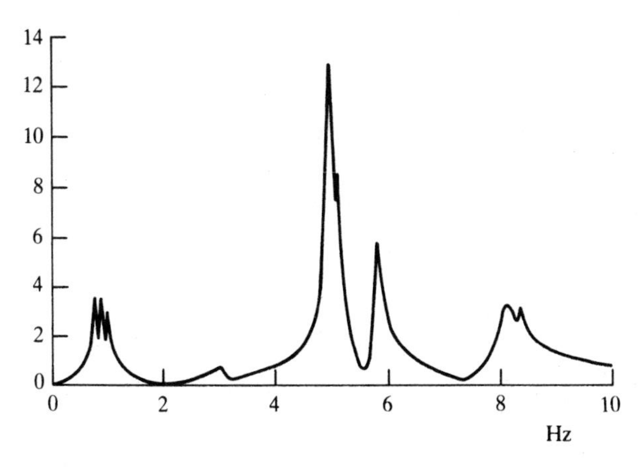

Figure 6.10 The PSD function of a tremorous speech signal represented by a 20th-order AR process. [From Gath and Yair (1987, IEEE), Figure 5, with permission.]

tremors [Gath and Yair (1987)]. The relative magnitude of the coefficients of the process determine whether the process is a low-pass, high-pass, or band-pass process. A direct method to derive autocorrelation functions that will produce a PSD with a certain form is to rewrite equation (6.9) as a cosine series:

$$S(f) = T\left(R(0) + 2\sum_{k=1}^{\infty} R(k)\cos(2\pi fT)\right). \tag{6.63}$$

Derivation of this equation is straightforward and is left as an exercise. Equation (6.63) is a Fourier cosine series with coefficients

$$[Z(0),\ldots,Z(k),\ldots] = [TR(0),\ldots,2TR(k),\ldots].$$

The integral equations are

$$Z(0) = \frac{1}{2f_N}\int_{-f_N}^{f_N} S(f)\,df, \tag{6.64}$$

$$Z(k) = \frac{1}{f_N}\int_{-f_N}^{f_N} S(f)\cos(2\pi fkT)\,df, \qquad k \geq 1. \tag{6.65}$$

As with any Fourier series, the accuracy of the approximation depends upon the number of coefficients $Z(k)$. Let us now derive some autocorrelation functions that are appropriate for a band-pass process.

Example 6.6

Develop the autocorrelation function of a second-order process that has the general properties of the band-pass spectrum in Figure 6.9. Using the fact that the PSD is an even function, the first term is

$$Z(0) = TR(0) = \frac{1}{2f_N}\int_{-f_N}^{f_N} S(f)\,df$$

$$= \frac{1}{f_N}\int_0^{f_N} S(f)\,df$$

$$= 2T\int_{f_l}^{f_u} \frac{\sigma^2}{2(f_u - f_l)}\,df = T\sigma^2,$$

or

$$R(0) = \sigma^2.$$

The other coefficients are

$$Z(k) = \frac{1}{f_N}\int_{-f_N}^{f_N} S(f)\cos(2\pi fkT)\,df$$

$$= \frac{2}{f_N}\int_0^{f_N} S(f)\cos(2\pi fkT)\,df$$

$$= 4T\int_{f_l}^{f_u} \frac{\sigma^2}{2(f_u - f_l)}\cos(2\pi fkT)\,df$$

$$= \frac{2T\sigma^2}{(f_u - f_l)}\int_{f_l}^{f_u} \cos(2\pi fkT)\,df$$

$$= \frac{2T\sigma^2}{(f_u - f_l)}\left(\frac{\sin(2\pi fkT)}{2\pi kT}\right)\Bigg|_{f_l}^{f_u}$$

$$= \frac{\sigma^2}{\pi k(f_u - f_l)}(\sin(2\pi f_u kT) - \sin(2\pi f_l kT)).$$

For this example, let the upper and lower frequency bounds be equal to 50% and 25% of the Nyquist frequency, respectively, as

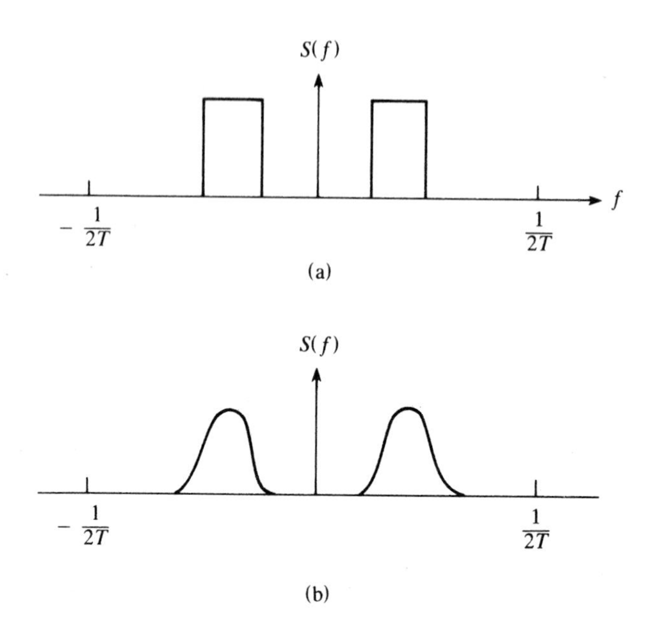

Figure 6.11 Ideal and model PSDs: (a) general band-pass and (b) second-order MA model with band-pass PSD and $T = 0.1$ s.

shown in Figure 6.11(a). Then

$$Z(k) = \frac{2\sigma^2}{\pi k f_N}(\sin(0.50\pi k - \sin(0.25\pi k))$$

and

$$R(k) = \frac{2\sigma^2}{\pi k}(\sin(0.50\pi k - \sin(0.25\pi k)), \qquad k \geq 1,$$

or

$$R(k) = [0.19\sigma^2, -0.32\sigma^2, -0.36\sigma^2, 0.0, 0.22\sigma^2, 0.11\sigma^2, \ldots].$$

If we had a second-order process with a variance of 1.0, the autocorrelation function would be

$$R(k) = [1.0, 0.19, -0.32, 0.0, \ldots].$$

The PSD is

$$S(f) = T\left(R(0) + 2\sum_{k=1}^{2} R(k)\cos(2\pi fkT)\right)$$

$$= T(1.0 + 2.0 \cdot 0.19\cos(2\pi fT) - 2.0 \cdot 0.32\cos(4\pi fT)),$$

shown in Figure 6.11(b) with $T = 0.1$ s. Notice that it only approximates the ideal model but nonetheless, it is a band-pass process.

References

Cadzow, J. (1987). *Foundations of Digital Signal Processing and Data Analysis*. Macmillan Publishing Co., New York.

Chatfield, C. (1975). *The Analysis of Time Series: Theory and Practice*. Halstead Press—John Wiley and Sons, New York.

Defatta, D., J. Lucas, and W. Hodgkiss (1988). *Digital Signal Processing: A System Design Approach*. John Wiley and Sons, New York.

Gath, I. and E. Yair (1987). Comparative evaluation of several pitch process models in the detection of vocal tremor. *IEEE Trans. Biomed. Engng.* 34:532–538.

Jenkins, G. and D. Watts (1968). *Spectral Analysis and its Applications*. Holden-Day, San Francisco.

Jong, M. (1982). *Methods of Discrete Signal and Systems Analysis*. McGraw-Hill Book Co., New York.

Koopmans, L. (1974). *The Spectral Analysis of Time Series*. Academic Press, New York.

Oppenheim, A. and A. Willsky (1983). *Signals and Systems*. Prentice-Hall, Englewood Cliffs, NJ.

Pandit, S. and S. Wu (1983). *Time Series Analysis with Applications*. John Wiley and Sons, New York.

Priestley, M. (1981). *Spectral Analysis and Time Series*, Vol. 1: *Univariate Series*. Academic Press, New York.

Rabiner, L. and R. Schafer (1978). *Digital Processing of Speech Signals*. Prentice-Hall, Englewood Cliffs, NJ.

Zeimer, R., W. Tranter, and D. Fannin (1987). *Signals and Systems—Continuous and Discrete*. Macmillan Publishing Co., New York.

Exercises

6.1 In Section 6.2 are listed properties of the PSD. Prove properties a and b, that the PSD is an even and real function.

6.2 Prove that $R_{yx}(k) = R_{xy}(-k)$.

6.3 Using

$$R_y(k) = E[y(n)y(n+k)]$$

$$= \sum_{i=0}^{q} \sum_{l=0}^{q} b(i)b(l)R_x(k+i-l),$$

show that $R_y(k) = 0$ for $|k| > q$.

6.4 For the following systems, find the transfer functions and express them in polar form:

a. $y(n) = x(n) - 1.4x(n-1) + 0.48x(n-2)$.

b. $y(n) + 0.25y(n-1) - 0.5y(n-2) + 0.7y(n-3) = x(n)$.

c. $y(n) - 0.89y(n-1) + 0.61y(n-2)$
$= x(n) + 0.54x(n-1)$.

6.5 Using the basic definition of an MA process [equation (6.36)], prove that the variance of the output is

$$\sigma_y^2 = \sigma_x^2 \sum_{l=0}^{q} b(l)^2$$

if $x(n)$ is zero-mean white noise.

6.6 Develop the power transfer function relationship [equation (6.35)] from the equation

$$S_y(f) = \sum_{k=-\infty}^{\infty} \sum_{i=-\infty}^{\infty} \sum_{l=-\infty}^{\infty} h(i)h(l)R_x(k+i-l)e^{-j2\pi fkT}.$$

[*Hint*: Sum over k first.]

6.7 Derive the relationship between the autocorrelation function of the system output and the cross correlation between the input and the output. [Start with the convolution relationship and multiply by $y(n-k)$.]

6.8 A boxcar, or rectangular, moving average is described by the MA system relationship,

$$y(n) = \frac{1}{q+1} \sum_{l=0}^{q} x(n+l).$$

For $q = 3$ and $R_x(k) = 4\delta(k)$, show that

$$R_y(k) = 4\left(1 - \frac{|k|}{4}\right) \quad \text{for } |k| \leq 3.$$

6.9 For a second-order MA process,

$$y(n) = x(n) - 1.4x(n-1) + 0.48x(n-2),$$

do the following:

a. Derive σ_y^2 and $R_y(k)$ when $\sigma_x^2 = 1, 5$;
b. Generate 50 points of $y(n)$ with $\sigma_x^2 = 5$ and plot them:
c. Make the scatter plots for $y(n)$ versus $y(n-2)$ and $y(n)$ versus $y(n-4)$; are these plots consistent with the theory?

6.10 Start with a general first-order AR system with a white noise input having a zero mean. Prove that $E[y(3)x(4)] = 0$.

6.11 Show in detail that, for a first-order AR system with a white-noise input, $\rho(2) = a(1)^2$.

6.12 Show that the second term on the right-hand side of equation (6.49) is zero.

6.13 For the third-order AR model

$$y(n) + 0.25y(n-1) - 0.5y(n-2) + 0.7y(n-3) = x(n),$$

show that

$$R_y(0) + 0.25R_y(1) - 0.5R_y(2) + 0.7R_y(3) = \sigma_x^2.$$

Remember that $R_y(k)$ is an even function.

6.14 For a general fourth-order AR system, find the equations that must be solved in order to use the recursion relationship for finding the values of the correlation function.

6.15 For the second-order AR system in Example 6.2, do the following:

a. Find the variance of the output signal if the variance of the input signal is 22.
b. Plot $R_y(k)$ for $-10 \le k \le 10$.

6.16 For the third-order system in Exercise 6.13, do the following:

a. Find the recursion relationship for the correlation function.
b. Using the equations for $k = 1$ and $k = 2$, solve for the initial conditions for $\rho_y(k)$.
c. Generate the values of $\rho_y(k)$ for $0 \le k \le 10$.

6.17 Take the second-order AR system

$$y(n) - 0.79y(n-1) + 0.22y(n-2) = x(n)$$

and let $\sigma_x^2 = 10$. Generate and plot 100 points of the output signal and label it $y_1(n)$. Use another seed number for generating another set of values for $x(n)$ and generate another output signal, $y_2(n)$. Are $y_1(n)$ and $y_2(n)$ exactly the same? Are their qualitative characteristics similar?

6.18 Prove that the normalized PSD $S_N(f) = S(f)/\sigma^2$ and the correlation function $\rho(k)$ are a Fourier transform pair.

6.19 Prove that

$$S(f) = R(0) + 2\sum_{k=1}^{\infty} R(k)\cos(2\pi fkT).$$

[*Hint*: Remember that $R(k)$ is an even function.]

6.20 Sketch Figure 6.5 with $T = 5.0$ s, 0.5 s, and 0.5 ms.

6.21 Derive the PSD, shown in Figure 6.8(d), of the first-order, high-frequency MA process in equation (6.62); $T = 2$ ms and $\sigma_x^2 = 0.25$.

6.22 For the high-pass process

$$y(n) = 0.5x(n) - 0.5x(n-1),$$

with $\sigma_x^2 - 0.25$, calculate and plot the PSD for $T = 1$ s and $T = 10$ s.

6.23 Calculate and plot the PSD for the following first-order MA processes:

a. $y(n) = 0.7x(n) + 0.3x(n-1)$, $T = 1$ s, $\sigma_x^2 = 2$;

b. $y(n) = 0.4x(n) - 0.6x(n-1)$, $T = 1$ ms, $\sigma_x^2 = 1$.

Are they high-pass or low-pass processes? Verify that the area under $S(f)$ equals σ_y^2.

6.24 Derive and sketch the PSD for the following second-order MA processes:

a. $y(n) = x(n) + x(n-1) + x(n-2)$, $T = 1$ s, $\sigma_x^2 = 1$;

b. $y(n) = x(n) + 0.5x(n-1) - 0.3x(n-2)$, $T = 1$ ms, $\sigma_x^2 = 5$.

6.25 Show that the process

$$y(n) = x(n) + 0.8x(n-1) + 0.5x(n-2),$$

$$T = 1 \text{ s}, \sigma_x^2 = 1,$$

has the normalized power density spectrum

$$S^*(f) = 1 + 1.27\cos(2\pi f) + 0.53\cos(4\pi f).$$

6.26 Develop the ACF for a third-order model of a low-pass MA process with $f_u = 0.75 f_N$ and $\sigma^2 = 25$.

a. What are the ACF and PSD?

b. Plot them for parametric T and $T = 20$ s.

6.27 Develop the ACF for a third-order model of a high-pass MA process with $f_l = 0.35 f_N$ and $\sigma^2 = 100$.

a. What are the ACF and PSD?

b. Plot them for parametric T and $T = 20$ ms.

Chapter 7 Spectral Analysis for Random Signals—Classical Estimation

7.1 Spectral Estimation Concepts

The determination of the frequency bands that contain energy or power in a sample function of a stationary random signal is called *spectral analysis*. The mode in which this information is presented is an energy or power spectrum. It is the counterpart of the Fourier spectrum for deterministic signals. The shape of the spectrum and the frequency components contain important information about the phenomena being studied. Two applications will be shown before the methodologies are studied. Vibrations are an intrinsic component of rotating machinery. Excessive levels indicate malfunction of some component. Consider Figure 7.1. It is the power spectrum of the vibration measurements from a 1/15-hp electric motor [Noori and Hakimmashhadi (1988)]. Vibrations at different frequencies are normal and are created by various moving parts. Some of the frequency components and their sources are indicated in the figure. A malfunction, such as worn-out bearings or loose components, is indicated when its corresponding frequency component has too large a magnitude. Thus, the spectrum is a valid tool for machine diagnosis. Another application is in the investigation of the behavior of the smooth muscle of the intestines by measuring the electrical activity, the electrogastrogram (EGG),

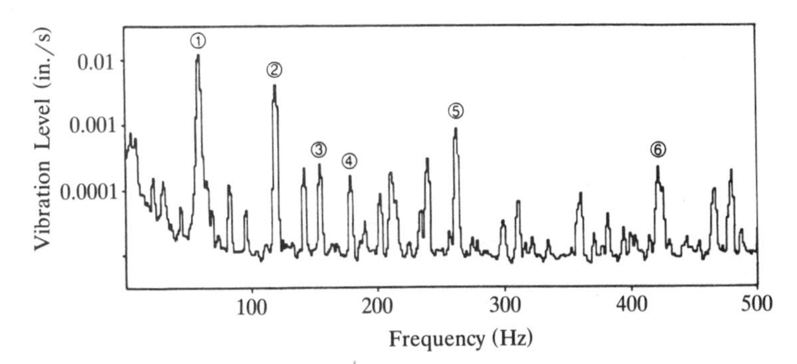

Figure 7.1 Vibration spectrum of a 1/15-hp electric motor. The frequencies of the peaks in the spectrum correspond to various motions in the motor: (1) 60 Hz, rotor imbalance; (2) 120 Hz, electrically induced vibration; (3) 155 Hz, bearing outer race; (4) 180 Hz, bent shaft; (5) 262 Hz, bearing inner race; (6) 420 Hz, bearing rolling elements. [Form Noori and Hakimmashhadi (1988), Figure 18.39, with permission.]

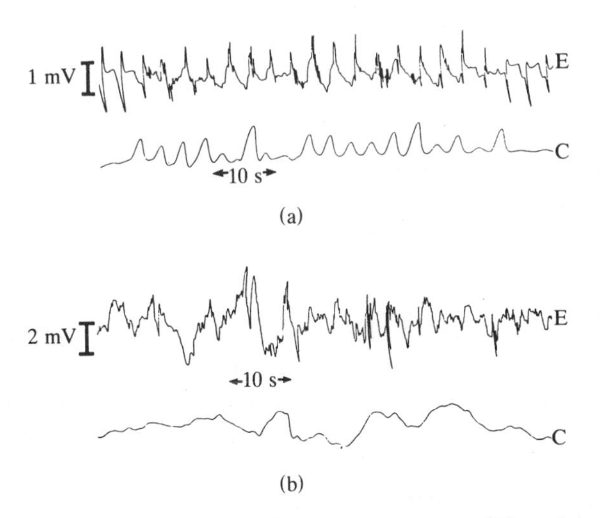

Figure 7.2 Typical electrogastrograms (E) and contractile activity (C) from (a) the duodenum and (b) the colon in a dog. [From Reddy, Collins, and Daniel (1987), Figure 1, with permission.]

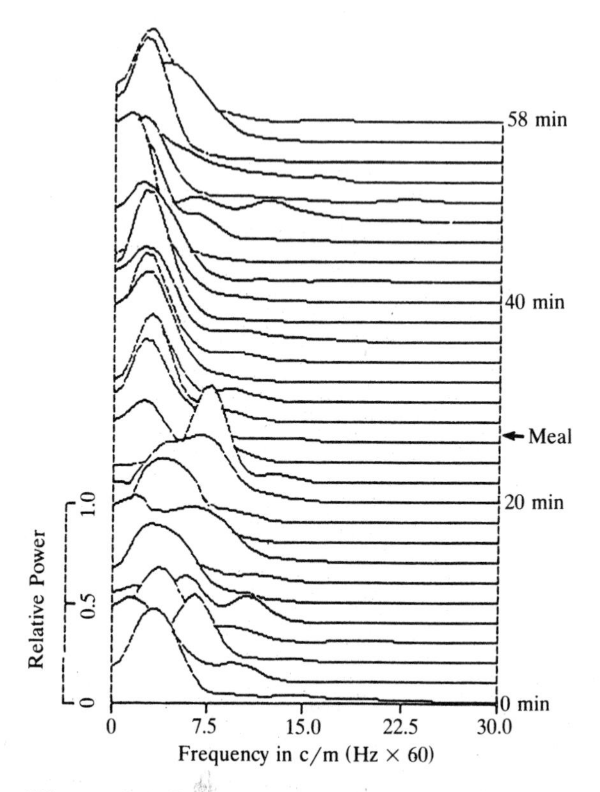

Figure 7.3 Power spectra from EGG of colon. Each spectrum is calculated from successive 2-min time periods. [From Reddy, Collins, and Daniel (1987), Figure 7b, with permission.]

generated during contraction. The smooth muscle plays an important role in the regulation of gastrointestinal motility and the EGG reflects the mode of muscular contraction. Figure 7.2 shows a measurement from the colon of an experimental dog [Reddy, Collins, and Daniel (1987)]. The activity is irregular. However, when a specific stimulus such as eating, is impressed, the activity pattern changes dramatically and becomes regular. Spectra from consecutive two-minute periods are shown in Figure 7.3. Before eating, the peaks of the power spectra occur at different frequencies over time. Notice that, after eating, the peaks occur at approximately the same frequency, indicating a change in function. These are only a sample of the many existing applications. The task now is to learn how to estimate these spectra accurately.

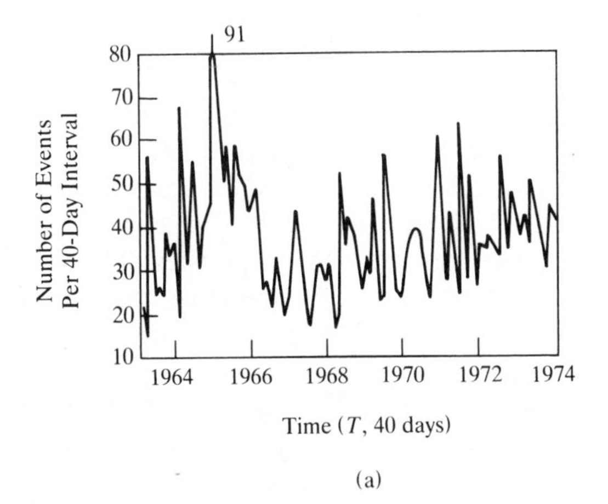

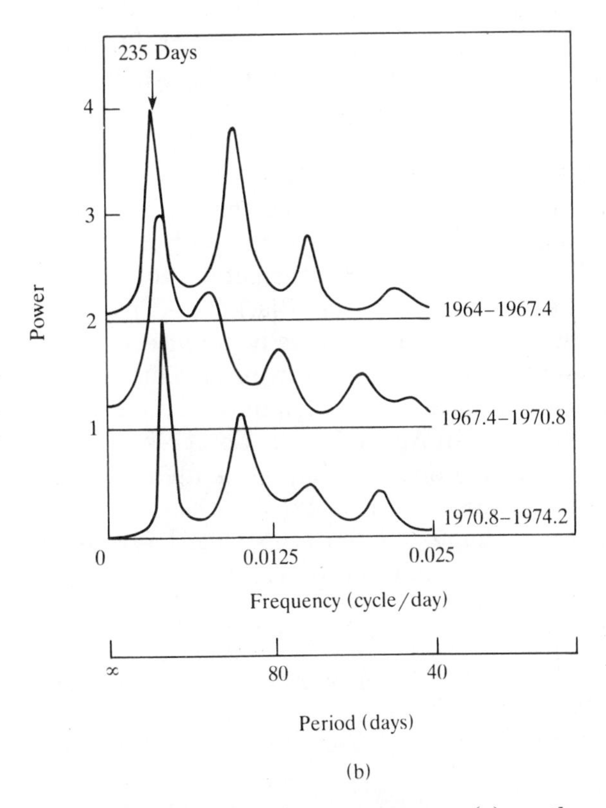

Figure 7.4 Earthquake occurrences: (a) earthquake occurrence rate from 1964–1974; (b) spectra for three nonoverlapping time periods. [From Landers and Lacoss (1977, IEEE), Figures 4 and 5, with permission.]

In general, there are two basic methods that are considered classical approaches. The modern methods will be studied in the next chapter. One classical approach is to develop an estimator from the Fourier transform of the sample function. This is called the direct, or *periodogram*, method and is based on Parseval's theorem. The name evolved in the context of the first applications of spectral analysis in astronomy and geophysics. The frequencies were very low, on the order of cycles per week, and it was more convenient to use the period, or $1/f$, as the independent variable. Figure 7.4 is an example and shows a signal of earthquake occurrences and the estimated spectra. More will be stated about these spectra in subsequent sections. The other approach is based on the fact that the spectral density function is the Fourier transform of the autocovariance function, as discussed in Section 6.2. With this method an estimator of the PSD is derived from the Fourier transform of the estimator of the autocorrelation function. This is called the indirect, or *Blackman–Tukey* (BT) method.

7.1.1 Developing Procedures

The classical estimators and their properties are developed directly from the definition of the PSD, and both methods have the same statistical properties. Their mathematical equivalence will be derived from the definition of the Blackman–Tukey estimator. Complete derivations are mathematically intense but are understandable for someone knowledgable in probability and statistics. Some properties are derived in this chapter and the expressions for the variance are derived in Appendix 7.1. Comprehensive presentations can be found in references such as Geckinli and Yavuz (1983), Jenkins and Watts (1968), and Priestley (1981).

Recall from Chapter 5 that the biased estimator of the autocorrelation function, equation (5.27), is

$$\hat{R}(k) = \begin{cases} \dfrac{1}{N} \displaystyle\sum_{n=0}^{N-|k|-1} x(n)x(n+k), & |k| \leq M, \\ 0, & |k| > M. \end{cases} \tag{7.1}$$

The bounds on the maximum lag M are typically $0.1N \leq M \leq$

$0.3N$. The BT estimator is

$$\hat{S}(f) = T \sum_{k=-M}^{M} \hat{R}(k) w(k) e^{-j2\pi fkT}, \tag{7.2}$$

where $w(k)$ is a window function. The other estimator, the periodogram, can be obtained without first estimating $\hat{R}(k)$. This is done by incorporating the rectangular data window $d_R(n)$ into equation (7.1). Recall from Section 3.4 that the rectangular data window is defined as

$$d_R(n) = \begin{cases} 1, & 0 \leq n \leq N-1, \\ 0, & \text{elsewhere.} \end{cases} \tag{7.3}$$

Now equation (7.1) is rewritten as

$$\hat{R}(k) = \frac{1}{N} \sum_{n=-\infty}^{\infty} x(n) d_R(n) x(n+k) d_R(n+k). \tag{7.4}$$

Its Fourier transform is

$$\hat{S}(f) = T \sum_{k=-\infty}^{\infty} \left(\frac{1}{N} \sum_{n=-\infty}^{\infty} x(n) d_R(n) x(n+k) d_R(n+k) \right) \times e^{-j2\pi fkT}. \tag{7.5}$$

The equation will be rearranged to sum over the index k first, or

$$\hat{S}(f) = \frac{T}{N} \sum_{n=-\infty}^{\infty} x(n) d_R(n) \sum_{k=-\infty}^{\infty} x(n+k) d_R(n+k) e^{-j2\pi fkT}. \tag{7.6}$$

Recognizing that the second summation will be a Fourier transform with the substitution $l = n + k$, equation (7.6) becomes

$$\hat{S}(f) = \frac{T}{N} \sum_{n=-\infty}^{\infty} x(n) d_R(n) e^{j2\pi fnT} \sum_{l=-\infty}^{\infty} x(l) d_R(l) e^{-j2\pi flT}. \tag{7.7}$$

The second summation of equation (7.7) is proportional to the Fourier transform of the sample function $x(l)d_R(l)$, a signal with finite energy because it is a truncated version of $x(n)$. Its DTFT is

$$\hat{X}(f) = T \sum_{l=-\infty}^{\infty} x(l) d_R(l) e^{-j2\pi flT}. \tag{7.8}$$

The first summation is also proportional to a Fourier transform, with $f = -(-f)$, or

$$\hat{S}(f) = \frac{1}{NT} \hat{X}(-f) \hat{X}(f) = \frac{1}{NT} \hat{X}^*(f) \hat{X}(f), \tag{7.9}$$

where $\hat{X}^*(f)$ is the complex conjugate of the estimate of the Fourier transform. Equation (7.9) defines the periodogram estimate. Because the periodogram is calculated with a different procedure than the BT estimator, it is also given a different symbol and is represented as

$$I(f) = \frac{1}{NT} \hat{X}^*(f) \hat{X}(f). \tag{7.10}$$

7.1.2 Sampling Moments of Estimators

Knowledge of the sampling properties of the PSD estimators is essential and provides the framework for additional steps that will improve the estimation procedure. The mean of the spectral estimate is defined as

$$E[\hat{S}(f)] = E\left[T \sum_{k=-\infty}^{\infty} \hat{R}(k) e^{-j2\pi fkT}\right]. \tag{7.11}$$

Substituting for $\hat{R}(k)$ from equation (7.4) yields

$$E[\hat{S}(f)] = \frac{T}{N}\sum_{k=-\infty}^{\infty}\left(E\left[\sum_{n=-\infty}^{\infty} x(n)d_R(n)x(n+k)d_R(n+k)\right]\right)$$

$$\times e^{-j2\pi fkT}$$

$$= \frac{T}{N}\sum_{k=-\infty}^{\infty}\left(\sum_{n=-\infty}^{\infty} E[x(n)x(n+k)]d_R(n)d_R(n+k)\right)$$

$$\times e^{-j2\pi fkT}$$

$$= \frac{T}{N}\sum_{k=-\infty}^{\infty} R(k)e^{-j2\pi fkT}\sum_{n=-\infty}^{\infty} d_R(n)d_R(n+k). \tag{7.12}$$

The second summation results from the implied data window and is a correlation of the rectangular data window with itself. It is defined separately as

$$w(k) = \frac{1}{N}\sum_{n=-\infty}^{\infty} d_R(n)d_R(n+k)$$

$$= \begin{cases} 1 - \dfrac{|k|}{N}, & |k| \le N-1, \\ 0, & \text{elsewhere.} \end{cases} \tag{7.13}$$

It has a triangular shape in the lag domain and is plotted in Figure 7.5(a). Equation (7.11) is rewritten as

$$E[\hat{S}(f)] = T\sum_{k=-\infty}^{\infty} w(k)R(k)e^{-j2\pi fkT}. \tag{7.14}$$

Thus, the mean value of the PSD differs from the actual PSD because $R(k)$ is multiplied by a window that is called a *lag*

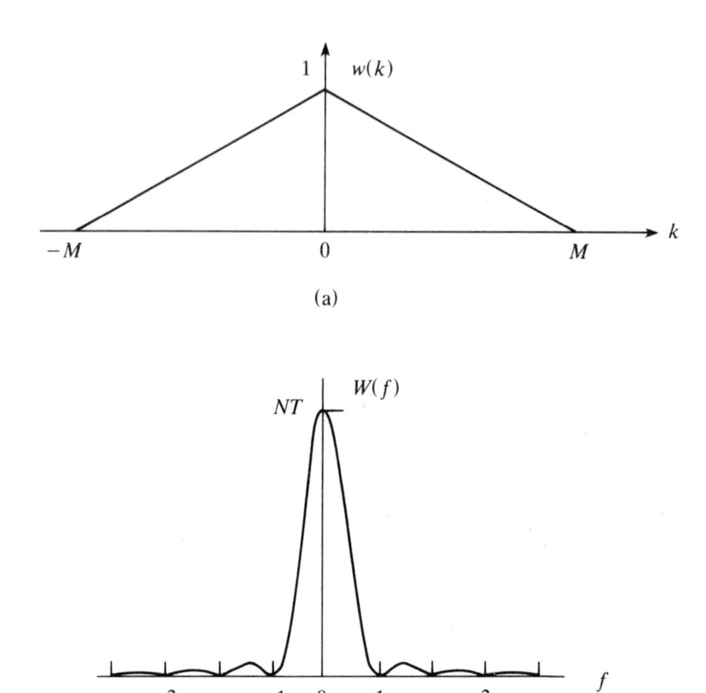

Figure 7.5 Implicit window function, Fejer's kernel: (a) triangular lag window; (b) corresponding spectral window.

window. Its effect is appreciated directly by invoking the convolution theorem. In the frequency domain, equation (7.14) becomes

$$E[\hat{S}(f)] = \int_{-1/2T}^{1/2T} S(g)W(f-g)\,dg. \tag{7.15}$$

The function $W(f)$, the Fourier transform of $w(k)$, is called the *spectral window* and is plotted in Figure 7.5(b). This particular spectral window is called *Fejer's kernel* and has the formula

$$W(f) = \frac{T}{N}\left(\frac{\sin(\pi fNT)}{\sin(\pi fT)}\right)^2. \tag{7.16}$$

It has a main lobe and side lobes similar to the other windows studied in Chapter 3. The convolution operation in equation (7.15) produces, in general, a biased estimate of $S(f)$. As N approaches infinity, Fejer's kernel approaches a delta function and the estimator becomes asymptotically unbiased.

Example 7.1

For the purpose of demonstrating the implicit effect of a finite signal sample on the bias, presume that the data window produces the spectral window shown in Figure 7.5 and that the actual PSD has a triangular shape, as shown in Figure 7.6. The convolution results in the spectrum plotted with the dashed line in Figure 7.6. Compare these spectra. Notice that some magnitudes are underestimated and others are overestimated.

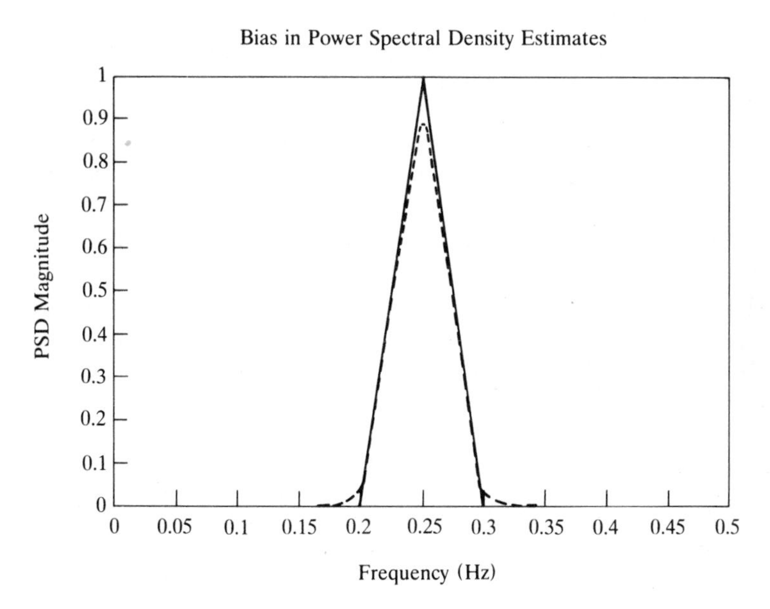

Figure 7.6 Schematic representation of bias caused by finite sample function: (—), actual PSD; (---), biased estimate of PSD.

More quantitative knowledge of the bias is available if advanced theory is considered. It is sufficient to state for now that the bias term is on the order of $(\log N)/N$ if the first derivative of $S(f)$ is continuous. Mathematically, this is expressed as

$$E[\hat{S}(f)] = S(f) + \mathcal{O}\left(\frac{\log N}{N}\right). \tag{7.17}$$

It is consistent with intuition that if the sample function has a large number of sample points, then the bias is small or negligible.

The derivation of the variance of the sample PSD is very complicated mathematically and is usually studied at advanced levels. A complete derivation for large N is described by Jenkins and Watts (1968) and is summarized in Appendix 7.1. Several sources derive the variance of a white noise PSD, e.g., Kay (1988). For now, let it suffice to state that the sample variance is

$$\operatorname{Var}[\hat{S}(f)] = S^2(f)\left(1 + \left(\frac{\sin(2\pi fNT)}{N\sin(2\pi fT)}\right)^2\right) \tag{7.18}$$

when N is large. Study equation (7.18) closely; notice that only the second term on the right-hand side is a function of N. The variance is inconsistent and equals the square of the magnitude of the actual PSD for large values of N. Essentially, this is not a good situation. The estimators developed from basic definitions are biased and inconsistent, both undesired properties. Techniques that improve the properties of these spectral estimators will be studied after considering their sampling distribution. The sampling distribution will be derived from the periodogram representation. From now on, the calculated version of the spectral estimators will also be used. Thus, the periodogram will be calculated at discrete frequencies $f = m/NT$. The change in notation is $I(m) = I(m/NT)$ and

$$I(m) = \frac{1}{NT}\hat{X}(m)\hat{X}^*(m). \tag{7.19}$$

Example 7.2

The sampling theory for power spectra shows that their estimates are inconsistent. This can be easily demonstrated through simulation. A random number generator was used to produce two

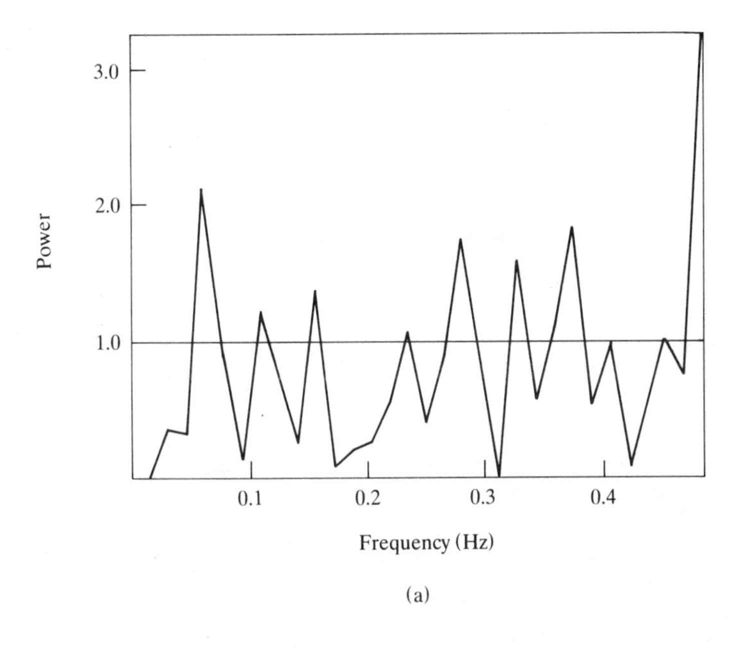

(a)

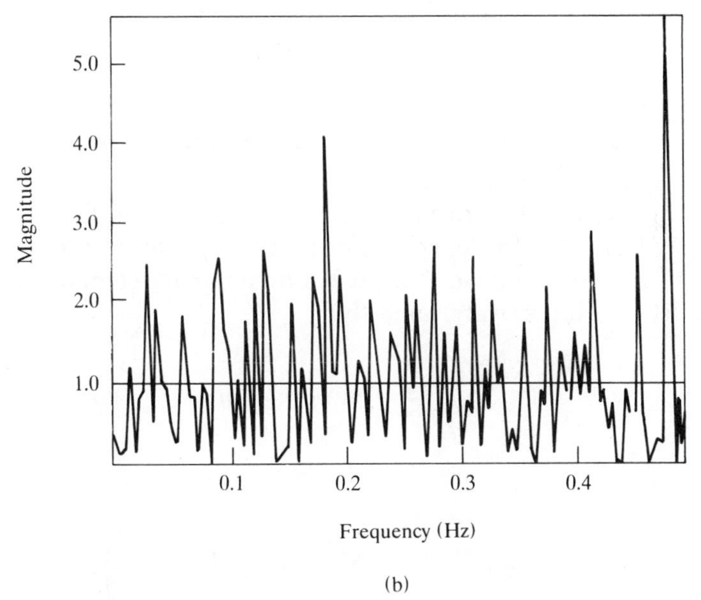

(b)

Figure 7.7 Estimates of the PSD of a uniform white noise process with $\sigma^2 = 1$, and $T = 1$ are plotted: (a) $N = 64$; (b) $N = 256$.

uniform white noise processes $x_1(n)$ and $x_2(n)$ with 64 and 256 points, respectively. The signal amplitudes were scaled so that $\sigma^2 = 1$ and the sampling interval T is 1. The periodograms were produced using equation (7.19). The theoretical spectrum and the periodograms' estimates are plotted in Figure 7.7. Notice that there is great variation in the magnitude of the estimated spectra. They do not resemble the white noise spectra and the increase in N does not improve the quality of the estimate. Only the frequency spacing has been reduced.

7.2 Sampling Distribution for Spectral Estimators

The sampling distribution of the spectral density function and the relationships among its harmonic components are very important for understanding the properties of PSD estimators. The derivations of some of these properties are presented and the derivations of others are left as exercises.

7.2.1 Spectral Estimate for White Noise

Let $x(n)$ represent an N-point sample of a Gaussian white noise process with zero mean and variance σ^2. Its DFT is defined as usual and for the purposes of this derivation is represented in complex Cartesian form with real and imaginary components $A(m)$ and $B(m)$, respectively, such that

$$\frac{X(m)}{\sqrt{NT}} = A(m) - jB(m),$$

with

$$\begin{aligned} A(m) &= \sqrt{\frac{T}{N}} \sum_{n=0}^{N-1} x(n)\cos\left(\frac{2\pi mn}{N}\right), \\ B(m) &= \sqrt{\frac{T}{N}} \sum_{n=0}^{N-1} x(n)\sin\left(\frac{2\pi mn}{N}\right). \end{aligned} \tag{7.20}$$

The periodogram as defined in equation (7.19) then becomes

$$I(m) = A^2(m) + B^2(m), \qquad m = 0, 1, \ldots, [N/2], \quad (7.21)$$

where $[N/2]$ represents the integer value of $N/2$. It can be shown that the estimates of the magnitudes of the PSD have a χ^2 distribution and that the harmonic components are uncorrelated with one another [Jenkins and Watts (1968)]. (They are also independent because they also have a Gaussian distribution.) The first- and second-order moments will be discussed first.

7.2.1.1 Moments

Because $A(m)$ and $B(m)$ are linear combinations of a Gaussian random variable [equations (7.20)], they are also Gaussian random variables. The mean value of the real part of the harmonic components is

$$E[A(m)] = E\left[\sqrt{\frac{T}{N}} \sum_{n=0}^{N-1} x(n) \cos\left(\frac{2\pi mn}{N}\right)\right]$$

$$= \sqrt{\frac{T}{N}} \sum_{n=0}^{N-1} E[x(n)] \cos\left(\frac{2\pi mn}{N}\right) = 0. \quad (7.22)$$

The mean value of the imaginary part $B(m)$ of the harmonic components is also zero. The variance is now written as the mean square and is, for the real part,

$$\operatorname{Var}[A(m)] = E[A^2(m)]$$

$$= E\left[\left(\sqrt{\frac{T}{N}} \sum_{n=0}^{N-1} x(n) \cos\left(\frac{2\pi mn}{N}\right)\right)^2\right]. \quad (7.23)$$

This operation requires evaluating the mean values of all the cross products. Because $x(n)$ is white noise, by definition they are zero and

$$E[A^2(m)] = \frac{T}{N}\sum_{n=0}^{N-1} E[x^2(n)]\left(\cos\left(\frac{2\pi mn}{N}\right)\right)^2$$

$$= \frac{T\sigma^2}{N}\sum_{n=0}^{N-1}\cos^2\left(\frac{2\pi mn}{N}\right). \tag{7.24}$$

After some extensive but straightforward algebra with complex variables, it can be shown that

$$E[A^2(m)] = \frac{T\sigma^2}{N}\left(\frac{N}{2} + \cos\left(\frac{2\pi m(N-1)}{N}\right)\frac{\sin(2\pi m)}{2\sin(2\pi m/N)}\right)$$

$$= \begin{cases} \dfrac{T\sigma^2}{2}, & m \neq 0 \text{ and } \left[\dfrac{N}{2}\right], \\ T\sigma^2, & m = 0 \text{ and } \left[\dfrac{N}{2}\right]. \end{cases} \tag{7.25}$$

The derivation is left as an exercise. The term $B^2(m)$ is also a random variable with the same properties as $A^2(m)$. The next task is to define the correlational properties. The covariance of the real and imaginary parts of the DFT can be derived using the trigonometric or complex exponential forms. The complex exponentials provide a shorter derivation. Start with the expression for the covariance of different harmonics of the DFT,

$$E[X(m)X^*(p)] = E\left[T\sum_{n=0}^{N-1} x(n)e^{-j2\pi mn/N} \cdot T\sum_{l=0}^{N-1} x(l)e^{j2\pi pl/N}\right]$$

$$= T^2\sum_{n=0}^{N-1}\sum_{l=0}^{N-1} E[x(n)x(l)]e^{-j2\pi mn/N}e^{j2\pi pl/N}$$

$$= T^2\sum_{n=0}^{N-1}\sigma^2 e^{-j2\pi(m-p)n/N}. \tag{7.26}$$

Using the geometric sum formula, equation (7.26) becomes

$$E[X(m)X^*(p)] = T^2\sigma^2 \frac{1 - e^{-j2\pi(m-p)}}{1 - e^{-j2\pi(m-p)/N}} = \begin{cases} NT^2\sigma^2, & m = p, \\ 0, & \text{otherwise.} \end{cases} \quad (7.27)$$

It can be similarly shown that

$$E[X(m)X(p)] = \begin{cases} NT^2\sigma^2, & m = p = 0 \text{ and } m = p = [N/2], \\ 0, & \text{otherwise.} \end{cases} \quad (7.28)$$

Thus, the harmonic components of the Fourier transform are uncorrelated with one another. Using the last two relationships, it can be shown that the real and imaginary components at different frequencies are also uncorrelated. First express them in real and imaginary parts, or

$$E[X(m)X(l)] = E[(A(m) - jB(m))(A(l) - jB(l))] \quad (7.29)$$

and

$$E[X(m)X^*(l)] = E[(A(m) - jB(m))(A(l) + jB(l))]. \quad (7.30)$$

Solving these two equations simultaneously yields

$$\begin{aligned} \text{Cov}[A(m), A(l)] &= \text{Cov}[B(m), B(l)] = 0, \quad m \neq l, \\ \text{Cov}[A(m), B(l)] &= \text{Cov}[B(m), A(l)] = 0, \quad m \neq l. \end{aligned} \quad (7.31)$$

7.2.1.2 Sample Distribution

The density function for the periodogram needs to be known so that confidence intervals can be developed for the estimators. The pdf can be defined because it is known that the sum of squared independent Gaussian variables with a unit variance form a chi-square (χ_ν^2) random variable. The number of degrees of freedom ν is equal to the number of independent terms summed. Through standardization a function of $I(m)$ can be made to have unit variance. Thus, standardizing and squaring $A(m)$ and $B(m)$ yields

$$\frac{A^2(m)}{T\sigma^2/2} + \frac{B^2(m)}{T\sigma^2/2} = \frac{I(m)}{T\sigma^2/2} = \chi_2^2, \qquad m \neq 0 \text{ and } \left[\frac{N}{2}\right]. \quad (7.32)$$

Because the zero and Nyquist frequency terms have only real parts,

$$\frac{I(m)}{T\sigma^2} = \frac{A^2(m)}{T\sigma^2} = \chi_1^2, \qquad m = 0 \text{ and } \left[\frac{N}{2}\right]. \quad (7.33)$$

The mean and variance of the periodogram can now be derived from the chi-square random variable. It is known that

$$E\left[\chi_\nu^2\right] = \nu \quad \text{and} \quad \text{Var}\left[\chi_\nu^2\right] = 2\nu. \quad (7.34)$$

Thus, for all frequency components except for $m = 0$ and $[N/2]$,

$$E\left[\frac{I(m)}{T\sigma^2/2}\right] = 2 \quad \text{or} \quad E[I(m)] = T\sigma^2 \quad (7.35)$$

and

$$\text{Var}\left[\frac{I(m)}{T\sigma^2/2}\right] = 4 \quad \text{or} \quad \text{Var}[I(m)] = T^2\sigma^4. \quad (7.36)$$

It can be similarly shown that, for the frequencies $m = 0$ and $[N/2]$,

$$E[I(m)] = T\sigma^2 \quad \text{and} \quad \text{Var}[I(m)] = 2T^2\sigma^4. \quad (7.37)$$

This verifies the general results observed in the previous section. The estimation procedure is inconsistent and the variance of the estimate is equal to the square of the actual PSD magnitude. Let us now study several examples of results from estimating the PSD of a known independent process.

7.2.2 Sampling Properties for General Random Processes

The theoretical derivations of the sampling properties of a white noise process can be extended to the general random process with the use of signal models. Any random process can be expressed as a weighted sum of white-noise values. If the sum has a finite number of terms, then the process is a moving-average one; if the sum is infinite, then the process is an AR or ARMA one. The general model is

$$y(n) = \sum_{l=-\infty}^{\infty} h(l)x(n-l), \tag{7.38}$$

where $x(n)$ is a white noise process. Referring back to Chapters 3 and 6, this also is the form of a convolution between a sequence of coefficients $[h(n)]$ and $x(n)$. The Fourier transform of equation (7.38) is

$$Y(m) = H(m)X(m), \tag{7.39}$$

with

$$H(m) = T\sum_{n=-\infty}^{\infty} h(n)e^{-j2\pi mn/N}. \tag{7.40}$$

For a sample function of finite duration for the process $y(n)$, the periodogram is

$$I_y(m) = \frac{Y(m)Y^*(m)}{NT} = \frac{H(m)X(m)H^*(m)X^*(m)}{NT}$$

$$= |H(m)|^2 I_x(m). \tag{7.41}$$

All of the statistical variations are contained in the $x(n)$ process. It is desired to have an expression that relates the periodogram of $y(n)$ to its actual PSD, $S_y(m)$. This can be accomplished by again using a model for the random process with $S_x(m) = T\sigma^2$. For a system model it is known from Chapter 6 that

$$S_y(m) = |H(m)|^2 S_x(m). \tag{7.42}$$

Substituting equations (7.42) and (7.41) into equation (7.32) yields

$$\frac{2I_y(m)}{S_y(m)} = \chi_2^2, \qquad m \neq 0 \text{ and } \left[\frac{N}{2}\right],$$

and

$$\frac{I_y(m)}{S_y(m)} = \chi_1^2, \qquad m = 0 \text{ and } \left[\frac{N}{2}\right]. \tag{7.43}$$

Equation (7.43) defines the relationship between the periodogram estimate of a PSD and the actual spectrum. Using the moments of the χ_2^2 random variable, it can be seen that the estimate is unbiased,

$$E\left[\frac{2I_y(m)}{S_y(m)}\right] = \nu = 2 \quad \text{or} \quad E\left[I_y(m)\right] = S_y(m), \tag{7.44}$$

and inconsistent

$$\text{Var}\left[\frac{2I_y(m)}{S_y(m)}\right] = 2\nu = 4 \quad \text{or} \quad \text{Var}\left[I_y(m)\right] = S_y^2(m). \tag{7.45}$$

The same results apply for $m = 0$ and $[N/2]$. At first these results seem to agree with the bias and variance results stated in Section 7.1.2 for large N. This is true because in the convolution operation of equation (7.38) it is presumed that enough points of $x(n)$ are available to represent accurately $y(n)$. For signals with small N, additional processing must be done to make $I_y(m)$ unbiased so that equation (7.45) is appropriate. This situation will be addressed in the next section.

Equation (7.45) shows that the periodogram is not only inconsistent but also has a very large variance similar to the situation with periodogram of the white noise process. Several examples will serve to illustrate this point and then methods to reduce the variance will be discussed.

Example 7.3

Let us consider the random process with the AR model

$$y(n) = y(n-1) - 0.5y(n-2) + x(n),$$

with $T = 1$ and $x(n)$ being zero-mean white noise with $\sigma^2 = 1$. The power transfer function was derived in Example 6.4 and the PSD for $y(n)$ is

$$S_y(m) = \frac{1}{2.25 - 3\cos(2\pi m/N) + \cos(4\pi m/N)} S_x(m).$$

For this process the spectrum $S_y(m)$ equals the power transfer function that is plotted in Figure 6.5. There is appreciable power in the range of 0.05 to 0.15 Hz. The periodogram for this process is estimated from sample functions with $N = 64$ and $N = 256$. These are plotted in Figure 7.8. Notice that these estimates are very erratic in magnitude and that increasing N does not improve the quality of the estimation. Again, this shows that the variance of the estimate is large and inconsistent.

Example 7.4

Very seldom in real applications does $T = 1$. If, for a white noise process, $T = 0.1$ s and $T\sigma^2 = 1$, the estimated PSD would only change in the frequency scale as plotted in Figure 7.9(a). Now, obviously, $\sigma^2 = 10$. If the variance remained 1, then the PSD would appear as in Figure 7.9(b). Notice the additional change in scale to keep the area equal to the variance.

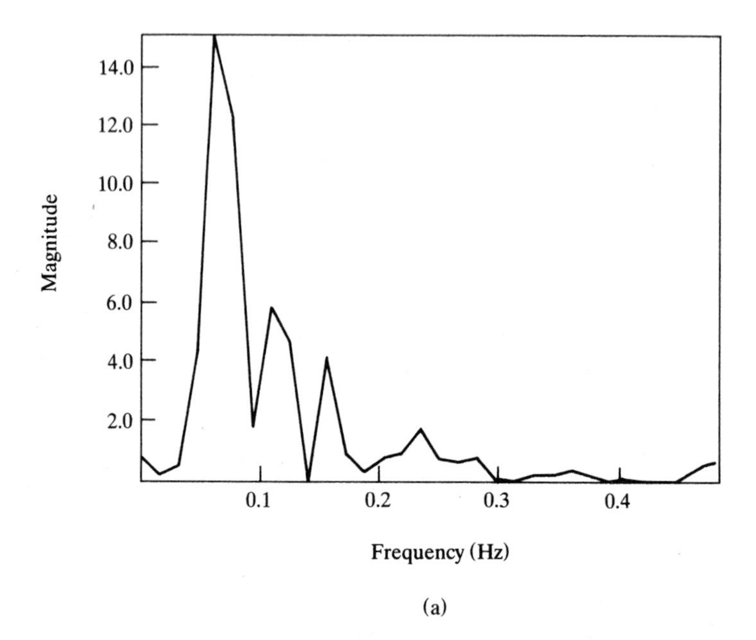

(a)

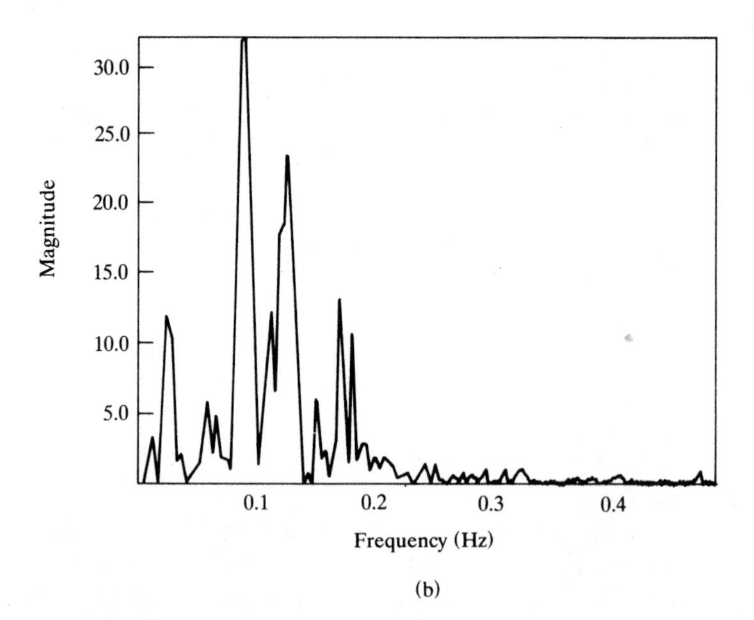

(b)

Figure 7.8 Estimates of the PSD of a Gaussian second-order AR process with $\sigma_x^2 = 1$, and $T = 1$ are plotted: (a) $N = 64$; (b) $N = 256$.

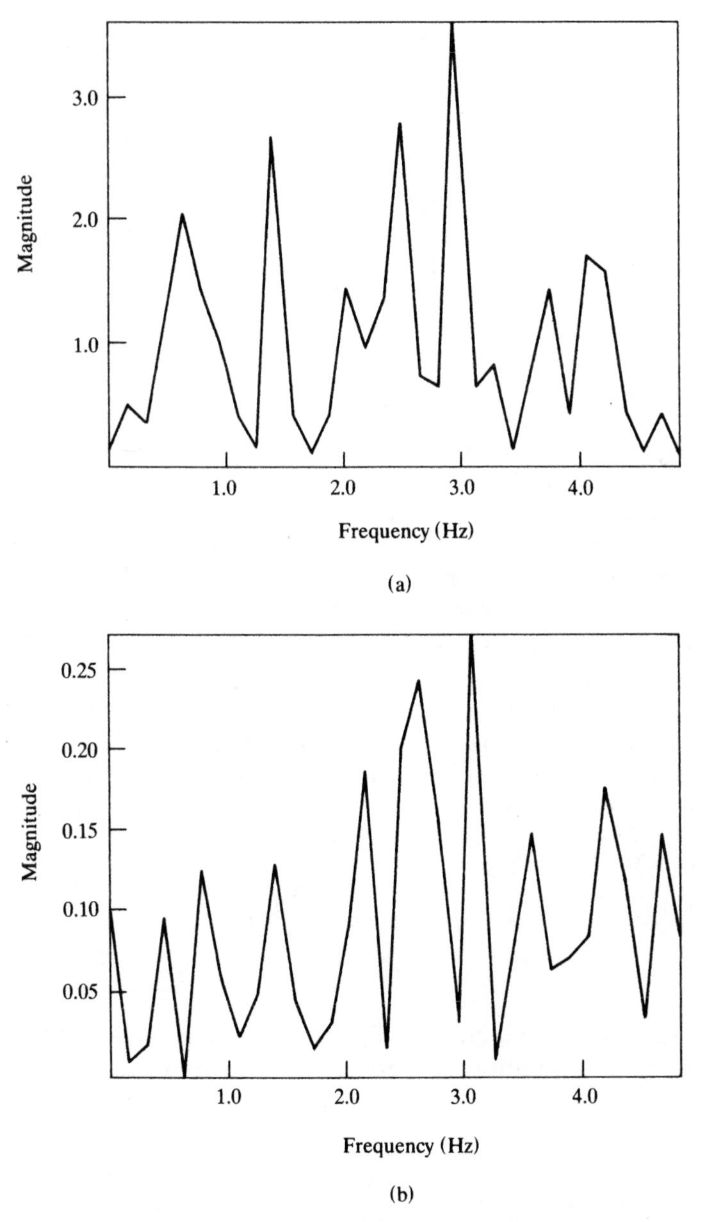

Figure 7.9 Estimates of the PSD of a uniform white noise process with 64 points; (a) $T = 0.1$, $\sigma^2 = 10$; (b) $T = 0.1$, $\sigma^2 = 1$.

7.3 Consistent Estimators — Direct Methods

During the 1950s and 1960s, much research was devoted to improving the properties of spectral estimators. The results were successful and produced useful techniques that are still used. As with most estimators there is a compromise. Any technique that reduces the variance of an estimate also increases its bias and vice versa. This trade-off will be discussed simultaneously with the explanation of the techniques.

7.3.1 Spectral Averaging

The simplest technique for developing a consistent estimator is called *periodogram averaging*, or the *Bartlett approach*. It is based upon the principle that the variance of a K-point sample average of independent random variables with variance σ^2 is σ^2/K. Thus, an ensemble average of K periodogram estimates produces an estimate whose variance is reduced by the factor K. In this situation, an N-point sample function is divided into K segments, each containing M points. This is illustrated in Figure 3.21(a); the signal is divided into 4 segments of 60 points. The periodogram for each segment $I_i(m)$ is estimated; the integer i is the index for the time series. The averaged periodogram is formed by averaging over all the periodograms at each frequency, or

$$I_K(m) = \frac{1}{K} \sum_{i=1}^{K} I_i(m). \tag{7.46}$$

Because the periodogram estimates are independent among each other, the summation in equation (7.46) is a summation of $2K$ squared terms if $m \neq 0$ or $M/2$; that is,

$$KI_K(m) = \sum_{i=1}^{K} A_i^2(m) + B_i^2(m), \tag{7.47}$$

and $I_K(m)$ has 2K degrees of freedom. Its relationship to the true PSD is still found by transforming it to be a chi-square random variable; thus,

$$\frac{2KI_K(m)}{S(m)} = \chi^2_{2K}, \qquad m \neq 0 \text{ and } \left[\frac{M}{2}\right] \tag{7.48}$$

and

$$E\left[\frac{2KI_K(m)}{S(m)}\right] = \nu = 2K \quad \text{or} \quad E[I_K(m)] = S(m) \tag{7.49}$$

and the estimator is unbiased. Remember that this bias property assumes a large N. The expectation of equation (7.46) is

$$E[I_K(m)] = \frac{1}{K}\sum_{i=1}^{K} E[I_i(m)] = E[I(m)], \qquad |m| \leq \left[\frac{M}{2}\right]. \tag{7.50}$$

The bias, which is essentially the result of leakage error (Section 3.4), is now defined exactly as in equations (7.14) and (7.15), with $M = N$. This bias is minimized in the same manner as the leakage was reduced for deterministic functions; that is, multiply the observed signal by a data window before the periodogram is produced. Now we have

$$y(n) = x(n)d(n), \qquad 0 \leq n \leq M - 1, \tag{7.51}$$

and

$$I_i(m) = \frac{1}{MT} Y_i(m) Y^*_i(m). \tag{7.52}$$

Either the Hamming or the Hanning data window is usually used for this purpose. The leakage error has been resolved but another source of constant bias has been introduced that is easily corrected. Remember that the variance is the area under the PSD. However, if equation (7.51) is considered together with the fact that the amplitude of a data window is between 0 and 1, the sample variance of $y(n)$ is

$$\hat{\sigma}_y^2 = E\left[\frac{1}{M}\sum_{n=0}^{M-1} x^2(n)d^2(n)\right] = \sigma_x^2 \frac{1}{M}\sum_{n=0}^{M-1} d^2(n) \le \sigma_x^2. \quad (7.53)$$

Thus, the variance of signal $y(n)$ is less than that of $x(n)$. The reduction factor is the average square value of the data window, as expressed in equation (7.53), and is called *process loss* (PL). To correct for PL, simply divide the periodogram estimate by the process loss factor. The PL factors of several windows are tabulated in Appendix 7.3. The corrected estimator with the windowed signal is approximately unbiased and is

$$I_i(m) = \frac{1}{\text{PL}}\frac{1}{MT} Y_i(m) Y_i^*(m). \quad (7.54)$$

Then it can be stated that

$$E[I_K(m)] = S(m), \qquad |m| \le \left[\frac{M}{2}\right] \quad (7.55)$$

and

$$\text{Var}\left[\frac{2KI_K(m)}{S(m)}\right] = 2\nu = 2\cdot 2K \quad \text{or} \quad \text{Var}[I_K(m)] = \frac{S^2(m)}{K}. \quad (7.56)$$

The periodogram averaging reduces the variance by a factor of K. This is not without a cost. Remember that only M points are used for each $I_i(m)$ and the frequency spacing becomes $1/MT$. Hence, the induced leakage before windowing is spread over a broader frequency range as K is increased.

Example 7.5

The estimate of the white-noise spectrum from Example 7.2 will be improved with spectral averaging. The time series with 256 points is subdivided into four segments and each segment is multi-

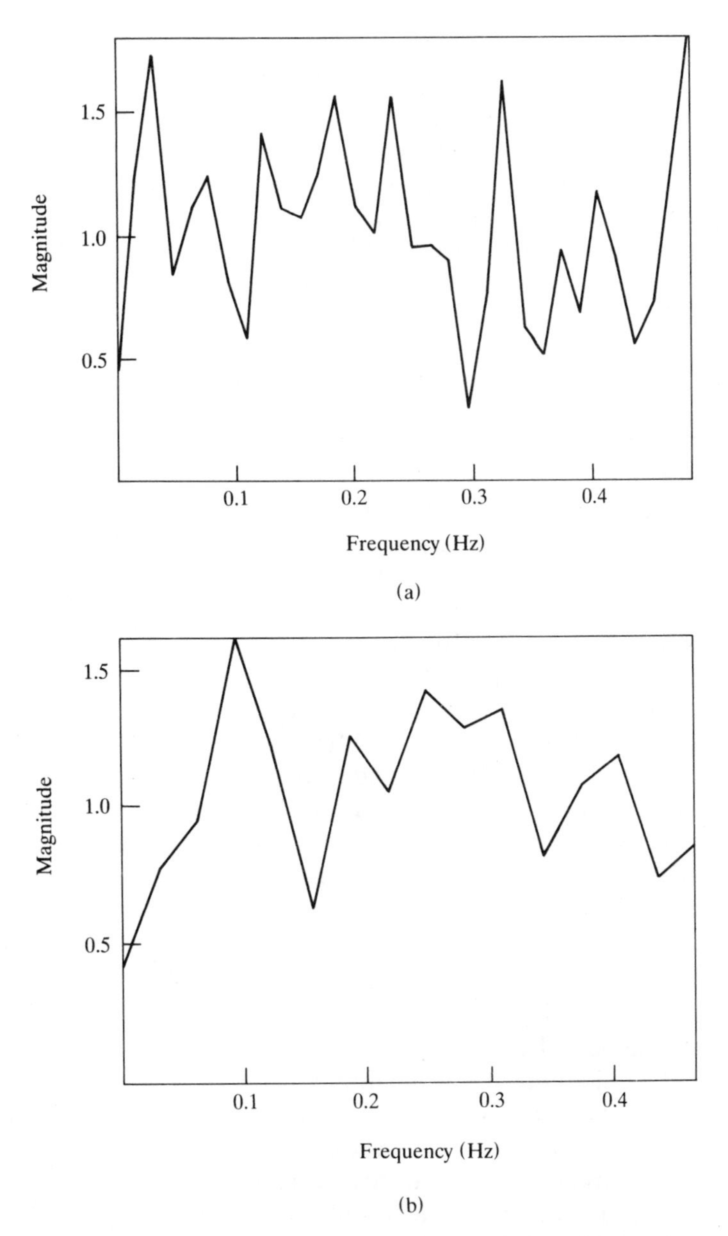

Figure 7.10 Estimates of a spectrum of the white noise process in Figure 7.7 using spectral averaging: (a) $N = 256$, $K = 4$; (b) $N = 256$, $K = 8$.

plied by a Hamming window. The spectral resolution becomes $1/MT = 1/64$ Hz. Each individual spectrum has properties similar to those shown in Figure 7.7(a). After correcting for process loss, the ensemble average is shown in Figure 7.10(a). It is much improved and resembles what is expected for a spectrum of white noise. The same procedure is repeated with $K = 8$ and the spectral estimate is shown in Figure 7.10(b).

Example 7.6

The spectral estimate for any random process can be improved with averaging. Consider a second-order AR Gaussian process with $a_1 = a_2 = 0.75$, $T = 0.1$, and $\sigma_x^2 = 20$. A sample function is generated with $N = 256$ and is divided into four segments. Figures 7.11(a) and 7.11(b) show the entire signal and the first segment after applying a Hamming data window. The four periodogram estimates, $I_1(m)$, $I_2(m)$, $I_3(m)$, and $I_4(m)$, are calculated and plotted in Figures 7.11(c)–7.11(f). The frequency resolution is $1/6.4$ Hz. Notice that the individual spectra are still very erratic. The ensemble average is plotted in Figure 7.11(g). It definitely resembles the theoretical spectrum in Figure 7.11(h). As an exercise, verify some of the values of the spectrum at several of the harmonics.

7.3.2 Confidence Limits

The chi-square relationship between the actual PSD and its estimate can be used to establish limits. These limits are boundaries for the actual PSD. Given the estimate, the actual PSD lies within the boundaries with a desired probability. This is a two-tailed test because an upper and a lower bound are necessary. Let the significance level be α and designate one limit as $L1$ and let it be the right-hand bound on the distribution. Then,

$$\text{Prob}\left[\chi_\nu^2 \geq L1\right] = \alpha/2. \tag{7.57}$$

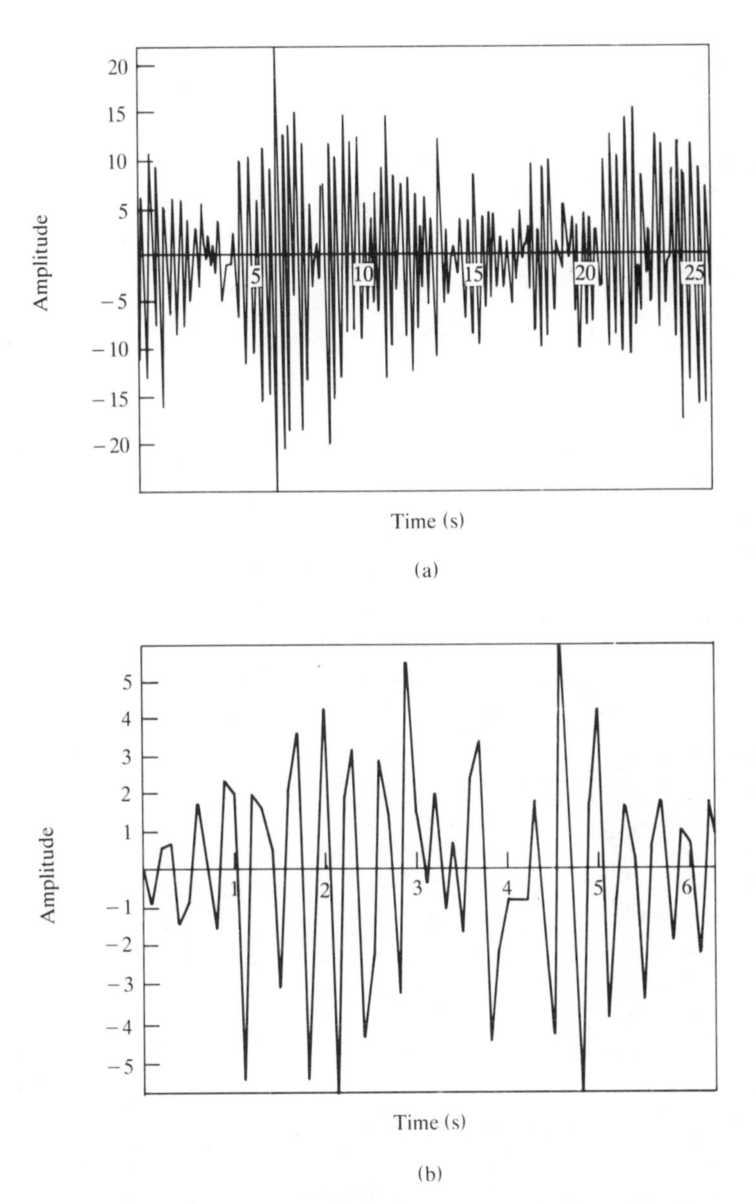

Figure 7.11 The spectra of a Gaussian second-order AR process estimated using segment averaging: (a) sample function with $N = 256$; (b) segment 1 multiplied by a Hamming window; (c)–(f) periodogram spectra of segments 1–4; (g) the average PSD estimate; (h) the ideal spectrum.

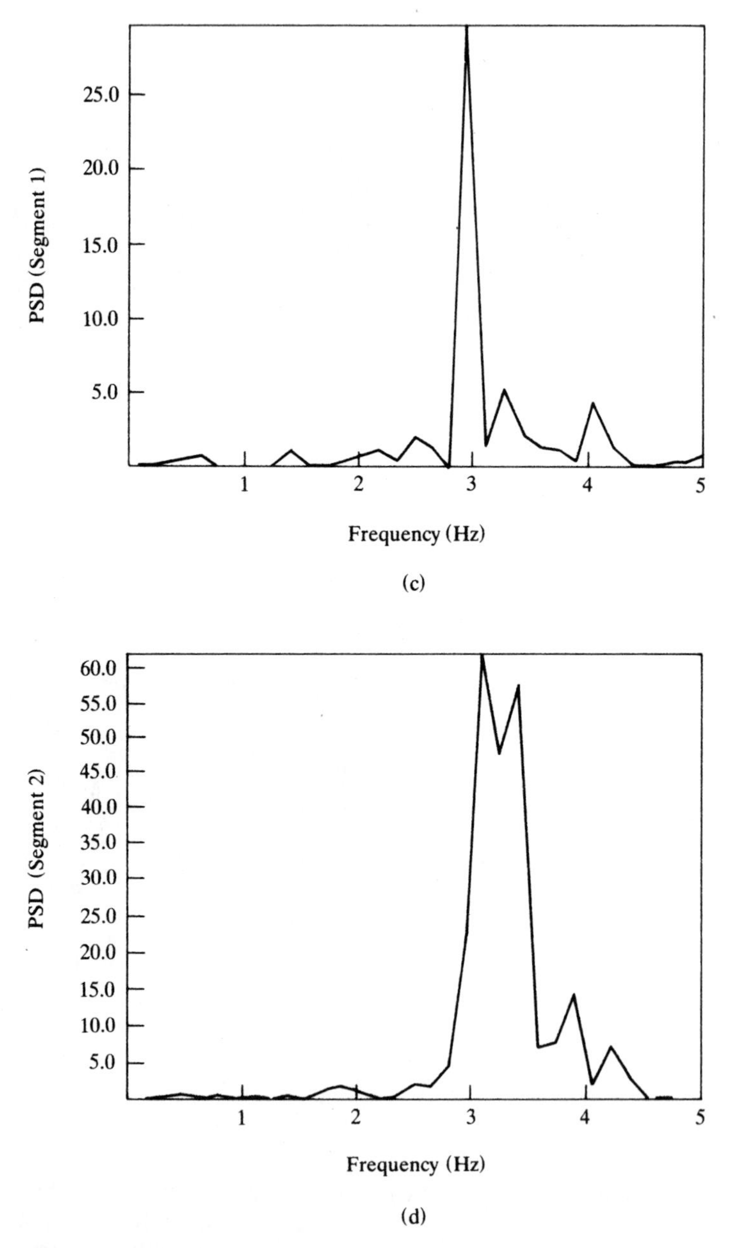

Figure 7.11 (*Continued*).

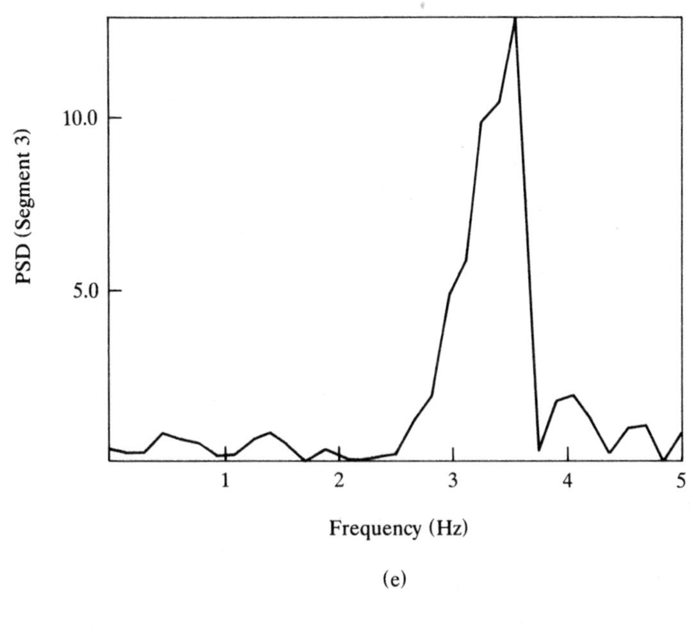

(e)

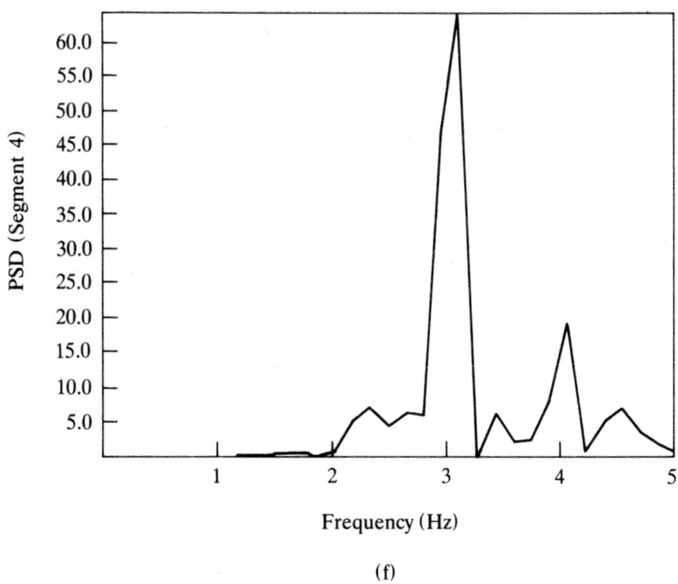

(f)

Figure 7.11 (*Continued*).

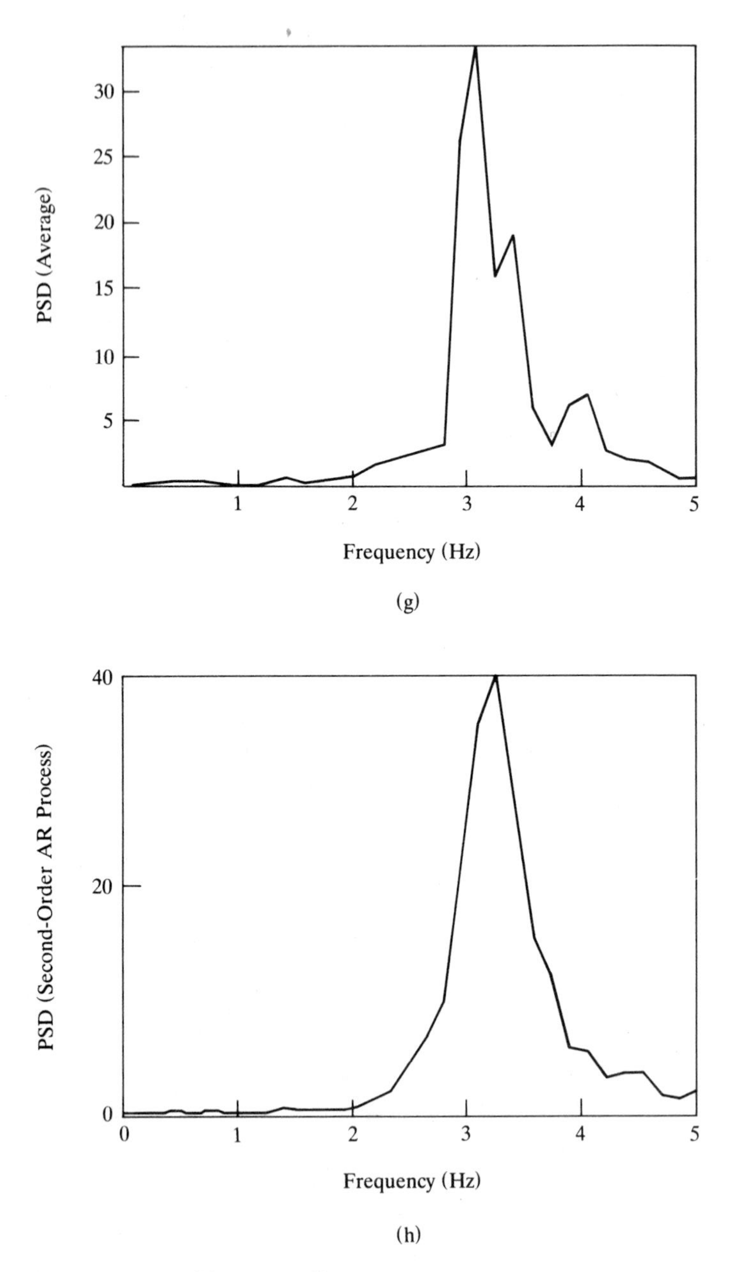

Figure 7.11 (*Continued*).

As is usual, the other limit is designated $L2$ and is the left-hand bound and

$$\text{Prob}\left[\chi_\nu^2 \geq L2\right] = 1 - \alpha/2 \quad \text{or} \quad \text{Prob}\left[\chi_\nu^2 \leq L2\right] = \alpha/2. \quad (7.58)$$

These establish the limits for a confidence level of $1 - \alpha$. These probabilities and boundaries are illustrated in Figure 7.12. For

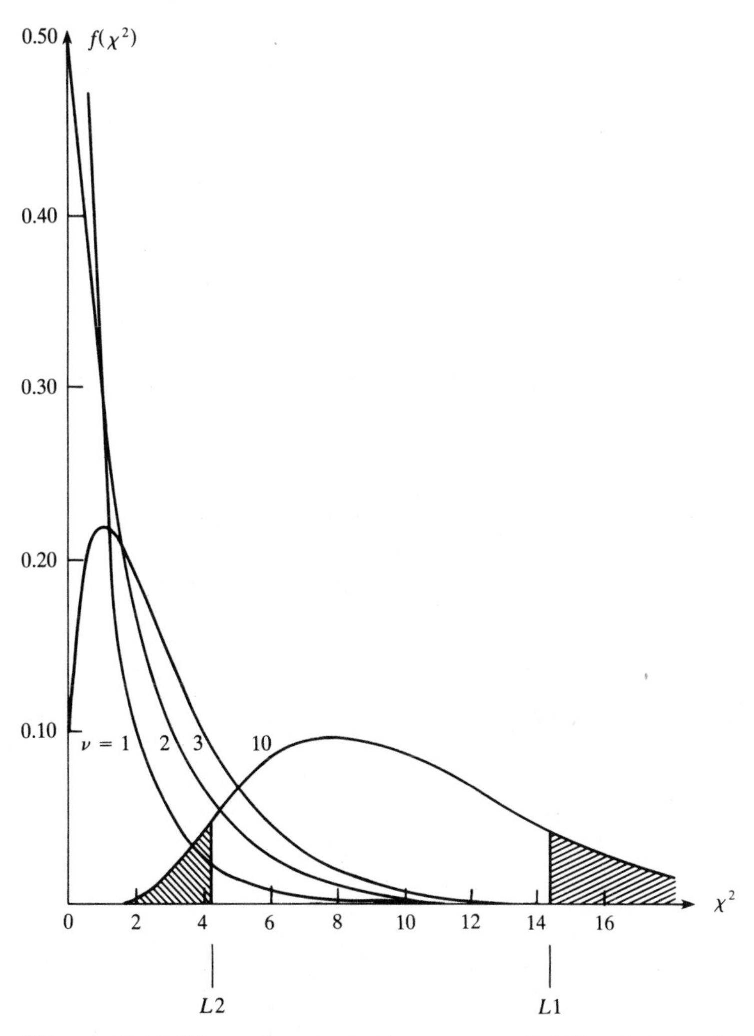

Figure 7.12 Plots of several chi-square probability density functions ($\nu = 1, 2, 3, 10$) and the confidence limits $L1$ and $L2$ for the pdf with $\nu = 10$.

equation (7.57) the lower bound is established with equation (7.48) and

$$\chi^2_{2K} = \frac{2KI_K(m)}{S(m)} \leq L1 \quad \text{or} \quad S(m) \geq \frac{2KI_K(m)}{L1}. \tag{7.59}$$

Similarly, the upper bound is established for equation (7.58) and is

$$\chi^2_{2K} = \frac{2KI_K(m)}{S(m)} \geq L2 \quad \text{or} \quad S(m) \leq \frac{2KI_k(m)}{L2}. \tag{7.60}$$

The values for $L1$ and $L2$ are read directly from the chi-square distribution function table for $\nu = 2K$ degrees of freedom. Often these limits are written

$$L1 = \chi^2_{2K,\alpha/2} \quad \text{and} \quad L2 = \chi^2_{2K,1-\alpha/2}. \tag{7.61}$$

Example 7.7

What is the relationship between the averaged periodogram and a signal's PSD for a 95% confidence interval if six segments, $K = 6$, are used. From each equation (7.61), with $\alpha = 0.05$,

$$L1 = \chi^2_{12,0.025} = 23.34 \quad \text{and} \quad L2 = \chi^2_{12,0.975} = 4.40.$$

From equations (7.59) and (7.60) the bounds on the spectrum are

$$\frac{2KI_K(m)}{L1} \leq S(m) \leq \frac{2KI_K(m)}{L2},$$

or

$$\frac{12I_K(m)}{23.34} = 0.51I_6(m) \leq S(m) \leq \frac{12I_K(m)}{4.40} = 2.73I_6(m).$$

Thus, with a probability of 0.95 the actual spectrum lies somewhere between a factor of 0.51 and 2.73 times estimated spectrum.

Example 7.8

What are the upper and lower bounds for estimating the PSD when 30 spectra are averaged and a 99% confidence interval is desired?

$$L1 = \chi^2_{60,\,0.005} = 91.95 \quad \text{and} \quad L2 = \chi^2_{60,\,0.995} = 35.53.$$

The bounds are

$$\frac{60 I_K(m)}{91.95} \leq S(m) \leq \frac{60 I_K(m)}{35.53}.$$

Example 7.9

Let us now determine the limits for the estimates of the PSD of a white noise process by determining the boundaries as shown in Examples 7.7 and 7.8. A sample function containing 960

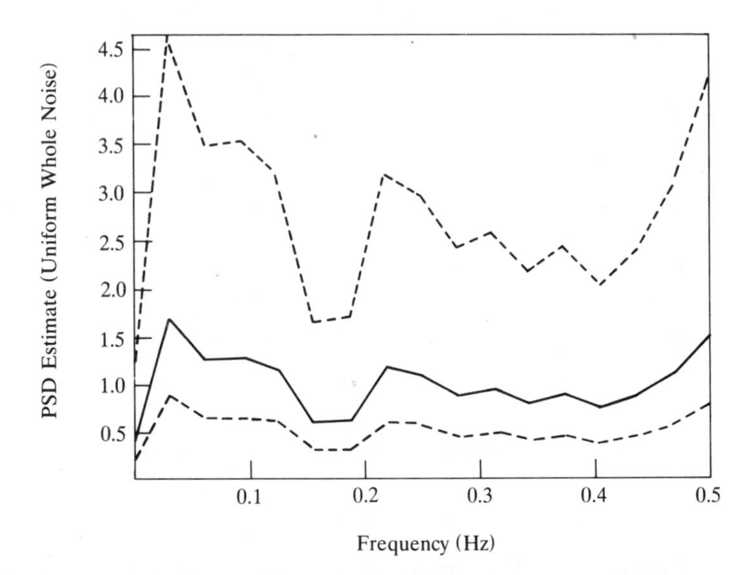

Figure 7.13 The estimated spectrum (—) from a white noise signal ($N = 960$), with $T = 1.0$, $\sigma^2 = 1.0$, and its confidence limits (---) for $K = 30$.

points was divided into 30 segments and the average periodogram calculated. The 95% confidence limits are

$$0.65 I_K(m) \le S(m) \le 1.69 I_K(m).$$

These bounds are plotted in Figure 7.13 along with the estimate. Again notice the asymmetry in the upper and lower bounds. The ideal spectrum lies within these bounds with a probability of 0.95.

Example 7.10

The confidence limits are established for the estimate of the PSD of the random process in Example 7.6, with $\sigma_x^2 = 5$. The result is plotted in Figure 7.14. The ideal spectrum is found in Figure 7.11(g) by correcting the power scale for a lower value of σ_x^2

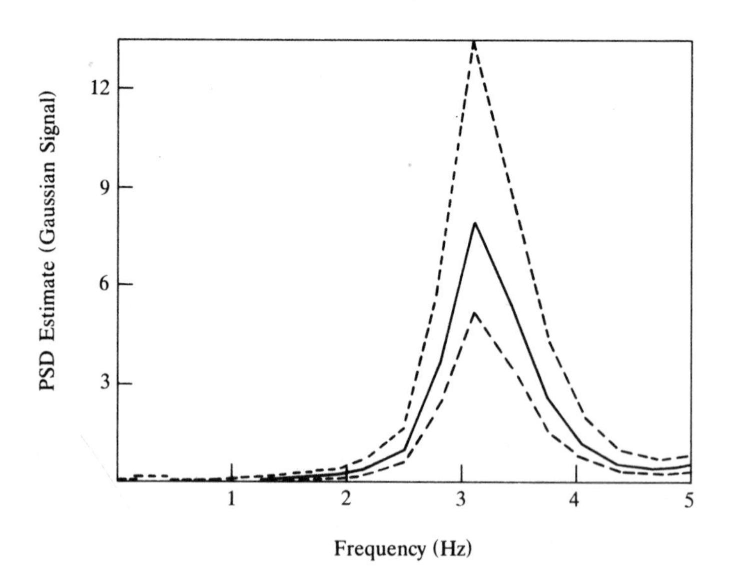

Figure 7.14 The estimated spectrum (—) from a second-order Gaussian process described in Example 7.10 ($a_1 = a_2 = 0.75$, $\sigma_x^2 = 5$, $T = 0.1$, $K = 30$, *and* $N = 960$) and its confidence limits (– – –).

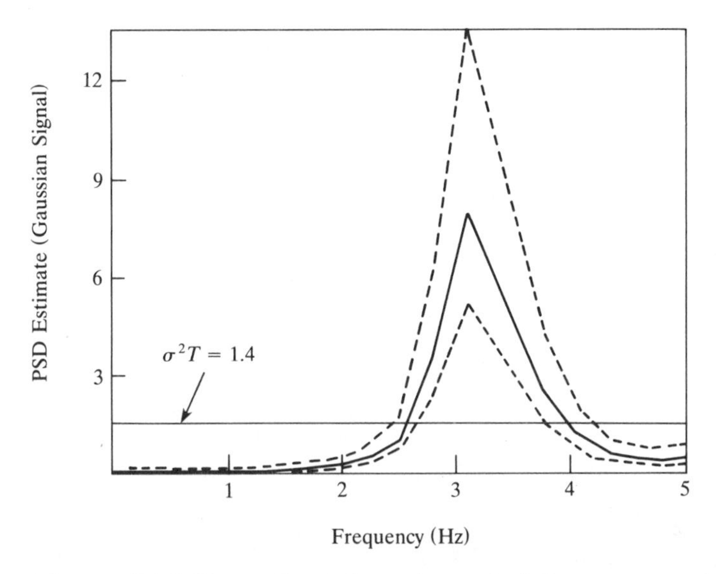

Figure 7.15 The estimated spectrum (—) from a second-order process its confidence limits (- - -), and an ideal white-noise spectrum with the sample variance.

(divide by 4). The ideal spectrum is well within the confidence limits.

Now let us suppose that we do not know anything about the characteristics of the random process. We hypothesize that it is white noise with the sample variance. This spectrum and the limits for the random process are shown in Figure 7.15. The hypothesized spectrum is outside of the designated boundaries. Thus we would reject the hypothesis and state definitely that the process is not white noise but has some structure. The next step is to propose several alternate hypotheses and test them also.

Example 7.11

The second-order process in Example 7.10 is divided into 60 segments. The averaged estimate and confidence limits are plotted in Figure 7.16. The confidence limits are much closer together and

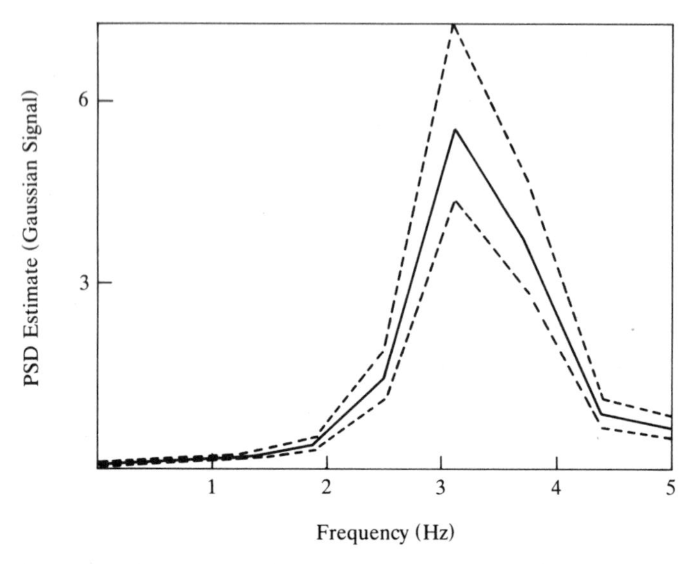

Figure 7.16 Spectral estimate and confidence limits for a second-order AR signal in Example 7.11. The estimate was obtained by averaging 60 periodogram segments.

permit more stringent judgement about possible ideal spectra that are consistent with the data. The frequency resolution has now been reduced substantially. Notice that the peak of the spectrum is broader and its magnitude is less.

7.3.3 Summary of Procedure for Spectral Averaging

The procedure for spectral averaging is sequential and is as follows:

1. Decide upon the spectral resolution needed and thus choose M.
2. Divide the sample function into K segments.
3. Detrend and apply data window to each segment.
4. Zero-pad signals in each segment in order to use the FFT algorithm.
5. Calculate the periodogram for each segment.
6. Correct for process loss.
7. Take the ensemble average.

8. Calculate chi-square confidence bounds.
9. Plot $I_K(m)$ and confidence bounds.
10. Test consistency with any hypothesized ideal spectrum.

If one decides that the frequency spacing is smaller then necessary and that closer confidence bounds are desired, then the number of segments can be increased and the entire procedure repeated. Detrending needs to be reemphasized here. It was defined in Section 3.4.5 as fitting each segment of this signal with a low-order polynomial and subtracting the values of the fitted curve. Detrending is necessary in order to remove bias in the low-frequency range of the spectral estimate.

7.3.4 Welch Method

A variation of the segmentation and averaging or Bartlett method is the *Welch method* [Welch (1967)]. In this method the sample function is also divided into K segments containing M sample points. However, overlapping of segments is allowed and the initial data points in each segment are separated by D time units. A schematic is shown in Figure 7.17. There is an overlap of $M - D$ sample points and $K = (N - M)/D + 1$ segments are formed. The *percentage overlap* is defined as $100(M - D)/M$. Overlapping allows more segments to be created with an acquired sample function. Alternatively, this allows more points in a seg-

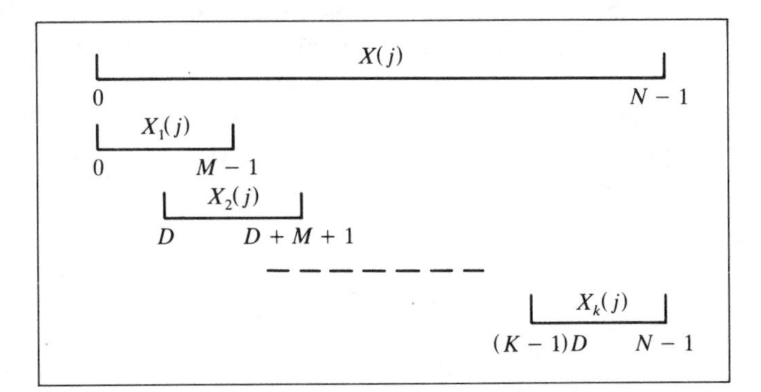

Figure 7.17 Illustration of the signal segments containing M points, with an overlap of $(M - D)$ points. [Adapted from Welch (1967, IEEE), Figure 1, with permission.]

ment for a given number of segments, thus reducing the bias. However, the segments are not strictly independent and the variance reduction factor VR is not K and depends on the data window. For any data window define the function $p(l)$ as

$$p(l) = \frac{\left(\sum_{n=0}^{M-1} d(n)d(n+lD)\right)^2}{\sum_{n=0}^{M-1} d^2(n)}. \tag{7.62}$$

The variance reduction factor is

$$\frac{K}{1 + 2\sum_{l=1}^{K-1} \frac{K-l}{K} p(l)}. \tag{7.63}$$

The equivalence number of degrees of freedom is twice this factor. Studying PSD estimates of simulated signals indicates that using a Hanning or Parzen data window with 50–65% overlap yields good results [Marple (1987)].

7.3.5 Spectral Smoothing

Perhaps the simplest method for reducing the variance of PSD estimates is *spectral smoothing*, or *the Daniell method*. It is also based on the notion of averaging independent spectral values; however, in this method the magnitudes are averaged over frequency domain. For *boxcar*, or *rectangular*, *smoothing*, the smoothed spectral estimate $\tilde{I}_K(m)$ is obtained by first calculating the periodogram or BT estimate and then averaging spectral magnitudes over a contiguous set of harmonic frequencies of $1/NT$, that is,

$$\tilde{I}_K(m) = \frac{1}{K} \sum_{j=-J}^{J} I(m-j), \qquad J = \frac{K-1}{2}. \tag{7.64}$$

The smoothing is symmetric and K is an odd integer. Again, the variance of the estimate is reduced by a factor of K and

$$\operatorname{Var}\left[\frac{2K\tilde{I}_K(m)}{S(m)}\right] = 2\nu = 2 \cdot 2K \quad \text{or} \quad \operatorname{Var}[\tilde{I}_K(m)] = \frac{S^2(m)}{K}. \tag{7.65}$$

As with averaging, the bias must be reduced by applying a data window to the signal and correcting $I(m)$ for process loss. There are some important differences from the results of the averaging technique. First, the frequency spacing is still $1/NT$; there are still $[N/2]$ harmonic values in $\tilde{I}_K(m)$. However, averaging makes them no longer independent; any value of $\tilde{I}_K(m)$ is composed of K consecutive harmonics of $I(m)$. Thus, the frequency resolution is decreased. The width in the frequency range over which the smoothing is done is called the *bandwidth* B. This is shown schematically in Figure 7.18. Another factor is that bias can be introduced. Consider the expectation of the smoothed estimate,

$$E[\tilde{I}_K(m)] = \frac{1}{K}\sum_{j=-J}^{J} E[I(m-j)] \approx \frac{1}{K}\sum_{j=-J}^{J} S(m-j). \tag{7.66}$$

If consecutive values of the actual spectra are not approximately

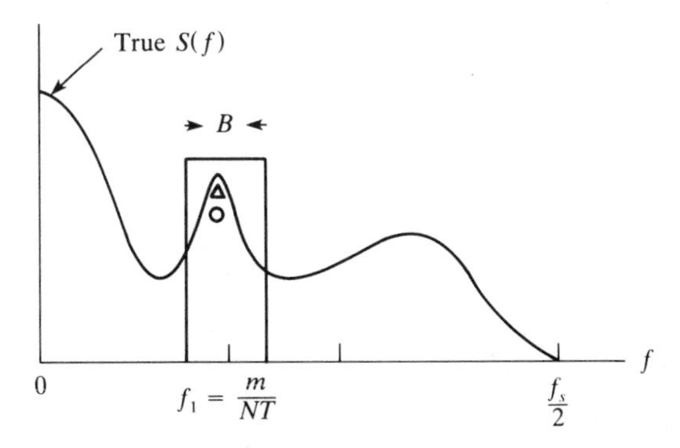

Figure 7.18 Schematic of rectangular smoothing; actual spectrum and rectangular spectral window of bandwidth B are shown. The triangle and circle represent peak values resulting from different smoothing bandwidths.

equal, then a bias can be caused. Obviously, local maxima will be underestimated and local minima will be overestimated. This only becomes a matter of concern when *narrow-band* signals are being analyzed.

Example 7.12

An AR(2) process is generated; $a(1) = a(2) = 0.75$, $\sigma_x^2 = 5$, $T = 0.02$, and $N = 64$. The periodogram is plotted in Figure 7.19(a) and is smoothed over three and seven harmonics, respectively. The resulting smoothed spectral estimates are shown in Figure 7.19(b) and 7.19(c) and are tabulated in Appendix 7.5. Verify some of these values. The band-widths of three- and seven-point smoothing are $B = 3/NT = 2.34$ Hz and $B = 7/NT = 5.47$ Hz, respectively. As

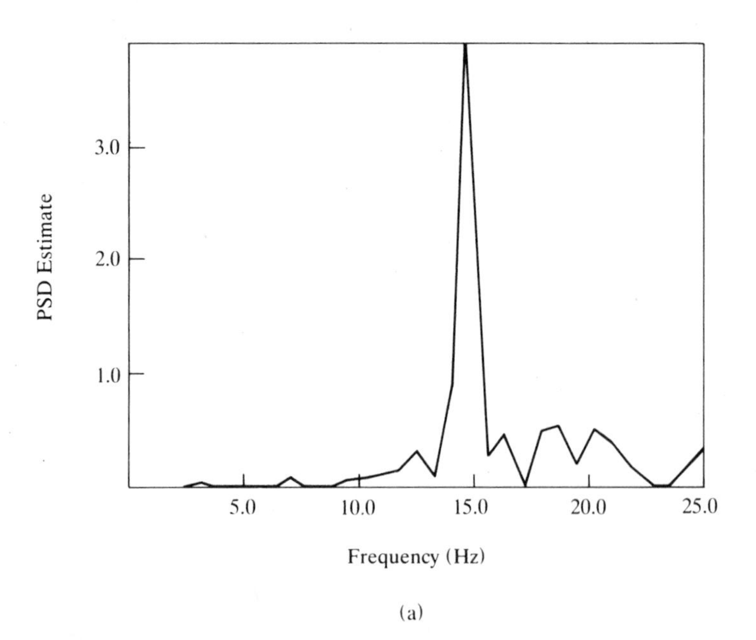

Figure 7.19 Spectral estimates with three- and seven-point smoothing for an AR signal with 64 points; a Hamming data window was used: (a) original estimate; (b) estimate with three-point smoothing; (c) estimate with seven-point smoothing.

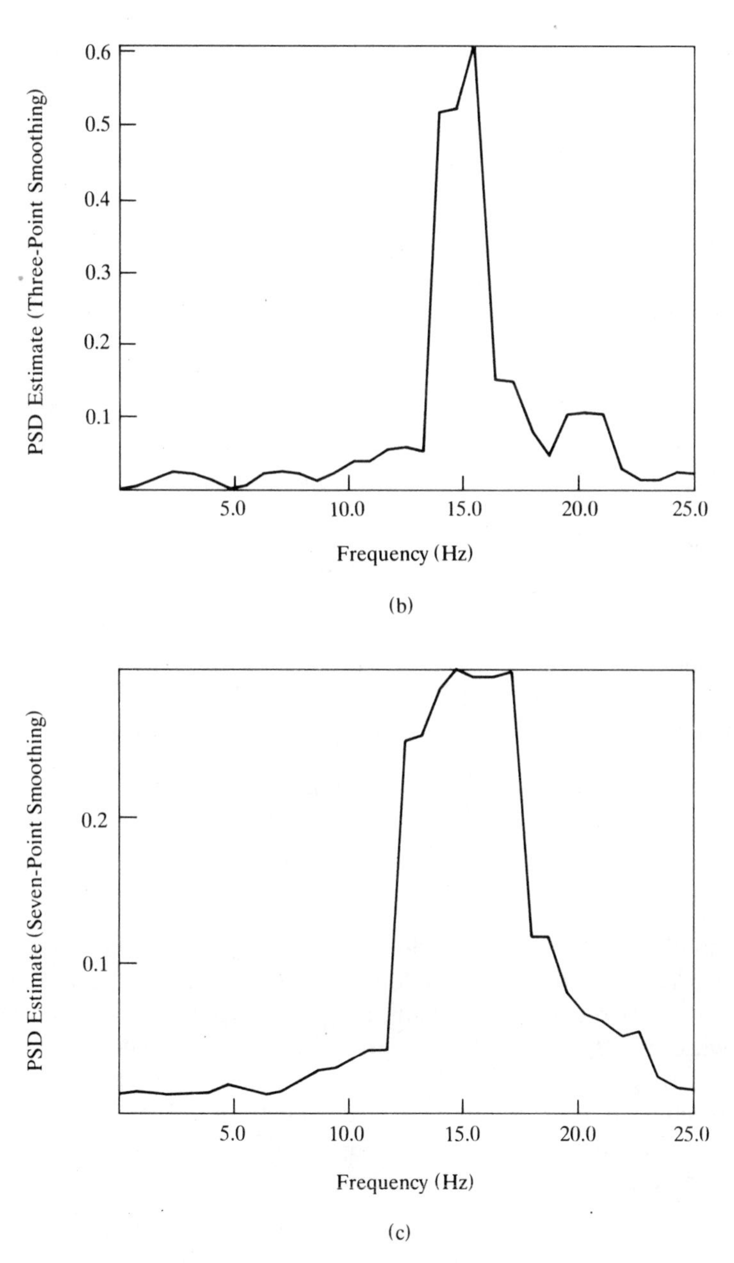

Figure 7.19 (*Continued*).

expected, the seven-point smoothing produces estimates with lesser variance and less-erratic characteristics. The penalty for this is that the resolution has decreased; that is, the bandwidth is wider or spectral values that are independent of one another are separated by seven harmonics. This is not a problem for the white noise process but is for a correlated random process. Observe in Figure 7.19(c) that the magnitude for $\tilde{I}_7(m)$ is less than that for $\tilde{I}_3(m)$. Thus, more of a negative bias is introduced because of the narrow spectral peak.

There are other smoothing weighting functions. The other one that is sometimes used is the triangular function [Schwartz and Shaw (1975)]. For smoothing over K terms,

$$\tilde{I}_K^t(m) = \frac{1}{J} \sum_{i=-J}^{J} \left(1 - \frac{|i|}{J}\right) I(m-i), \qquad J = \frac{K-1}{2}. \quad (7.67)$$

The benefit is that less bias is introduced for smoothing over the same number of harmonics. The cost is that the variance is not reduced as much. The number of degrees of freedom is

$$\nu = 2 \frac{2J+1}{\dfrac{4}{3} + \dfrac{2J^2+2J+1}{3J^3}}. \quad (7.68)$$

Bandwidth is difficult to define for nonrectangular smoothing. *Equivalent bandwidth* B_e is defined as the width of the rectangular window that has the same peak magnitude and area as the triangular window, $B_e = J/NT$. The proof is left as an exercise.

In general, it has been shown that bias is proportional to the second derivative of the true spectrum [Jenkins and Watts (1968)]. The proofs are quite complicated and a computer exercise is suggested to emphasize this concept.

7.3.6 Additional Applications

Several additional applications are briefly summarized in order to demonstrate how some of the principles that have been explained in the previous sections have been implemented.

In the development of magnetic storage media for computer disks, the size of the magnetic particles seems to influence the amount of noise generated during the reading of the stored signal. In particular, it was suspected that the amount of noise is dependent upon the frequency. In order to test this phenomenon, periodic square-wave signals were sampled and stored on a disk, with the frequency being changed at repeated trials. The recorded signals were read and their PSD estimated using periodogram averaging. The fundamental and harmonic frequency components of the square wave were removed from the spectra in order to study the noise spectra. They are plotted in Figure 7.20. The shapes of the PSDs are the same for all the signals. However, it is evident that the higher-frequency signals produce less noise.

Several biological parameters are monitored during maternal pregnancy and labor. One of these is the electrical activity of the uterus, called the electrohysterogram (EHG), recorded from a site on the abdomen. The characteristics of the EHG change as parturition becomes imminent. Figures 7.21(a) and 7.21(b) show that, as labor progresses, the EHG has more high-frequency components. The signals are monitored over an eight-hour period and are sampled at a rate of 20 Hz. Each sample must be at least 50 s long in order to capture the results of an entire contraction. For convenience, $N = 1024$ for each sample function. The periodogram is calculated for each sample function and is smoothed using a

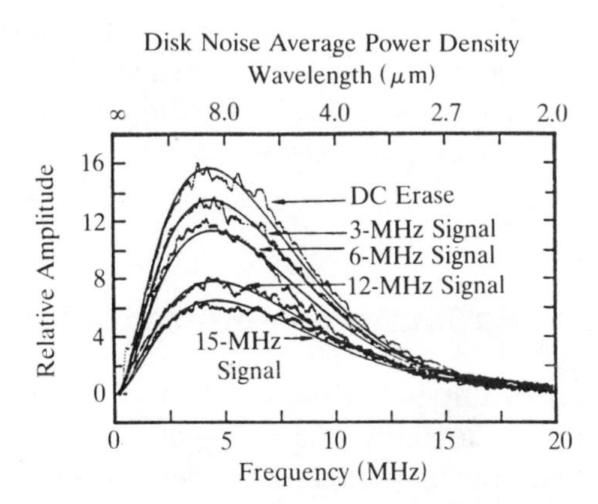

Figure 7.20 Noise spectra from a recording medium as a function of signal frequency. [From Anzaloni and Barbosa (1984, IEEE), Figure 3, with permission.]

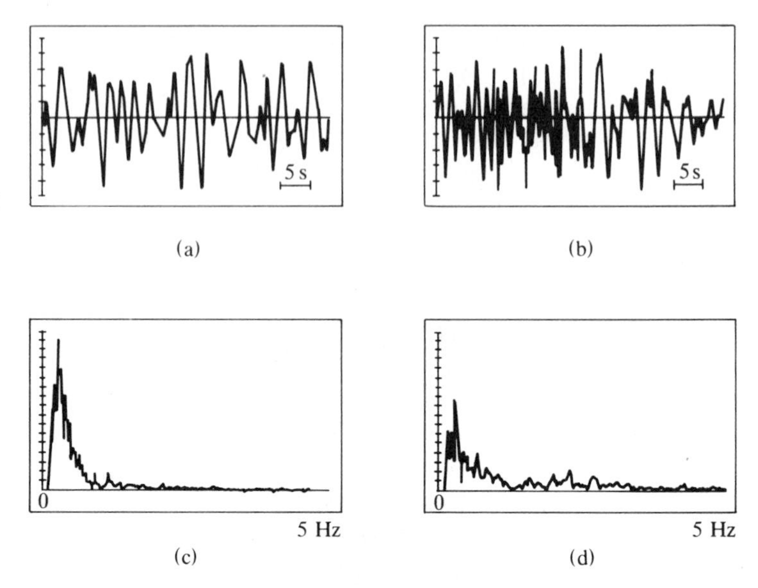

Figure 7.21 The EHG during pregnancy and during labor contraction: (a) EHG during pregnancy; (b) EHG during labor; (c) PSD of (a); (d) PSD of (b). [From Marque et al. (1986), Figure 2, with permission.]

Hanning spectral window. Representative spectra are shown in Figures 7.21(c) and 7.21(d). The actual changes in the spectra are monitored by calculating the energy in different frequency bands. The frequency bands low (L, 0.2–0.45 Hz), high (H, 0.8–3.0 Hz), and total (T, 0.2–3.0 Hz) are stipulated. An increase in the H/T ratio with a corresponding decrease in the L/T ratio indicates that the uterus is in its labor mode of contraction.

7.4 Consistent Estimators — Indirect Methods

7.4.1 Spectral and Lag Windows

Spectral smoothing is actually one mode of a very general method for improving spectral estimates. Because the smoothing functions are symmetric, the smoothing operation is mathematically equivalent to the convolution of the periodogram with an even function that is now represented by $W(f)$ or $W(m)$, depending upon whether the continuous or discrete frequency domain is being used. Again, because of the convolution in the frequency domain,

$$\tilde{I}_K(m) = I(m) * W(m), \tag{7.69}$$

the lag domain multiplication

$$\tilde{R}_K(k) = \hat{R}(k)w(k) \tag{7.70}$$

is implied. The functions $W(m)$ and $w(k)$ are Fourier transform pairs and are called the *spectral window* and the *lag window*, respectively. The difference with respect to smoothing is that emphasis is placed on operating in the lag domain because the spectral windows usually have a more complicated mathematical representation. The procedure is as follows:

a. $\hat{R}(k)$ is estimated;
b. The lag window is selected and the operation in equation (7.70) is performed;
c. The Fourier transform of $\tilde{R}_K(k)$ is calculated.

This is called the *Blackman–Tukey* (BT) *approach* and is named after two researchers who developed this technique. To distinguish the windowing approach from smoothing approach, the notation will be changed slightly: Equations (7.69) and (7.70) become

$$\tilde{S}_M(m) = I(m) * W(m) \tag{7.71}$$

and

$$\tilde{R}_M(k) = \hat{R}(k)w(k). \tag{7.72}$$

The use of the subscript M will become apparent in this section. Many spectral–lag window pairs have been developed, and most have a specialized usage. The ones most often used will be discussed. The article by Harris (1978) presents a comprehensive discussion of the characteristics of all the windows. Appendix 7.4 contains the mathematical formulations for the lag and spectral windows. Notice that all these windows have the same mathematical form as the data windows listed in Appendix 3.5. The only significant difference is that the lag windows range from $-MT$ to $+MT$ and the data windows range from 0 to $(N-1)T$. The triangular, or Bartlett, lag window is plotted in Figure 7.5. Consequently, the lag spectral windows do not have a phase component but the magnitude components are the same as those of the data spectral windows. They are plotted in Figure 3.15. Again, the basic concept is for the windowing operation to increase the number of degrees of freedom and hence the variance reduction factor. Focusing on the representation in equation (7.2), the PSD estimate from a truncated ACF estimate is

$$\hat{S}_M(f) = T\left(\hat{R}(0) + 2\sum_{k=1}^{M} \hat{R}(k)\cos(2\pi fkT)\right). \qquad (7.73)$$

This truncation represents the function of the rectangular lag window. For a weighted window, equation (7.73) becomes

$$\tilde{S}_M(f) = T\left(\hat{R}(0) + 2\sum_{k=1}^{M} w(k)\hat{R}(k)\cos(2\pi fkT)\right). \qquad (7.74)$$

The research into effective windows was to eliminate the leakage and bias produced in $\tilde{S}_M(f)$ through the implied convolution operation. Thus, the goal was to develop lag windows whose spectral windows had small side-lobe magnitudes, which minimized leakage, and narrow main-lobe bandwidths, which minimized bias. The design of various windows to optimize certain characteristics of the estimation procedure is called *window carpentry* in the older literature. It was discovered that optimizing one property precluded optimizing the other. These characteristics of windows were discussed in detail in Chapter 3. It can be understood from the discussion on smoothing that a wide main lobe tends to induce more bias, whereas a small side lobe tends to minimize leakage. Under the supposition that most spectra have a small first derivative, then the concentration on small side lobes is necessary. Thus the rectangular and Bartlett windows will not be good and the Hanning–Tukey, Hamming, and Parzen windows will be effective.

Before continuing the discussion on additional properties of windows needed for spectral analysis, let us examine a few examples of spectral estimation using the BT approach.

Example 7.13

The time series of Example 7.3 is used and the estimate of its ACF with $N = 256$ and $M = 25$ is shown in Figure 7.22(a). Simply calculating the Fourier transform after applying the rectangular window with $M = 25$ produces the estimate in Figure 7.22(b). This is calculating equation (7.74) with $f = m/2MT$ and $0 \leq m \leq 25$. Compare this estimate with the original periodogram estimate and the true spectrum. This is a more stable estimate than the periodogram because some smoothing was applied but it still does not resemble the true spectrum very much. In actuality, equation

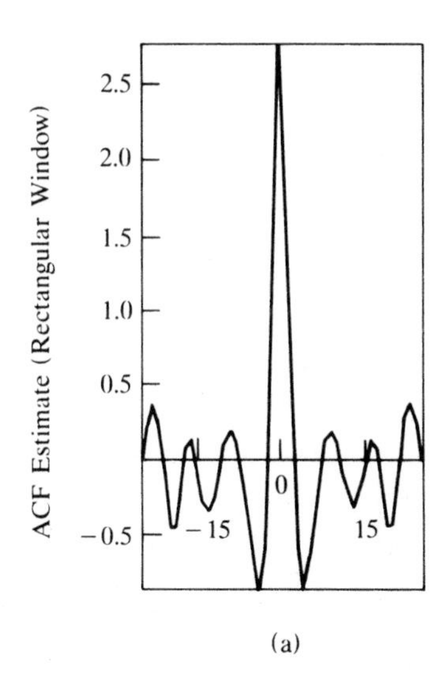

Figure 7.22 Estimates plotted from the time series in Example 7.13: (a) autocorrelation function; (b) spectrum from using rectangular lag window, $M = 25$; (c) autocorrelation function shifted and zero-padded; (d) product of ACF in (a) and Parzen lag window, $M = 25$; (e) spectrum using Parzen lag window, $M = 25$.

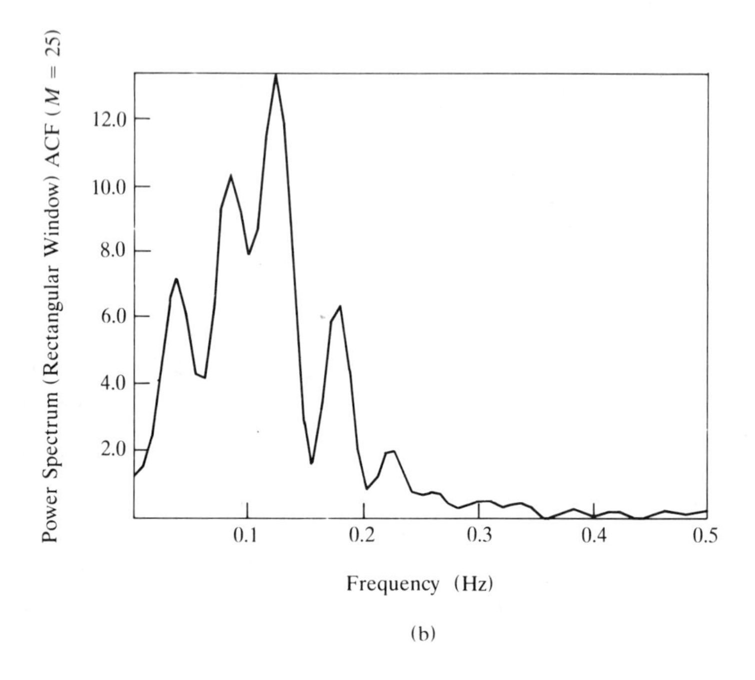

(b)

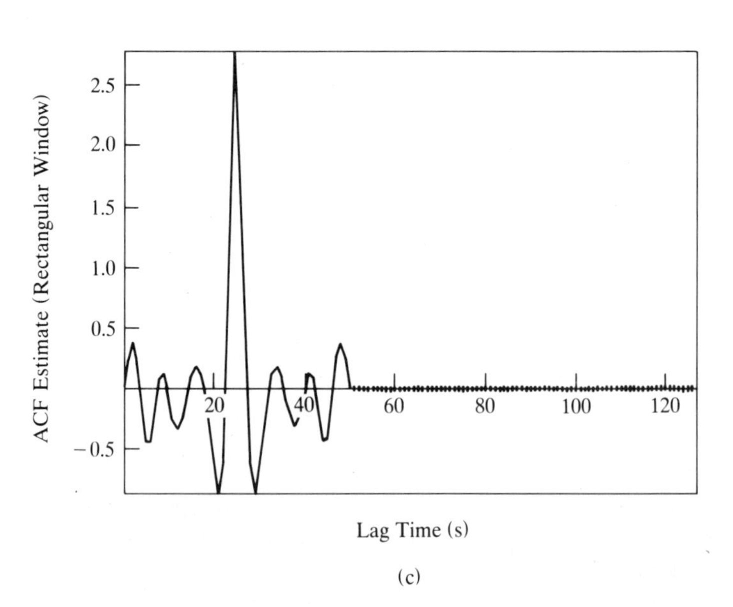

(c)

Figure 7.22 (*Continued*).

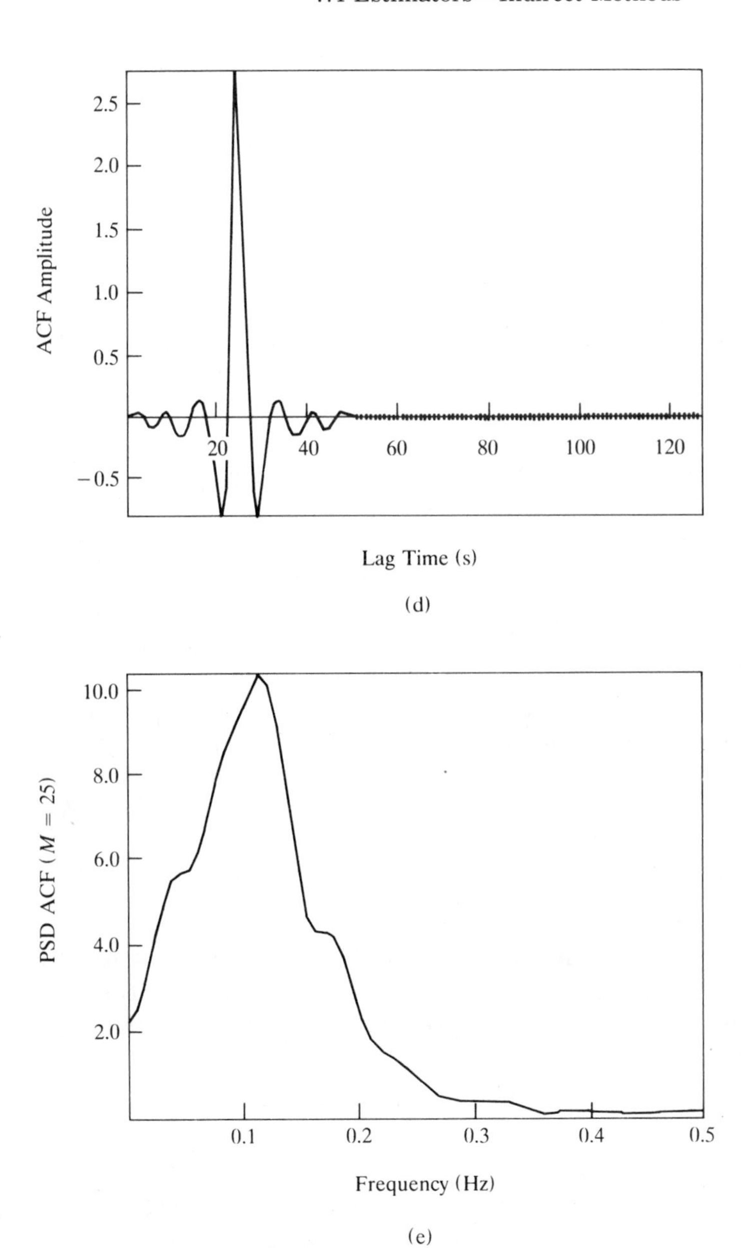

Figure 7.22 (*Continued*).

(7.74) is calculated indirectly using the FFT algorithm. $\tilde{R}_M(k)$ is shifted 25 times units and zeros added to give a total of 128 points, as shown in Figure 7.22(c). Then the FFT algorithm is used to calculate the DFT. The magnitude component of the DFT is the spectral estimate $\tilde{S}_M(m)$.

If a Parzen window is applied with $M = 25$, it results in $\tilde{R}_M(k)$ as shown in Figure 7.22(d) after shifting and zero padding. The resulting estimate, $\tilde{S}_M(m)$, is much better and is shown in Figure 7.22(e). A discussion of the variance reduction will make this result expected.

7.4.2 Important Details for Using FFT Algorithms

Before proceeding there must be a short discussion of the procedures for calculating the BT estimate. If equation (7.74) is calculated directly, then the results are straightforward. However, one finds that this requires a large computation time and the FFT algorithm is used for calculating the DFT. A statement of caution is needed. The FFT algorithms usually assume that the time function begins at $n = 0$. However, $\tilde{R}_M(k)$ begins at $k = -M$ and the total number of points is $2M + 1$. This means that, in order to use the FFT algorithm, $\tilde{R}_M(k)$ must be shifted M time units and that it must be zero padded. This is shown in Figure 7.22(c). What does the time shift imply? Will the resulting $\tilde{S}_M(m)$ be real or complex? It is known that the estimate must be a real function!

7.4.3 Statistical Characteristics of BT Approach

Examination of equation (7.71) shows that the initial estimated spectrum is convolved with the spectral window. How the bias, consistency, and confidence limits are evaluated in this situation must be examined.

7.4.3.1 Bias

Fortunately, for bias considerations the width of the spectral window dominates and the effect of the Fejer kernel can be neglected. Let us examine this briefly. The effect of applying a lag window produces a spectral estimate such that

$$\tilde{S}(f) = W(f) * \hat{S}(f) = W(f) * \int_{-1/2T}^{1/2T} S(g) D(f-g)\, dg$$

$$= \int_{-1/2T}^{1/2T} W(f-h) \int_{-1/2T}^{1/2T} S(g) D(h-g)\, dg\, dh. \quad (7.75)$$

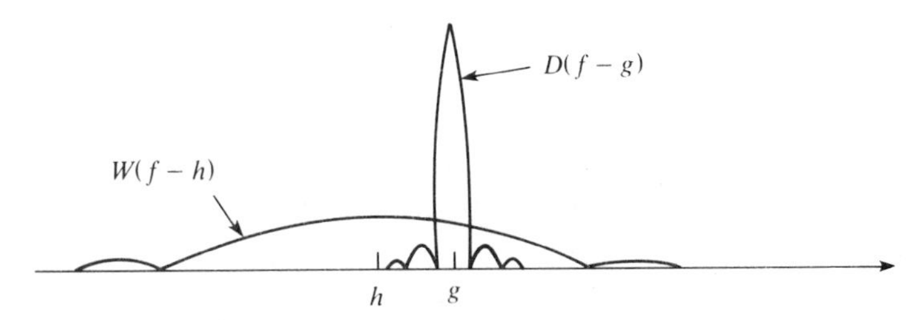

Figure 7.23 Sketch of spectral windows $D(f)$ and $W(f)$. [Adapted from Fante (1988), Figure 9.23, with permission.]

Interchanging the order of the independent variables for integrating produces

$$\tilde{S}(f) = \int_{-1/2T}^{1/2T} S(g)\left(\int_{-1/2T}^{1/2T} W(f-h)D(h-g)\,dh\right)dg. \quad (7.76)$$

Concentrate on the integration within the parentheses. The two windows are sketched in Figure 7.23. Because $M \le 0.3N$, $D(f)$ is much narrower than $W(f)$ and can be considered as a delta function. Therefore,

$$\tilde{S}(f) = \int_{-1/2T}^{1/2T} S(g)\left(\int_{-1/2T}^{1/2T} W(f-h)\delta(h-g)\,dh\right)dg$$

and

$$\tilde{S}(f) \approx \int_{-1/2T}^{1/2T} S(g)W(f-g)\,dg, \quad (7.77)$$

and the lag spectral window dominates the smoothing operation. The bias that can be induced is expressed as an expectation of equation (7.77),

$$E[\tilde{S}(f)] = E\left[\int_{-1/2T}^{1/2T} S(g)W(f-g)\,dg\right].$$

If the spectrum varies slightly over the width of the window, then

$$E[\tilde{S}(f)] = S(f)\int_{-1/2T}^{1/2T} W(g)\,dg, \quad (7.78)$$

because $S(f)$ and $W(f)$ are deterministic. In order to maintain an unbiased estimate, the spectral window must have an area of 1. If the spectrum does not vary smoothly relative to the window, then there is a bias that is proportional to the second derivative of the spectrum. Formulae for approximation can be found in Jenkins and Watts (1968).

7.4.3.2 Variance

The variance of the Blackman–Tukey procedure can be shown to be

$$\operatorname{Var}[\tilde{S}_M] = \frac{S^2(f)}{NT}\int_{-1/2T}^{1/2T} W^2(f)\,df = \frac{S^2(f)}{N}\sum_{k=-M}^{M} w^2(k). \tag{7.79}$$

The proof is given in Appendix 7.2. In order to obtain a measure of the amount of smoothing that occurs when using the various windows, the notion of equivalent bandwidth is used again. B_e is the bandwidth of the rectangular spectral window $W_e(f)$ that produces the same variance as given in equation (7.79). Using the previous definition and knowing that the area must be 1,

$$W_e(f) = \begin{cases} \dfrac{1}{B_e}, & -\dfrac{B_e}{2} \le f \le \dfrac{B_e}{2}, \\ 0, & \text{elsewhere,} \end{cases} \tag{7.80}$$

then

$$\int_{-1/2T}^{1/2T} W_e^2(f)\,df = \frac{1}{B_e}. \tag{7.81}$$

The spectral variance in terms of the equivalent bandwidth is

$$\operatorname{Var}[\tilde{S}_M(f)] = \frac{S^2(f)}{NTB_e}. \tag{7.82}$$

The point of reference for the BT approach is the rectangular lag window. That is, $\hat{R}(k)$ is simply truncated at lags $\pm M$. The PSD is estimated with equation (7.73). The frequency spacing is $1/2MT$. The variance reduction is simply calculated with equation (7.79). Then

$$\operatorname{Var}\left[\tilde{S}_M(f)\right] = \frac{S^2(f)}{N} \sum_{k=-M}^{M} w^2(k) \approx S^2(f)\frac{2M}{N} \quad (7.83)$$

and $\nu = N/M$. The equivalent bandwidth is simply found by equating equation (7.83) to equation (7.82). Thus,

$$\frac{2M}{N} = \frac{1}{NTB_e} \quad \text{or} \quad B_e = \frac{1}{2MT}$$

and B_e is equal to the frequency spacing. All of the other lag windows have some shaping and have greater variance reduction at the cost of a greater equivalent bandwidth for a given value of M. That is, there is an effective smoothing of the initial estimate caused by shaping the lag window. All of the lag windows have certain properties:

a. $w(0) = 1$, preserves variance;
b. $w(k) = w(-k)$, symmetry;
c. $w(k) = 0$, $|k| > M$ and $M < N$, truncation.

Because of these properties, the spectral windows have corresponding general properties:

a. $\int_{-1/2T}^{1/2T} W(f)\,df = 1$, no additional bias is induced;
b. $W(f) = W(-f)$, symmetry;
c. frequency spacing is $1/2MT$.

Example 7.14

For the Bartlett window, find the variance reduction factor, ν, and B_e. Because all these terms interrelated, B_e will be found first using equation (7.82). Now,

$$B_e\left(\int_{-1/2T}^{1/2T} W_M^2(f)\,df\right)^{-1} = \left(T\sum_{k=-M}^{M} w^2(k)\right)^{-1}.$$

B_e will be calculated form the lag window because the mathematical forms are simpler. To be more general, the integral approximation of the summation will be implemented with $\tau = kT$. Hence, the total energy is

$$\sum_{k=-M}^{M} w^2(k) \approx \frac{1}{T}\int_{-MT}^{MT} w^2(\tau)\, d\tau = \frac{1}{T}\int_{-MT}^{MT}\left(1 - \frac{|\tau|}{MT}\right)^2 d\tau$$

$$= \frac{2}{T}\int_0^{MT}\left(1 - \frac{2\tau}{MT} + \frac{\tau^2}{M^2T^2}\right) d\tau$$

$$= \frac{2}{T}\left(\tau - \frac{\tau^2}{MT} + \frac{\tau^3}{3M^2T^2}\right)\Bigg|_0^{MT} = \frac{2M}{3}$$

and

$$B_e = \frac{3}{2MT}.$$

The equivalent bandwidth is wider than that for the rectangular window. The variance is easily found from equation (7.82),

$$\mathrm{Var}\left[\tilde{S}(f)\right] = \frac{S^2(f)}{NTB_e} = \frac{S^2(f)2MT}{NT3} = \frac{S^2(f)M2}{N3}.$$

The VR is $3N/2M$ and $\nu = 3N/M$. Thus, with respect to the rectangular window, the variance has been reduced by a factor of 1/3 and the number of degrees of freedom increased by a factor of 3.

7.4.3.3 Confidence Limits

The other windows that are often used are the Tukey and Parzen windows. The choice of the window and the truncation point are important factors. It is a difficult choice because there are no obvious criteria, only general guidelines. The first important factor is the number of degrees of freedom. The general formula for any window is

$$\nu = \frac{2N}{\sum_{k=-M}^{M} w^2(k)}. \tag{7.84}$$

These are tabulated in Appendix 7.3. For the Tukey and Parzen windows in particular they are $2.67N/M$ and $3.71N/M$. For a given truncation lag on the same sample function, the Parzen window will yield a lower variance. Now comes the trade-off. The measure of the main-lobe width that quantitates the amount of smoothing accomplished over $I(m)$ is the effective bandwidth. It indicates the narrowest separation between frequency bands that can be detected is a measure of resolution. These are also given in Appendix 7.3 and are $2.67/MT$ and $3.71/MT$, respectively, for these two windows. As can be seen, these measures are consistent with our qualitative notions that the Tukey window would have better resolution. The other advantage for the Parzen function is that its spectral window is never negative and therefore a negative component is not possible in $\tilde{S}_M(m)$. However, in general it has been found that these two window functions produce comparable estimates.

The establishment of confidence limits for any spectral estimate is the same procedure as for periodogram estimates. The only difference is that equations (7.59)–(7.61) are written slightly differently, to reflect the different functions used. They become

$$\chi_\nu^2 = \frac{\nu \tilde{S}_M(m)}{S(m)} \le L1 \quad \text{or} \quad S(m) \ge \frac{\nu \tilde{S}_M(m)}{L1}, \tag{7.85}$$

$$\chi_\nu^2 = \frac{\tilde{S}_M(m)}{S(m)} \ge L2 \quad \text{or} \quad S(m) \le \frac{\nu \tilde{S}_M(m)}{L2}, \tag{7.86}$$

$$L1 = \chi_{\nu,\alpha/2}^2 \quad \text{and} \quad L2 = \chi_{\nu,1-\alpha/2}^2. \tag{7.87}$$

Example 7.15

For Example 7.13 and Figure 7.22 the confidence limits are calculated for the BT estimates using the Hamming and rectangular spectral windows. For the Hamming window with 95%

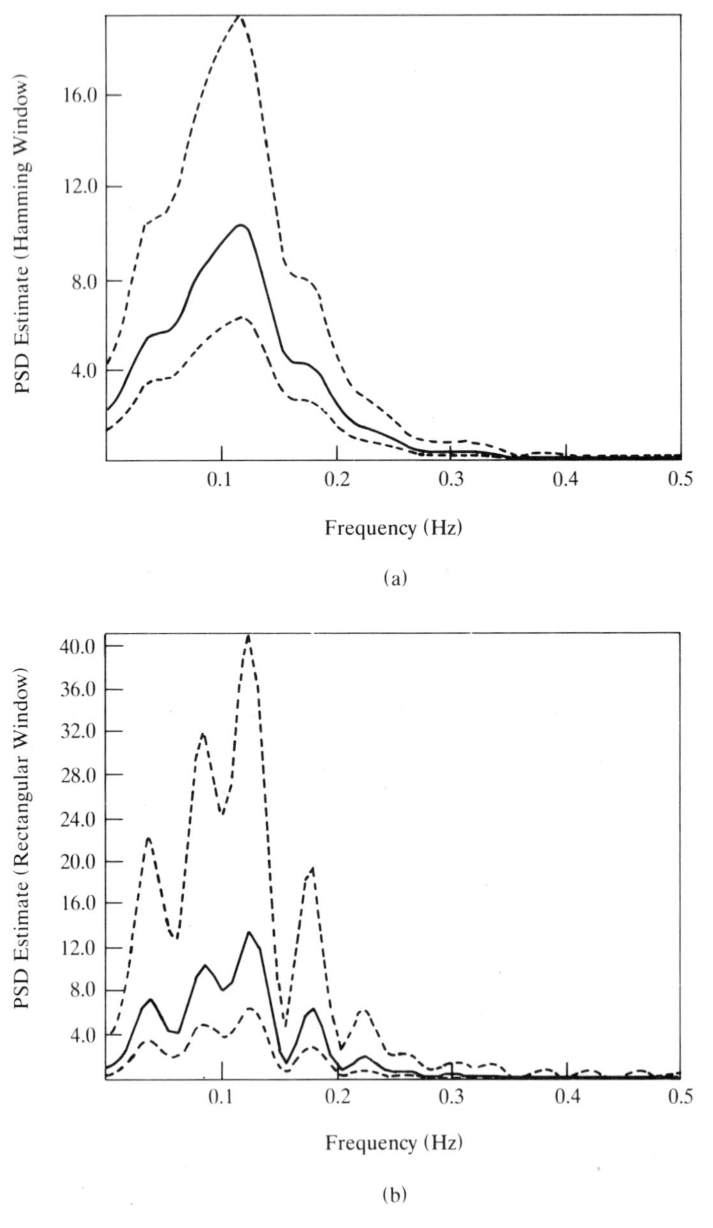

Figure 7.24 BT spectral estimates and 95% confidence limits for example 7.15, (a) Hamming and (b) rectangular lag windows.

confidence limits, $\nu = 2.51N/M = 2.51 \times 256/25 = 25.68 \approx 26$. Therefore,

$$L1 = \chi^2_{26,0.025} = 41.92 \quad \text{and} \quad L2 = \chi^2_{26,0.0975} = 13.84,$$

$$\frac{26\tilde{S}_M(m)}{41.92} \le S(m) \le \frac{26\tilde{S}_M(m)}{13.84}.$$

The PSD estimate and the confidence limits are plotted in Figure 7.24(a). The limits bound the actual PSD very well. The bounds for the rectangular lag window are plotted in Figure 7.24(b). Note that they create a much larger confidence interval and reflect the erratic nature of the estimate. For a PSD that does not contain narrow-band peaks, the Hamming lag window yields a much better PSD estimate than the rectangular lag window.

Several general principles hold for all window functions. B_e and ν are both inversely proportional to M. Thus, when M is made smaller to increase the number of degrees of freedom (reduce variance), the bandwidth will also increase (reduce resolution). Hence, when trying to discover if any narrow-band peaks occur, a larger variance must be tolerated. The tactic for approaching this dilemma is called *window closing*. Several estimates of a PSD are made with different values of M. These spectral estimates are assessed to judge whether the measured signal has a narrow-band or broadband spectrum. An example will help illustrate this point.

Example 7.16

Figure 7.25(a) shows a segment of a radar return signal, the sampling interval is 1 s and $N = 448$. It is necessary to determine its frequency components. The BT method of estimation using the Bartlett window will be implemented. The NACF is estimated for a maximum lag of 60 and is plotted in Figure 7.25(b). Several lags are used: $M = 16$, 48, and 60. With $T = 1$, the frequency spacings are 0.03125, 0.01042, and 0.00833 and the equivalent bandwidths are 0.09375, 0.03126, and 0.02499, respectively. It is seen in Figure 7.26 that the spectral estimate with $M = 16$ is smooth and does not reveal the peaks at frequencies $f = 0.07$ and 0.25 Hz that are

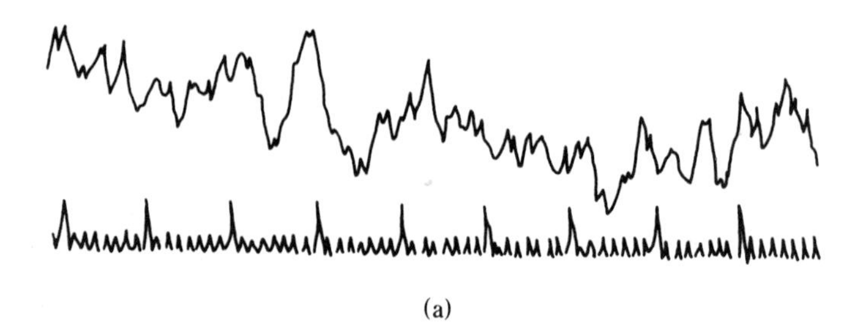

(a)

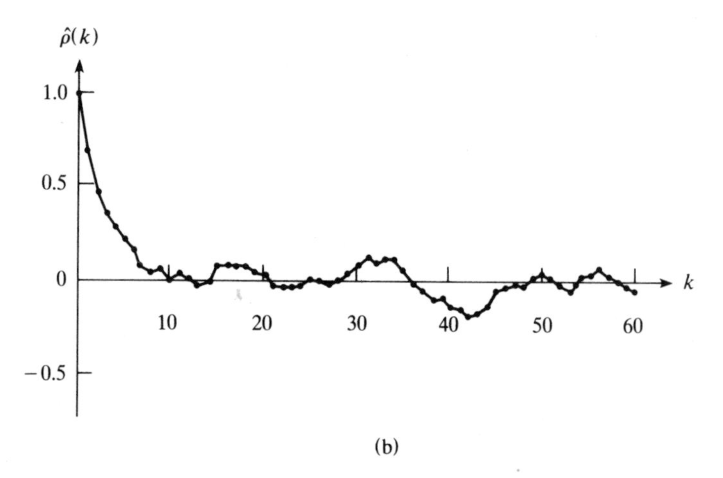

(b)

Figure 7.25 (a) A radar return signal and (b) its sample correlation function.

produced with the other two estimates. This is because B_e is wider than the local valleys in the PSD. Because the estimates with $M = 48$ and 60 lags are very similar, these can be considered as reasonable estimates. Naturally, the one to choose is the one with the greater degree of freedom, that is, the estimate with lag = 48. The number of degrees of freedom is 28.

This example, in general, illustrates the fact that larger lags are necessary to reveal the possibility of narrow spectral bands. Smaller lags are adequate if the true spectrum is very smooth. The only way to determine a reasonable lag value is to perform window closing and essentially search for ranges of lag values that produce similar estimates. Once a range is determined, the smallest maximum lag is used because it will give the largest number of degrees of freedom.

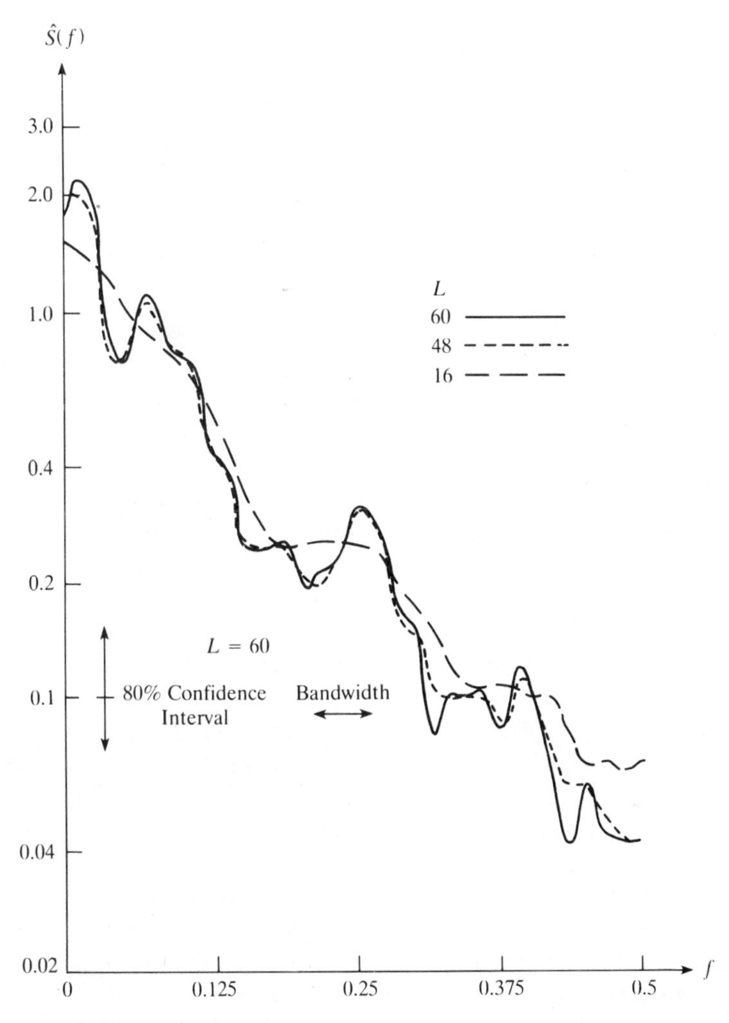

Figure 7.26 Estimates of the power spectral density function of the radar return signal in Figure 7.25. BT method is used and M = maximum lag. See Example 7.16 for details. [Adapted from Jenkins and Watts (1968), Figure 7.17, with permission.]

A slight variation in steps can be made with the Tukey window. Its spectral function is

$$W(m) = 0.25\delta(m-1) + 0.5\delta(m) + 0.25\delta(m+1) \quad (7.88)$$

and a convolution can be easily performed in the frequency domain.

The alternative approach is the following:

a. $\hat{R}(k)$ is estimated.
b. The Fourier transform of $\hat{R}(k)$ is calculated with the selected value of maximum lag.
c. $\hat{S}(m)$ is smoothed with equation (7.88).

7.5 Autocorrelation Estimation

The sum of the lag products is the traditional and direct method for estimating the autocorrelation function. With the advent of the FFT an indirect but much faster method is commonly used. It is based on the periodogram and the Fourier transform relationship between the PSD and the ACF. Now the estimate of the ACF is

$$\hat{R}(k) = \text{IDFT}[I(m)]. \tag{7.89}$$

The difficulty arises because discretization in the frequency domain induces periodicity in the time domain. Refer to Chapter 3. Thus equation (7.89) actually forms the circular ACF. In order to avoid this error, the original data must be padded with at least N zeros to form the ordinary ACF with equation (7.89). The steps in this alternate method are the following:

a. Zero-pad the measured data with at least N zeros.
b. Calculate the periodogram with an FFT algorithm.
c. Estimate $\hat{R}(k)$ through an inverse transform of $I(m)$.
d. Multiply $\hat{R}(k)$ by a selected lag window.
e. Calculate the Fourier transform of $\tilde{R}_M(k)$ to obtain $\tilde{S}_M(m)$.

The estimate produced using this method is not different from the direct BT approach. The only difference that will occur is that the frequency spacing will differ.

References

Anzaloni, A. and L. Barbosa (1984). The average power density spectrum of the readback voltage from particulate media. *IEEE Trans. Magnetics* 20:693–698.

Bendat, J. and A. Piersol (1986). *Random Data, Analysis and Measurement Procedures*. Wiley-Interscience, New York.

Bloomfield, P. (1976). *Fourier Analysis of Time Series—An Introduction*. John Wiley and Sons, New York.

Fante, R. (1988). *Signal Analysis and Estimation*. John Wiley and Sons; New York.

Geckinli, N. and D. Yavuz (1983). *Discrete Fourier Transformation and Its Applications to Power Spectra Estimation*. Elsevier Scientific Publishing Co., Amsterdam.

Haggan, V. and O. Oyetunji (1984). On the selection of subset autoregressive time series models. *J. Time Series Anal*. 5:103–114.

Harris, F. (1978). On the use of windows for harmonic analysis with the discrete Fourier transform. *Proc. IEEE* 66:51–83.

Jenkins, G. and D. Watts (1968). *Spectral Analysis and Its Applications*. Holden-Day, San Francisco.

Kay, S. (1988). *Modern Spectral Estimation*. Prentice-Hall, Englewood Cliffs, NJ.

Landers, T. and R. Lacoss (1977). Geophysical applications of spectral estimates. *IEEE Trans. Geoscience Electronics* 15:26–32.

Marple, S. (1987). *Digital Spectral Analysis with Applications*. Prentice-Hall, Englewood Cliffs, NJ.

Marque, C., J. Duchene, S. LeClercq, G. Panczer, and J. Chaumont (1986). Uterine EHG processing for obstetrical monitoring. *IEEE Trans. Biomed. Engrg*. 33:1182–1187.

Noori, M. and H. Hakimmashhadi (1988). Vibration analysis. In *Signal Processing Handbook* (C. Chen, ed.). Marcel Dekker, New York.

Priestley, M. (1981). *Spectral Analysis and Time Series*, Vol. 1: *Univariate Series*. Academic Press, New York.

Reddy, S., S. Collins, and E. Daniel (1987). Frequency analysis of gut EMG. *CRC Critical Reviews in Biomed. Engrg*. 15(2):95–116.

Schwartz, M. and L. Shaw (1975). *Signal Processing: Discrete Spectral Analysis, Detection, and Estimation*. McGraw-Hill Book co., New York.

Welch, P. (1967). The use of fast Fourier transform for the estimation of power spectra: A method based on time averaging over short, modified periodograms. *IEEE Trans. Audio and Electroacoustics* 15:70–73.

Exercises

7.1 Prove that the correlation of the rectangular data window with itself yields the triangular lag window.

7.2 Derive Fejer's kernel from the properties of the rectangular window.

7.3 A signal has the triangular PSD in Figure 7.6. Assume that the data acquisition and signal conditioning produce a rectangular spectral window of unity magnitude over the frequency range $-0.05 \leq f \leq 0.05$ Hz.
a. Sketch the spectral window.
b. Convolve the spectral window with the PSD and show the bias that will be produced in the estimated spectrum.

7.4 In the derivation of the variance for the variance of the periodogram in Appendix 7.1, some summations are simplified. Derive equation (A7.11) from equation (A7.10).

7.5 Prove that the mean of the variance of the real part of the estimate of the PSD for white noise is

$$E[A^2(m)] = \begin{cases} \dfrac{T\sigma^2}{2}, & m \neq 0 \text{ and } \left[\dfrac{N}{2}\right], \\ T\sigma^2, & m = 0 \text{ and } \left[\dfrac{N}{2}\right]. \end{cases}$$

Start with the defining equation [equation (7.24)],

$$E[A^2(m)] = \frac{T\sigma^2}{N} \sum_{n=0}^{N-1} \cos^2\left(\frac{2\pi mn}{N}\right),$$

expand it using Euler's formula, and simplify using the geometric sum formulae.

7.6 In Section 7.2.1 the real part of the DFT of a white noise process is $A(m)$. Derive the expression for the variance of $A(m)$ that is found in equation (7.25). [A possible approach to the derivation is to expand the squared cosine term found in equation (7.24) by using a double angle formula and then to express the cosine term as a summation of complex exponentials using Euler's formula. Find the resultant sum using the geometric sum formula.] If there are any difficulties, consult Bloomfield [(1976), page 15].

7.7 Prove that the harmonic components in a white noise process are uncorrelated, as stated in equation (7.28); that is,

$$E[X(m)X(p)] = \begin{cases} N\sigma^2, & m = p = 0 \text{ and } m = p = [N/2], \\ 0, & \text{otherwise.} \end{cases}$$

7.8 Prove that the real and imaginary parts of the harmonics of the DFT are uncorrelated. Start with equation (7.29).

7.9 Using the covariance expression in Appendix 7.1, prove that the PSD values evaluated at integer multiples of $1/NT$ are uncorrelated; that is, for $f_1 = m_1/NT$, $f_2 = m_2/NT$, and $m_1 \neq m_2$, prove that $\text{Cov}[I(f_1)I(f_2)] = 0$.

7.10 Use the sample spectrum plotted in Figure 7.8(a):

a. Sketch the spectrum if $T = 1$ ms and $T\sigma^2 = 1$.
b. Sketch the spectrum if $T = 1$ ms and $\sigma^2 = 1$.
c. Sketch the spectrum if $T = 50$ ms and $\sigma^2 = 1$.

7.11 A short time series $x(n)$ has the values 6, 4, 2, -1, -5, -2, 7, 5.

a. What is the sample variance?
b. Create a Hamming data window, $d(n)$, for $x(n)$.
c. Multiply $x(n)$ by $d(n)$. What are the resulting values?
d. What is the new sample variance? Is it consistent with the process loss?

7.12 Derive the process loss for the Hanning data window.

7.13 In Example 7.6 and Figure 7.11(g), verify the magnitudes of the averaged spectral estimate at the frequencies 2, 3, and 4 Hz.

7.14 Verify the confidence bounds on the spectral averaging used in Example 7.8.

7.15 What are the confidence bounds on the procedure performed in Example 7.6?

7.16 In the derivation of the variance of the spectral estimates that are found in Appendices 7.1 and 7.2, a form of the triangular spectral window is encountered,

$$G(f) = \left(\frac{\sin(\pi f T N)}{N \sin(\pi f T)} \right)^2 .$$

a. Show that $G(0) = 1$.
b. For $G(f)$ to approximate a delta function as $N \to \infty$, its area must be 1. Show that $NTG(f)$ could be a good approximation to $\delta(f)$ for large N.

7.17 For the smoothed spectral estimate in Figure 7.19(b), verify the magnitudes at the frequencies nearest 13, 15, and 20 Hz in the unsmoothed spectrum. The values are tabulated in Appendix 7.5. Repeat this for Figure 7.19(c).

7.18 What are the confidence limits on the seven-point spectral smoothing used in Example 7.11 for the following cases:
a. A 95% confidence level is desired.
b. A 99% confidence level is desired.

7.19 What are the bandwidths and degrees of freedom for nine-point rectangular and triangular spectral smoothing? How do they compare?

7.20 Derive the variance reduction factor, $\nu/2$, for triangular smoothing, as given in equation (7.68). Assume that the PSD is constant over the interval of smoothing. The following summation formula may be useful:

$$\sum_{i=1}^{N} i^2 = \frac{N(N+1)(2N+1)}{6} .$$

7.21 Prove that $B_e = J$ for triangular smoothing.

7.22 What type of bias will be produced in the variance reduction procedure if, with respect to a smoothing window, the following hold:
a. The actual PSD has a peak.
b. The actual PSD has a trough.

7.23 In Example 7.13, verify the frequency spacings, equivalent bandwidths, and degrees of freedom.

7.24 For the Tukey window, calculate the degrees of freedom and the equivalent bandwidth from equation (7.84).

7.25 For the BT estimate of the PSD using the rectangular lag window and plotted in Figure 7.24(b), verify the confidence limits also shown in the figure.

Computer Exercises

7.26 Generate a white noise process with a variance of 2 and a sampling frequency of 50 Hz.
a. Generate at least 1000 signal points.
b. Divide the signal into at least four segments and calculate the individual spectra and plot them.
c. What are the ranges of magnitudes of the individual periodograms?
d. Produce a spectral estimate through averaging.
e. What is the range of magnitudes? Did it decrease from those found in part c?
f. Find the 95% confidence limits and plot them with the spectral estimate.
g. Does the spectrum of the ideal process lie between the confidence limits?

7.27 The annual trapping of Canadian lynx form 1821 to 1924 has been modeled extensively [Haggan and Oyetunji (1984)]. Two models are

model a:

$$y(n) - 1.48y(n-2) + 0.54y(n-4) = x(n), \qquad \sigma_x^2 = 1.083,$$

model b:

$$y(n) - 1.492y(n-1) + 1.324y(n-2) = x(n), \qquad \sigma_x^2 = 0.0505,$$

after the data have been logarithmically transformed and detrended, and $x(n)$ is white noise. It is claimed that the trappings have a period of 9.5 years.

a. Generate a 200-point random process for each model.
b. Plot the true spectra for each model.
c. Estimate the spectra for each model using the same procedure.
d. Which model more accurately reflects the claimed period?

7.28 Do the following for an MA process with an autocorrelation function estimate

$$\hat{R}(0) = 25, \qquad \hat{R}(\pm 1) = 15, \qquad \hat{R}(k) = 0 \quad \text{for } |k| \geq 2.$$

a. Calculate $\hat{S}_1(m)$ directly using equation (7.73).
b. Calculate $\hat{S}_2(m)$ using an FFT algorithm with $N = 8$.
   1. Is $\hat{S}_2(m)$ real or complex?
   2. What part of $\hat{S}_2(m)$ is equal to $\hat{S}_1(m)$?
   3. How do the frequency spacings differ?

7.29 A daily census for hospital inpatients is tabulated in Appendix 7.6.

a. Plot the time series.
b. Estimate the ACF with $M = 20$.
c. What periodicities are present as ascertained from parts a and b?
d. Estimate the PSD using the BT technique with a suitable lag window.
e. What information is present in this particular $\tilde{S}_M(m)$?
f. Repeat steps a through e with $M = 10$ and $M = 30$.
g. How do these last two PSD estimates differ from the first one?

Appendices

Appendix 7.1 Variance of Periodogram

The bias of the periodogram was discussed in Section 7.1.2. The variance was also stated and it is important to have its derivation available. The system model will be used so that the variation is embedded in white-noise input, which has a zero mean and a variance σ_x^2. Direct derivation of the variance under different conditions can be found in several references, such as Jenkins and Watts (1968), Kay (1988), or Fante (1988). The periodogram estimate is

$$I_y(f) = \frac{T}{N}\hat{Y}(-f)\hat{Y}(f) = \frac{T}{N}\hat{Y}^*(f)\hat{Y}(f), \qquad \text{(A7.1)}$$

where

$$Y(f) = \sum_{n=0}^{N-1} y(n)e^{-j2\pi fnT} \quad \text{and} \quad Y(f) = H(f)X(f). \quad \text{(A7.2)}$$

When N is large, the covariance at different frequencies is written

$$\text{Cov}[I_y(f_1)I_y(f_2)]$$
$$= \text{Cov}[H(f_1)H^*(f_1)I_x(f_1)H(f_2)H^*(f_2)I_x(f_2)]. \quad \text{(A7.3)}$$

Because the variability is restricted to $x(n)$, equation (A7.3) becomes

$$\text{Cov}[I_y(f_1)I_y(f_2)]$$
$$= H(f_1)H^*(f_1)H(f_2)H^*(f_2)\text{Cov}[I_x(f_1)I_x(f_2)]. \quad \text{(A7.4)}$$

and

$$\text{Cov}[I_x(f_1)I_x(f_2)] = E[I_x(f_1)I_x(f_2)] - E[I_x(f_1)]E[I_x(f_2)]. \quad \text{(A7.5)}$$

Examine the second term on the right-hand side of equation (A7.5). Each factor is the mean of the periodogram of a white noise process and it is easily determined from equation (7.22) to be unbiased and, therefore,

$$E[I_x(f_1)] = E[I_x(f_2)] = T\sigma_x^2. \qquad \text{(A7.6)}$$

The mean product is much more complicated and becomes the

mean of four terms. Let the letters s, t, k, and l represent the summing indices. Now,

$$I_x(f_1) = \frac{T}{N} \sum_{s=0}^{N-1} \sum_{t=0}^{N-1} x(s)x(t)\exp(-j(s-t)2\pi f_1 T) \quad \text{(A7.7)}$$

$I_x(f_2)$ is similarly expressed and

$$E[I_x(f_1)I_x(f_2)] = \frac{T^2}{N^2} \sum_{s=0}^{N-1} \sum_{t=0}^{N-1} \sum_{k=0}^{N-1} \sum_{l=0}^{N-1} E[x(s)x(t)x(k)x(l)]$$
$$\times \exp(-j(s-t)2\pi f_1 T - j(k-l)2\pi f_2 T). \quad \text{(A7.8)}$$

As with the proofs of stationarity, it will be assumed that the fourth moment behaves approximately as that of a Gaussian process. Therefore,

$$E[x(s)x(t)x(k)x(l)]$$
$$= R_x(s-t)R_x(k-l) + R_x(s-k)R_x(t-l)$$
$$+ R_x(s-l)R_x(t-k). \quad \text{(A7.9)}$$

Equation (A7.8) will be evaluated on a term by term basis using equation (A7.9). Examine the sums produced from individual terms on the right of equation (A7.9). Remember that $R_x(s-t) = \sigma_x^2\delta(s-t)$ and $R_x(k-l) = \sigma_x^2\delta(k-l)$. The first term has exponents of zero when the autocorrelation are nonzero and summation produces the term $\Sigma_1 = N^2\sigma_x^4$. The second term has nonzero values only when $s = k$ and $t = l$, and its summation reduces to

$$\Sigma_2 = \sum_{s=0}^{N-1} \sum_{t=0}^{N-1} \sigma_x^4 \exp(-js2\pi(f_1+f_2)T + jt2\pi(f_1-f_2)T)$$
$$= \sigma_x^4 \sum_{s=0}^{N-1} \exp(-js2\pi(f_1+f_2)T) \sum_{t=0}^{N-1} \exp(+jt2\pi(f_1+f_2)T). \quad \text{(A7.10)}$$

Using the geometric sum formula and then Euler's formula,

$$\Sigma_2 = \sigma_x^4 \left(\frac{1 - \exp(-j2\pi(f_1 + f_2)TN)}{1 - \exp(-j2\pi(f_1 + f_2)T)} \times \frac{1 - \exp(+j2\pi(f_1 + f_2)TN)}{1 - \exp(+j2\pi(f_1 + f_2)T)} \right)$$

$$= \sigma_x^4 = \left(\frac{\sin(\pi(f_1 + f_2)TN)}{\sin(\pi(f_1 + f_2)T)} \right)^2. \qquad \text{(A7.11)}$$

The third term has properties similar to Σ_2 when $s = l$ and $t = k$, and its summation is

$$\Sigma_3 = \sigma_x^4 \left(\frac{\sin(\pi(f_1 - f_2)TN)}{\sin(\pi(f_1 - f_2)T)} \right)^2. \qquad \text{(A7.12)}$$

Summing Σ_1, Σ_2, and Σ_3 produces

$$E[I_x(f_1) I_x(f_2)] = \frac{T^2 \sigma_x^4}{N^2} \left(N^2 + \left(\frac{\sin(\pi(f_1 + f_2)TN)}{\sin(\pi(f_1 + f_2)T)} \right)^2 + \left(\frac{\sin(\pi(f_1 - f_2)TN)}{\sin(\pi(f_1 - f_2)T)} \right)^2 \right). \qquad \text{(A7.13)}$$

Subtracting the product of the means of the periodograms from equation (A7.13), yields the covariance between periodogram values of white noise,

$$\text{Cov}[I_x(f_1) I_x(f_2)] = \sigma_x^4 T^2 \left(\left(\frac{\sin(\pi(f_1 + f_2)TN)}{N \sin(\pi(f_1 + f_2)T)} \right)^2 + \left(\frac{\sin(\pi(f_1 - f_2)TN)}{N \sin(\pi(f_1 - f_2)T)} \right)^2 \right). \qquad \text{(A7.14)}$$

From system theory it is known that

$$S_y(f) = H(f)H^*(f)S_x(f). \tag{A7.15}$$

Substituting equations (A7.15) and (A7.14) into equation (A7.3) produces

$$\mathrm{Cov}\left[I_y(f_1)I_y(f_2)\right] = S_y(f_1)S_y(f_2)\left(\left(\frac{\sin(\pi(f_1+f_2)TN)}{N\sin(\pi(f_1+f_2)T)}\right)^2 + \left(\frac{\sin(\pi(f_1-f_2)TN)}{N\sin(\pi(f_1-f_2)T)}\right)^2\right) \tag{A7.16}$$

This is a complicated expression that has a very important result. Because in discrete-time processing the spectral values are evaluated at integer multiples of the resolution $1/NT$, the periodogram magnitude values are uncorrelated. This proof is left as an exercise. The variance is found when $f_1 = f_2$,

$$\mathrm{Var}\left[I_y(f)\right] = S_y^2(f)\left(1 + \left(\frac{\sin(2\pi fTN)}{N'\sin(2\pi fT)}\right)^2\right). \tag{A7.17}$$

Thus, even when N approaches infinity,

$$\mathrm{Var}\left[I_y(f)\right] \to S_y^2(f) \tag{A7.18}$$

and the periodogram is an inconsistent estimator.

Appendix 7.2 Proof of Variance of BT Spectral Smoothing

From Section 7.4.3 the BT estimator is

$$\tilde{S}(f) = T\sum_{k=-M}^{M} w(k)\hat{R}(k)e^{-j2\pi fkT} = \int_{-1/2T}^{1/2T}\hat{S}(g)W(f-g)\,dg. \tag{A7.19}$$

its variance is then

$$\operatorname{Var}[\tilde{S}(f)] = E\left[\left(\int_{-1/2T}^{1/2T} \hat{S}(g)W(f-g)\,dg\right)^2\right] - E[\hat{S}(f)]^2$$

$$= E\left[\left(\int_{-1/2T}^{1/2T} \hat{S}(g)W(f-g)\,dg\right)\right.$$

$$\left.\times\left(\int_{-1/2T}^{1/2T} \hat{S}(h)W(f-h)\,dh\right)^2\right] - E[\tilde{S}(f)]^2. \quad \text{(A7.20)}$$

Expanding the squared mean and collecting terms, equation (A7.20) becomes

$$\operatorname{Var}[\tilde{S}(f)] = \int_{-1/2T}^{1/2T}\int_{-1/2T}^{1/2T} W(f-g)W(f-h)$$
$$\times \operatorname{Cov}[\hat{S}(g)\hat{S}(h)]\,dg\,dh. \quad \text{(A7.21)}$$

Because $\hat{S}(f)$ is a periodogram, its covariance is given by equation (A7.16) in Appendix 7.1. For large N each squared term approaches $\delta(\cdot)/NT)$ and equation (A7.21) becomes

$$\operatorname{Var}[\tilde{S}(f)] \approx \int_{-1/2T}^{1/2T}\int_{-1/2T}^{1/2T} W(f-g)W(f-h)$$
$$\times \frac{S(g)S(h)}{NT}(\delta(g+h)+\delta(g-h))\,dg\,dh \quad \text{(A7.22)}$$

$$\approx \frac{1}{NT}\int_{-1/2T}^{1/2T} (W(f+h)W(f-h)$$
$$+W(f-h)^2)S(h)^2\,dh. \quad \text{(A7.23)}$$

Equation (A7.23) is again simplified by assuming N is large and, consequently, the spectral window is narrow. Then there will not be appreciables overlap of windows located at frequencies $(f+h)$ and $(f-h)$. Thus, the first term in the parentheses of the integrand can be considered zero. Also the PSD can be considered constant over the window width. Thus,

$$\mathrm{Var}[\tilde{S}(f)] \approx \frac{S(f)^2}{NT} \int_{-1/2T}^{1/2T} W(h)^2 \, dh. \qquad \text{(A7.24)}$$

Appendix 7.3 Window Characteristics

Window	Equivalent Bandwidth (bins)	Ratio (Highest Sidelobe Level to Peak)	Degrees of Freedom	Process Loss
Rectangular	1.00	0.220	$\frac{N}{M}$	1.00
Bartlett	1.33	0.056	$3\frac{N}{M}$	0.33
Hanning–Tukey	1.50	0.026	$2.67\frac{N}{M}$	0.36
Hamming	1.36	0.0089	$2.51\frac{N}{M}$	0.40
Parzen	1.86	0.0024	$3.71\frac{N}{M}$	0.28

Appendix 7.4 Lag Window Functions

Lag Window*	Spectral Window
Rectangular	
$w_R(k) = 1, \quad \lvert k \rvert \le M$	$W_R(f) = T\frac{\sin(2\pi fMT)}{\sin(\pi fT)}$
Triangular–Bartlett	
$w_B(k) = 1 - \frac{\lvert k \rvert}{M}, \quad \lvert k \rvert \le M$	$W_B(f) = \frac{T}{M}\left(\frac{\sin(\pi fMT)}{\sin(\pi fT)}\right)^2$
Hanning–Tukey	
$w_T = \frac{1}{2}\left(1 + \cos\left(\frac{\pi k}{M}\right)\right)$	$W_T(f) = 0.5W_R(f) + 0.25W_R\left(f + \frac{1}{2MT}\right) + 0.25W_R\left(f - \frac{1}{2MT}\right)$

Lag Window*	Spectral Window
Hamming	
$w_H(k) = 0.54 + 0.46\cos\left(\frac{\pi k}{M}\right)$	$W_H(f) = 0.54W_R(f) + 0.23W_R\left(f + \frac{1}{2MT}\right) + 0.23W_R\left(f - \frac{1}{2MT}\right)$
Parzen	
$w_p(k) = \begin{cases} 1 - 6\left(\frac{k}{M}\right)^2 + 6\left(\frac{\|k\|}{M}\right)^3, & \|k\| \le \frac{M}{2}, \\ 2\left(1 - \frac{\|k\|}{M}\right)^3, & \frac{M}{2} < \|k\| \le M \end{cases}$	$W_P(f) = \frac{8T}{M^3}\left(\frac{3}{2}\frac{\sin^4(\pi fMT/2)}{\sin^4(\pi fT)} - \frac{\sin^4(\pi fMT/2)}{\sin^2(\pi fT)}\right)$

*All lag windows have $w(k) = 0$ for $|k| > M$.

Appendix 7.5 Spectral Estimates from Smoothing

Spectral Estimates from Three- and Seven-Point Smoothing In Figures 7.19(b) and 7.19(c)

Frequency	Original Periodogram	Periodogram with Three-Point Smoothing	Periodogram with Seven-Point Smoothing
0.00000	0.00003	0.00454	0.01344
0.78125	0.00680	0.00667	0.01491
1.56250	0.01318	0.01568	0.01384
2.34375	0.02705	0.02586	0.01309
3.12500	0.03735	0.02336	0.01369
3.90625	0.00567	0.01487	0.01406
4.68750	0.00158	0.00382	0.01992
5.46875	0.00420	0.00507	0.01762
6.25000	0.00941	0.02260	0.01321
7.03125	0.05418	0.02484	0.01545

Spectral Estimates from Three- and Seven-Point Smoothing In Figures 7.19(b) and 7.19(c) (Continued).

Frequency	Original Periodogram	Periodogram with Three-Point Smoothing	Periodogram with Seven-Point Smoothing
7.81250	0.01094	0.02386	0.02194
8.59375	0.00646	0.01293	0.02909
9.37500	0.02139	0.02495	0.03059
10.15625	0.04699	0.04088	0.03673
10.93750	0.05427	0.04039	0.04386
11.71875	0.01991	0.05711	0.04353
12.50000	0.09714	0.05930	0.25210
13.28125	0.06086	0.05404	0.25618
14.06250	0.00413	0.51547	0.28618
14.84375	1.48142	0.52037	0.29994
15.62500	0.07555	0.60708	0.29561
16.40625	0.26426	0.15202	0.29617
17.18750	0.11624	0.14909	0.29847
17.96875	0.06678	0.08262	0.11871
18.75000	0.06482	0.05061	0.11793
19.53125	0.02024	0.10270	0.08260
20.31250	0.22305	0.10446	0.06660
21.09375	0.07008	0.10338	0.06000
21.87500	0.01701	0.03044	0.05273
22.65625	0.00423	0.01394	0.05513
23.43750	0.02058	0.01291	0.02525
24.21875	0.01392	0.02384	0.01818
25.00000	0.03702	0.02162	0.01635

Appendix 7.6 Hospital Census Data

Hospital Inpatient Daily Census, April 21 to July 19*

397	462	486	483	477	438	407	421	480	484
486	479	415	400	419	477	510	503	500	435
408	417	478	497	500	512	450	421	423	471
496	478	463	413	396	366	375	444	469	480
439	402	442	492	507	518	493	439	399	428
476	499	488	460	419	380	406	472	502	495
490	443	398	417	490	505	499	484	430	384
392	452	455	426	414	405	379	410	485	514
525	511	461	436	444	488	494	510	493	429
392	420	466	476	494	484	423	388	411	472

*Total 100 observation (read row-wise).

Chapter 8 Random Signal Modeling and Modern Spectral Estimation

8.1 Introduction

The preceding analyses of properties of random signals are focused upon estimating the first- and second-order stochastic characteristics of the time series and their spectral density composition. More detailed information about a time series can be obtained if the time series itself can be modeled. In addition, its power spectrum can be obtained from the model. This approach has been given increased emphasis in engineering during the last decade, with new algorithms being developed and many useful applications appearing in books and journal publications.

Basically, the concept is to model the signal using the principles of casual and stable discrete-time systems, as studied in Chapter 6. Recall that in the time domain, the input $x(n)$ and output $y(n)$ of a system are related by

$$y(n) = -\sum_{i=1}^{p} a(i)y(n-i) + \sum_{l=0}^{q} b(l)x(n-l). \qquad (8.1)$$

In signal modeling, the output function is known. From what principle does the input function arise? As was learned in Chapter 6, the system can, when the input is white noise, create an output

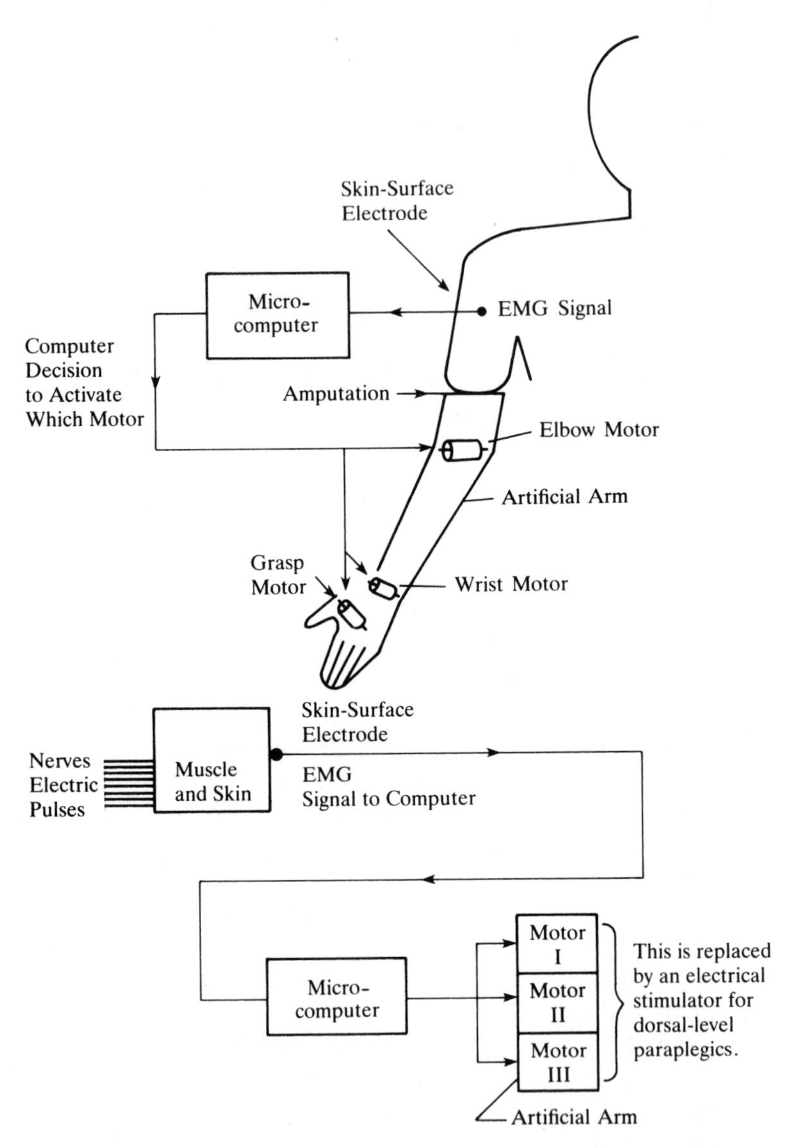

Figure 8.1 Schematic diagram of (a) arm prosthesis and (b) signal identification and control process; $f_s = 2$ kHz. [Adapted from Graupe (1984), Figure B2, with permission.]

with properties of other random signals. Thus the properties of $y(n)$ depend upon the variance of the white noise and on the parameters $a(i)$, $b(l)$, p, and q. This approach is called *parametric signal modeling*. Because equation (8.1) also defines an autoregressive moving average model, the approach is synonymously called *ARMA modeling*. Also recall that if the system's parameters and input are known, the power spectral density function of the output can be determined through the power transfer function. With $x(n)$ being white noise, then

$$S_y(f) = |H(f)|^2 S_x(f) = |H(f)|^2 \sigma_x^2 T. \tag{8.2}$$

Thus, an alternative approach to determining the PSD of a signal can be defined upon a parameteric, or ARMA, model. Collectively, this approach to signal processing is called modern signal processing.

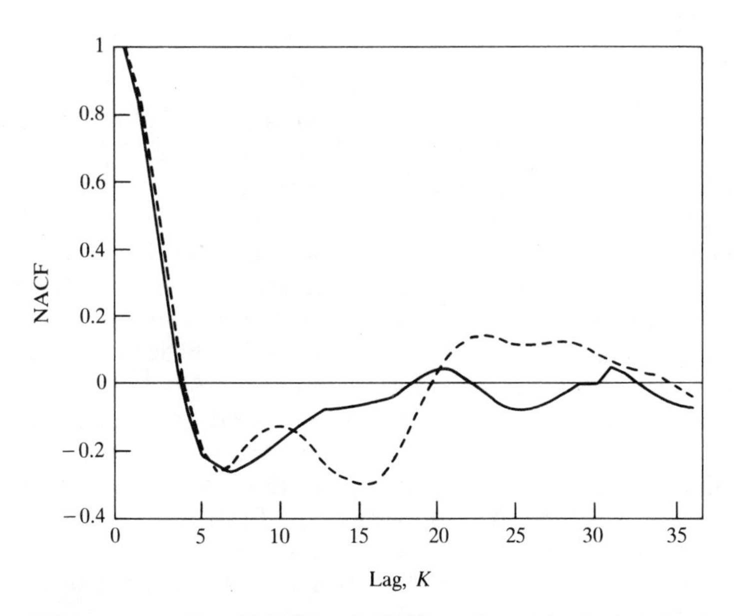

Figure 8.2 The NACFs of EMG and residual signals from a contracting muscle (500 signal points): (—) NACF at 25% MVC, (---) NACF at 50% MVC. [Adapted from Triolo, Nash, and Moskowitz (1988, IEEE), Figures 6 and 8, with permission.]

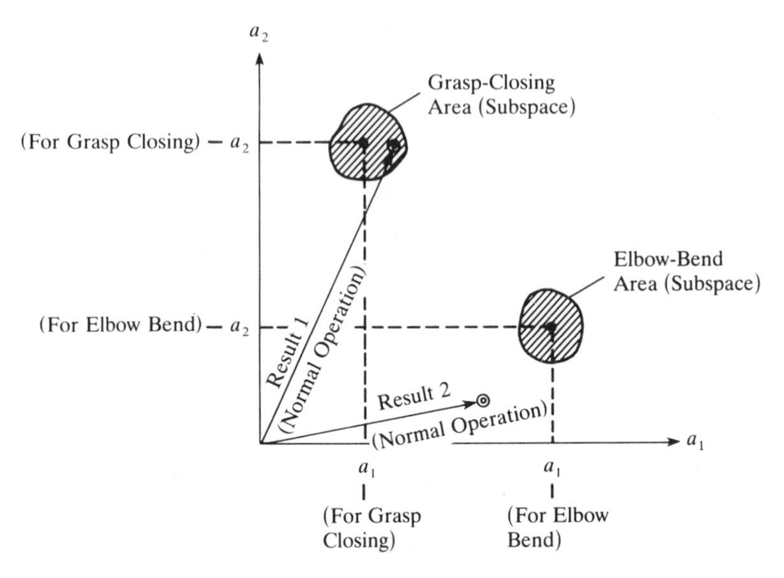

Figure 8.3 Parameter spaces for two movements based on the first two AR parameters. [Adapted from Graupe (1984), Figure B1, with permission.]

An application in the time domain will help illustrate the utility of this approach. If a signal containing 1000 points can be modeled by equation (8.1) with the orders p and q being 20 or less, then the information in this signal can be represented with relatively few parameters. This condition is true for most signals [Jansen (1985)]. Thus, further analyses and interpretation are facilitated. One such application is the use of the electromyogram (EMG) from a leg or arm muscle to control a device for a paralyzed person or an artificial limb for an amputee. A schematic diagram for controlling a limb prosthesis is shown in Figure 8.1. The concept is that different intensities of contraction will produce a signal with different structure and thus different parameter sets. These sets will make the prosthesis perform a different function. The change in structure is indicated by the change in NACF, as shown in Figure 8.2. When the parameter sets caused by different levels of contraction are plotted (see Figure 8.3) one can see that the parameter sets are disjoint. It is then straightforward to have a multiple-task controller with this scheme. AR(4) models have been found to be successful [Trilolo, Nash, and Moskowitz (1988) and Hefftner, Zucchini, and Jaros (1988)].

8.2 Model Development

Model development is based upon the fact that signal points are interrelated across time, as expressed by equation (8.1). This interrelatedness can be expressed by the autocorrelation function. Thus, there is information contained in several signal values that can be used to estimate future values. For instance, what is the estimate of the present value of $y(n)$ based upon two previously measured values? Mathematically this is expressed as

$$\hat{y}(n) = h(1)y(n-1) + h(2)y(n-2). \tag{8.3}$$

This weighted sum is called a *linear prediction*. Naturally, the desire is to estimate the value of $y(n)$ with the smallest error possible. Define the error as

$$\epsilon(n) = y(n) - \hat{y}(n). \tag{8.4}$$

Notice that the error is also a time series that is stationary because $y(n)$ is stationary. Again, the mean squared error (MSE) will be minimized. Notice that the MSE is also the variance σ_ϵ^2 of the error time series. By definition,

$$\begin{aligned} \text{MSE} &= E\left[\epsilon^2(n)\right] = E\left[(y(n) - \hat{y}(n))^2\right] \\ &= E\left[(y(n) - h(1)y(n-1) - h(2)y(n-2))^2\right]. \end{aligned} \tag{8.5}$$

The coefficients will be those that minimize the MSE. To find $h(1)$,

$$\begin{aligned} &\frac{\partial}{\partial h(1)}\text{MSE} \\ &= \frac{\partial}{\partial h(1)} E\left[(y(n) - h(1)y(n-1) - h(2)y(n-2))^2\right] \\ &= E\left[\frac{\partial}{\partial h(1)}(y(n) - h(1)y(n-1) - h(2)y(n-2))^2\right] \\ &= -2E\left[(y(n) - h(1)y(n-1) - h(2)y(n-2))y(n-1)\right] \\ &= 0. \end{aligned} \tag{8.6}$$

The factor -2 can be disregarded and the terms within the square brackets must be expanded to simplify the expectations,

$$\begin{aligned}(y(n) - h(1)y(n-1) - h(2)y(n-2))y(n-1) \\ = y(n)y(n-1) - h(1)y(n-1)y(n-1) \\ - h(2)y(n-2)y(n-1).\end{aligned} \tag{8.7}$$

The terms in equation (8.7) are cross products and its expectation will yield an equation containing some terms of the autocorrelation function of $y(n)$. The expectation is

$$\begin{aligned}E[y(n)y(n-1) - h(1)y(n-1)y(n-1) \\ -h(2)y(n-2)y(n-1)] \\ = E[y(n)y(n-1)] - E[h(1)y(n-1)y(n-1)] \\ - E[h(2)y(n-2)y(n-1)] \\ = R_y(-1) - h(1)R_y(0) - h(2)R_y(1) = 0.\end{aligned}$$

Because all ACFs are even functions, then

$$R_y(1) - h(1)R_y(0) - h(2)R_y(1) = 0. \tag{8.8}$$

Similarly, another equation is obtained by minimizing equation (8.5) with respect to $h(2)$,

$$R_y(2) - h(1)R_y(1) - h(2)R_y(0) = 0. \tag{8.9}$$

As can be appreciated, if some values of the ACF of $y(n)$ are known, then the weighting coefficients can be found. The value of one more quantity must be found. Because the MSE was minimized, its magnitude is important in order to judge the accuracy of the model. It can be directly obtained from equations (8.5), (8.8), and (8.9) and is derived in Appendix 8.1. It is

$$\sigma_\epsilon^2 = R_y(0) - h(1)R_y(1) - h(2)R_y(2). \tag{8.10}$$

The last three equations are usually represented in matrix form as

$$\begin{bmatrix} R(0) & R(1) & R(2) \\ R(1) & R(0) & R(1) \\ R(2) & R(1) & R(0) \end{bmatrix} \begin{bmatrix} 1 \\ -h(1) \\ -h(2) \end{bmatrix} = \begin{bmatrix} \sigma_\epsilon^2 \\ 0 \\ 0 \end{bmatrix}. \tag{8.11}$$

This solution form is very prevalent for parameteric modeling and is called the *Yule–Walker* (YW) equations. The 3×3 matrix also has a special form. Notice that all the terms on the diagonal, subdiagonal, and supradiagonal are the same. This matrix form is called a *Toeplitz matrix*. Now combine the prediction equations (8.3) and (8.4) and write them as

$$y(n) - h(1)y(n-1) - h(2)y(n-2) = \epsilon(n). \tag{8.12}$$

This is exactly the form of a second-order AR system with input $\epsilon(n)$. The nature of $\epsilon(n)$ determines the adequacy of the model and we stipulate that not only must the variance of $\epsilon(n)$ be minimum but it also must be a white noise sequence. It is known from Chapter 6 that an AR system can produce an output with structure when its input is uncorrelated. Thus, knowing the autocorrelation function of a signal permits the development of models for it.

The AR model is used most often because the solution equations for its parameters are simpler and more developed than those for either MA or ARMA models. Fortunately, this approach is valid as well as convenient because of the *Wold decomposition theorem* [Fante (1988) and Priestley (1981)]. It states that the random component of any stationary process can be modeled by a stable causal linear system with a white noise input. The impulse response is essentially an MA process that can have infinite order. This has been demonstrated in Chapter 5. The AR and ARMA process models also produce such impulse responses. Thus, an ARMA or MA process can be adequately represented by an equivalent AR process of suitable, possibly infinite, order. Typically, one uses different criteria to determine the model order. These criteria are functions of estimates of σ_ϵ^2. However, in most signal processing situations, the values of $R(k)$ are not known either. Thus the modeling must proceed from a different viewpoint. Within the next several sections, methods will be derived to develop signal models. The derivations are very similar to the foregoing procedure and parallel the philosophy of other estimating procedures. That is, the first attempt for implementation is to use the theoretical procedure with the theoretical functions being replaced by their estimates.

8.3 Random Data Modeling Approach

8.3.1 Basic Concepts

The *random data model* is the name given to the approach for modeling signals based upon the notion of linear prediction when the autocorrelation function is unknown. Again, it is desired to select model parameters that minimize the squared error. Start with the general equation for linear prediction using the previous p terms,

$$\hat{y}(n) = h(1)y(n-1) + h(2)y(n-2) + \cdots + h(p)y(n-p), \tag{8.13}$$

with $\hat{y}(n)$ representing the predicted value of the time series at time n. The prediction error is

$$\epsilon(n) = y(n) - \hat{y}(n) \tag{8.14}$$

and the error is a random time sequence itself. Combining equations (8.14) and (8.13) to express the signal and error sequence in a system form produces

$$\epsilon(n) = y(n) - h(1)y(n-1) - h(2)y(n-2) - \cdots - h(p)y(n-p). \tag{8.15}$$

The error sequence is the output of a pth-order MA system with the signal as the input and weighting coefficients $b(i) = -h(i)$, $0 \le i \le p$. If the sequence $\epsilon(n)$ is a white-noise sequence, the process defined by equation (8.15) is called *prewhitening* the signal $y(n)$.

Another aspect to this situation is that usually the mean m of the process is also unknown. Thus, to be all inclusive,

$$y(n) = z(n) - m, \tag{8.16}$$

where $z(n)$ is the signal to be modeled. The error is then explicitly given as

$$\epsilon(n) = (z(n) - m) + a(1)(z(n-1) - m) + \cdots + a(p)(z(n-p) - m). \tag{8.17}$$

Now the error is based not only upon the signal points available but also the mean value. So a slightly different approach must be considered. Let the total squared error (TSE) be the error criterion. Because $p + 1$ points are used in equation (8.17), the error sequence begins at time point p and

$$\begin{aligned} \text{TSE} &= \sum_{n=p}^{N-1} \epsilon^2(n) \\ &= \sum_{n=p}^{N-1} ((z(n) - m) + a(1)(z(n-1) - m) \\ &\qquad + \cdots + a(p)(z(n-p) - m))^2. \end{aligned} \tag{8.18}$$

Examination of equation (8.18) reveals that because of the different time indices in each term, the resulting summations are over different time ranges of the process $z(n)$. This can have a significant effect upon the estimation of the parameters when N is small. Let us develop the procedure for estimating m and $a(1)$ in a first-order model in order to appreciate the potential complexities. This is done in the following example.

Example 8.1

For any point in time, the error sequence and the TSE for a first-order model are

$$\epsilon(n) = (z(n) - m) + a(1)(z(n-1) - m),$$

$$\text{TSE} = \sum_{n=1}^{N-1} ((z(n) - m) + a(1)(z(n-1) - m))^2.$$

The parameters m and $a(1)$ are found through the standard minimization procedure. Thus,

$$\frac{\partial}{\partial a(1)}\text{TSE} = \sum_{n=1}^{N-1} ((z(n) - m) + a(1)(z(n-1) - m)) \times (z(n-1) - m) = 0$$

and

$$\frac{\partial}{\partial m}\text{TSE} = \sum_{n=1}^{N-1} ((z(n) - m) + a(1)(z(n-1) - m)) \times (1 + a(1)) = 0.$$

Rearranging the last equation and indicating the estimates of m and $a(1)$ with circumflexes yields

$$\sum_{n=1}^{N-1} (z(n) - \hat{m}) + \hat{a}(1) \sum_{n=1}^{N-1} (z(n-1) - \hat{m}) = 0.$$

The subtle point is to notice that the summation in the preceding equation is over different time spans of $z(n)$. The first summation involves the last $N - 1$ points and the second summation involves the first $N - 1$ points. Dividing the equation by $N - 1$ and designating the sums as different estimates of the mean, $\hat{m}_2$ and $\hat{m}_1$, respectively, gives

$$(\hat{m}_2 - \hat{m}) + \hat{a}(1)(\hat{m}_1 - \hat{m}) = 0.$$

One can easily ascertain that if N is large, then

$$\hat{m}_2 \approx \hat{m}_1 \approx \hat{m}$$

and the conventional estimate of the mean is valid. Notice very clearly that this is not true if N is small.

The solution for the other equation is more complicated and contains cross products. Expanding produces

$$\sum_{n=1}^{N-1} (z(n) - \hat{m})(z(n-1) - \hat{m}) + \hat{a}(1)(z(n-1) - \hat{m})^2 = 0,$$

or

$$\hat{a}(1) = \frac{-\sum_{n=1}^{N-1}(z(n) - \hat{m})(z(n-1) - \hat{m})}{\sum_{n=1}^{N-1}(z(n-1) - \hat{m})^2}.$$

Again, if the numerator and denominator are divided by $N - 1$, they are estimates of the covariance of lag 1 and the variance, respectively. Thus,

$$\hat{a}(1) = -\frac{\hat{C}(1)}{\hat{C}(0)} = -\hat{\rho}(1),$$

the value of the negative of the sample NACF at lag 1. Notice that the estimator for the variance did not utilize all of the available sample points of $z(n)$. The total squared error is

$$\text{TSE} = (N-1)\big(\hat{C}(0) + \hat{a}(1)\hat{C}(1)\big).$$

The development of the models for signals will be based on the knowledge gained in Example 8.1. The signals will be considered to be long enough so that summations that contain the same number of points, but differ by their ranges of summations, can be considered as approximately equal. This means that *the mean value can be estimated immediately and subtracted from the signal*. That is, before being analyzed, the signal will be detrended as is necessary in conventional spectral analysis. The models will be developed on the zero-average sample signal. The criterion that will be used to derive the solutions for the models' parameters will be the minimization of the total squared error for linear prediction. Also remember that the estimate for the parameter of the first-order model is the negative value of the sample NACF at lag 1. It shall be seen that the NACF values are essential to this model development. The models derived in this manner are called the linear prediction coefficient (LPC) models.

The development of signal modeling techniques for small-sample signals requires studying in detail the effects of summing over different time ranges of the signal. These estimating techniques are a current topic of research interest and can be studied in more advanced textbooks and in the current literature. For more advanced detail, one can begin by studying the relevant material in Jenkins and Watts (1968) and Kay (1988).

Example 8.2

One signal that seems to be modeled well by a first-order AR process is the surface of a grinding wheel. The roughness of the surface dictates the texture of the surface of the object being ground. The surface is quantized by measuring the height of the

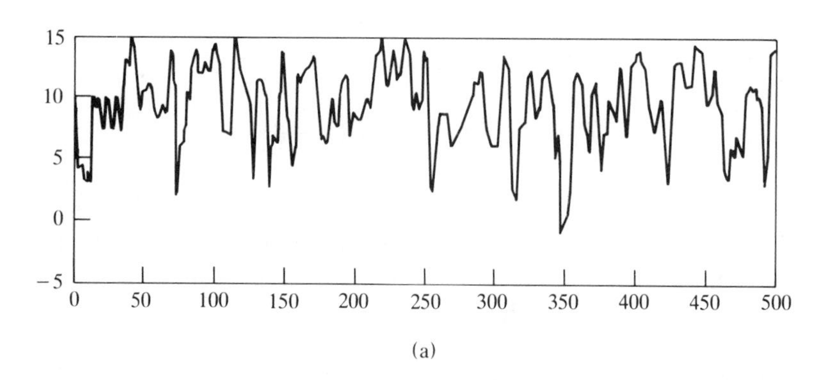

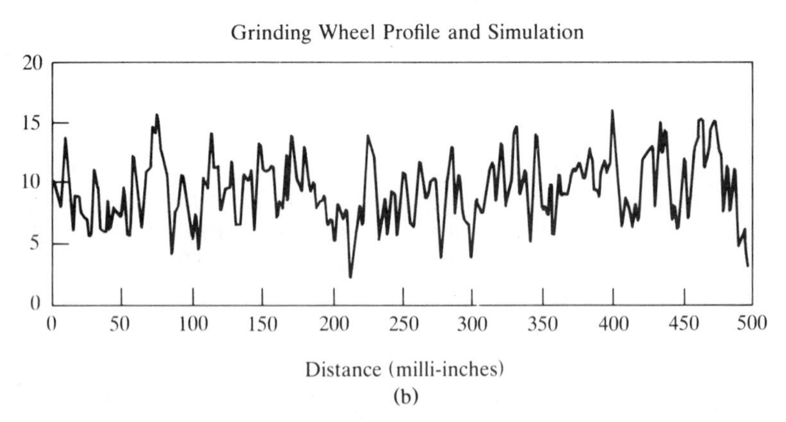

Figure 8.4 Grinding wheel profile: (a) surface height in milli-inches measured every 2 milli-inches; (b) simulated profile with first-order model.

wheel above a designated reference point. The sampling interval is 0.002 inches. A sample measurement is shown in Figure 8.4(a) and tabulated in Appendix 8.3 (where height is in milli-inches). The measurements have a nonzero mean of 9.419, which had to be subtracted from the signal before the parameter $a(1)$ could be calculated. The model is

$$y(n) = 0.627Y(n-1) + \epsilon(n) \quad \text{with } \sigma_\epsilon^2 = 6.55.$$

The sample standard derivation is 3.293. If one simulates the process with a generated $\epsilon(n)$, a resulting signal $s(n)$, such as that in Figure 8.4(b), is produced. Notice that $y(n)$ and $s(n)$ are not identical but that the qualitative appearances are similar. Because the pdf for $y(n)$ is approximately Gaussian, the pdf for $\epsilon(n)$ was chosen to be Gaussian also.

8.3.2 Solution of General Model

After detrending, the general signal model is equation (8.15) and the parameters are found through the minimization of the TSE, where

$$\begin{aligned}\text{TSE} &= \sum_{n=p}^{N-1} \epsilon^2(n) \\ &= \sum_{n=p}^{N-1} (y(n) + a(1)y(n-1) + a(2)y(n-2) \\ &\qquad + \cdots + a(p)y(n-p))^2. \quad (8.19)\end{aligned}$$

The general solution becomes

$$\begin{aligned}\frac{\partial}{\partial a(i)}\text{TSE} &= 0 \\ &= \sum_{n=p}^{N-1} (y(n) + a(1)y(n-1) + a(2)y(n-2) \\ &\qquad + \cdots + a(p)y(n-p))y(n-i), \quad (8.20)\end{aligned}$$

for $1 \leq i \leq p$. Thus, there are p equations of the form

$$-\sum_{n=p}^{N-1} y(n)y(n-i)$$

$$= \sum_{n=p}^{n-1} a(1)y(n-1)y(n-i) + a(2)y(n-2)y(n-i)$$

$$+ \cdots + a(p)y(n-p)y(n-i). \qquad (8.21)$$

If each term is divided by N, the summations become the biased estimates of the autocovariance and autocorrelation function, or

$$-\hat{R}(i) = a(1)\hat{R}(i-1) + a(2)\hat{R}(i-2) + \cdots + a(p)\hat{R}(i-p). \qquad (8.22)$$

The unbiased estimates could have also been used by dividing the summations by $N - p$. The lowercase symbol $r(k)$ is usually used as the symbol for the sample autocorrelation function $\hat{R}(k)$. This will be used to make the writing simpler. Equation (8.22) then becomes

$$a(1)r(i-1) + a(2)r(i-2) + \cdots + a(p)r(i-p) = -r(i). \qquad (8.23)$$

Examine the arguments of $r(k)$. Notice that some are negative. In rewriting these p equations, their even property will be utilized; thus, in matrix form they can be written as

$$\begin{bmatrix} r(0) & r(1) & \cdots & r(p-1) \\ r(1) & r(0) & \cdots & r(p-2) \\ \vdots & \vdots & & \vdots \\ r(p-1) & r(p-2) & \cdots & r(0) \end{bmatrix} \begin{bmatrix} a(1) \\ a(2) \\ \vdots \\ a(p) \end{bmatrix} = \begin{bmatrix} -r(1) \\ -r(2) \\ \vdots \\ -r(p) \end{bmatrix}. \qquad (8.24)$$

Notice that this matrix equation has the same form as equation (8.11) and contains the Yule–Walker equations for modeling measured signals. The square matrix of dimension $p \times p$ is called the *autocorrelation matrix*. The solution can be found with any technique that solves simultaneous linear equations, such as the Gaussian elimination method. The total squared error then becomes

$$\mathrm{TSE} = N\left(r(0) + \sum_{i=1}^{p} \hat{a}(i) r(i)\right); \tag{8.25}$$

TSE is also called the *sum of squared residuals*. Because of its correspondence to the theoretical model, the variance of the error sequence is estimated to be

$$s_p^2 = \frac{\mathrm{TSE}}{N} = r(0) + \sum_{i=1}^{p} \hat{a}(i) r(i). \tag{8.26}$$

Now that the signal can be modeled, it would be good to be assured that the system model is a stable one. Analytical derivations and research using simulated random signals has shown that the use of the biased autocorrelation estimates in the Yule–Walker equations always produces an ACF matrix that is positive definite. This in turn insures that the signal models will be stable. This will not be the situation if the unbiased estimates for $\hat{R}(k)$ are used [Kay (1988) and Marple (1987)]. As a reminder, a matrix is *positive definite* if its determinant and those of all its principal minors are positive. For a formal definition, consult any introductory textbook on matrices, such as Ayres (1962).

Example 8.3

Develop a second-order model for a temperature signal. Twelve measurements of the temperature signal and the estimates of its autocovariance function for five lags are tabulated in Table 8.1. The sample mean and variance are 9.467 and 4.68, respectively. The equations for a second-order model of a signal are

$$\hat{a}(1) r(0) + \hat{a}(2) r(1) = -r(1)$$

Table 8.1

Month	Temperature (°C)	$\hat{C}(k)$
1	3.4	15.1
2	4.5	7.05
3	4.3	−3.5
4	8.7	−9.79
5	13.3	−12.0
6	13.8	
7	16.1	
8	15.5	
9	14.1	
10	8.9	
11	7.4	
12	3.6	

and

$$\hat{a}(1)r(1) + \hat{a}(2)r(0) = -r(2).$$

From Table 8.1 and the variance, the sample NACF values are [1, 0.687, 0.322, − 0.16, − 0.447, − 0.547]. The solution for the model's parameters are

$$\hat{a}(1) = \frac{r(1)(r(2) - r(0))}{r^2(0) - r^2(1)} = \frac{\hat{\rho}(1)(\hat{\rho}(2) - 1)}{1 - \hat{\rho}^2(1)} = -0.883,$$

$$\hat{a}(2) = \frac{(r^2(1) - r(0)r(2))}{r^2(0) - r^2(1)} = \frac{(\hat{\rho}^2(1) - \hat{\rho}(2))}{1 - \hat{\rho}^2(1)} = 0.285.$$

The variance of the residual is found from equation (8.26),

$$s_p^2 = r(0)[1 + \hat{a}(1)\hat{\rho}(1) + \hat{a}(2)\hat{\rho}(2)]$$

$$= 21.9 \cdot 0.48 = 10.6.$$

8.3.3 Model Order

As one can surmise, the model for a signal can be any order that is desired. However, it should be as accurate as possible. After the parameters have been estimated, the actual error process can be generated from the signal, using the MA system equation

$$\epsilon(n) = y(n) + \hat{a}(1)y(n-1) + \hat{a}(2)y(n-2) + \cdots + \hat{a}(p)y(n-p). \tag{8.27}$$

Our knowledge of fitting regression equations to curves provides us with some general intuition. A model with too low an order will not represent the properties of the signals, whereas one with too high an order will also represent any measurement noise or inaccuracies and not be a reliable representation of the true signal. Thus, methods that will determine model order must be used. Some methods can be ascertained from the nature of the modeling process and some of these methods depend on the same concepts that are used in regression. The squared error is being minimized, so it seems that a plot of TSE or s_p^2 against model order p should reveal the optimal value of p. The plot should be monotonically decreasing and asymptotically approach zero. The lowest value of p associated with no appreciable decrease in s_p^2 should indicate the model order. Consider the error plot in Figure 8.5. The residual error does

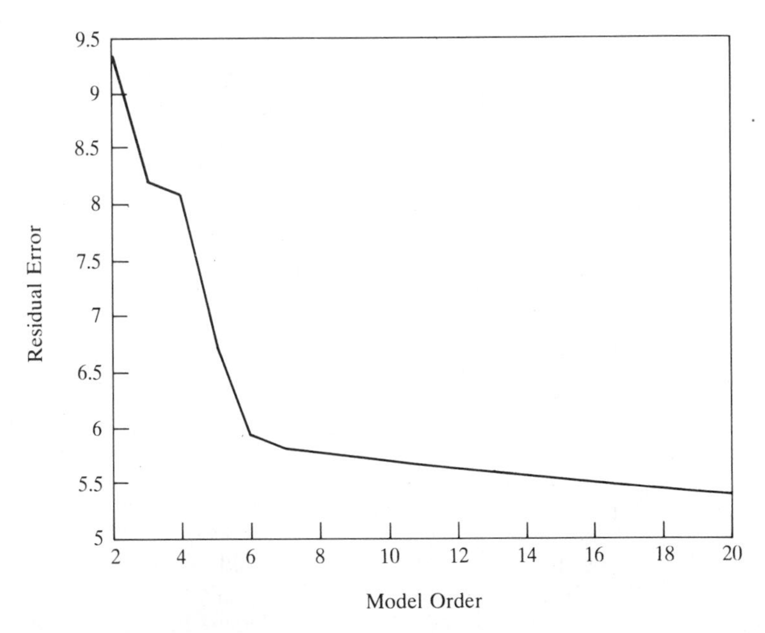

Figure 8.5 Plot of residual error versus model order for signal analyzed in Example 8.4.

decrease as expected but there is no definitive slope change to indicate clearly any asymptotic behavior. The largest incremental decrease occurs from orders 2 to 3 and the error does not seem to decrease much with orders 4 or 5. Thus, one could judge that the order for the signal is approximately 3. This is typical of the characteristics of residual error curves. As can be seen, there is no definitive curve characteristic and, in general, the residual variance is only an approximate indicator of model order.

Another approach based upon the principles of prediction is that one simply increases the model order until the residual process $\epsilon(n)$ in equation (8.27) becomes white noise. This is the approach used by Triolo, Nash, and Moskowitz (1988) and Hefftner, Zucchini, and Jaros (1988) for modeling the electromyogram from skeletal muscle. The NACF of the residuals is calculated from fitting models of various orders. Representative results are shown in Figure 8.6. The majority of signals studied in this manner produce white-noise residuals with fourth-order models and hence

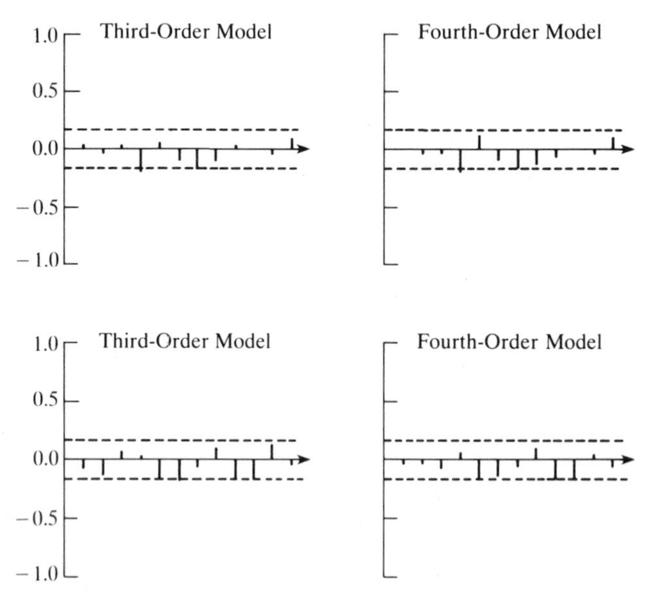

Figure 8.6 The NACFs of the residuals of some third- and fourth-order models fitted to EMG signals ($T = 2$ ms). The dashed lines are the $\pm 1.96/\sqrt{N}$ confidence limits. [Adapted from Hefftner, Zucchini, and Jaros (1988, IEEE), Figure 7, with permission.]

this order is used. A typical model found is

$$y(n) - 0.190y(n-1) + 0.184y(n-2) + 0.192y(n-3) + 0.125y(n-4) = \epsilon(n), \quad (8.28)$$

with $\hat{\sigma}_y^2 = 0.016$ (in volts).

Other criteria have been developed that are functions of the residual variance and have a well-defined minimum. These criteria are based upon concepts in mathematical statistics but are only approximate solutions; their validity is tested with simulations. These developments have been pioneered by Akaike and he has developed two criteria whose properties are summarized in recent advanced treatments. The first of these is the *final prediction error* (FPE), where

$$\text{FPE} = s_p^2 \frac{N+p+1}{N-p-1} \quad (8.29)$$

and the mean of the signal is zero. The fractional portion of FPE increases with p and accounts for the inaccuracies in estimating $a(i)$. A study of the characteristics of this criterion show that it tends to have a minimum at values of p that are less than the model order. The other criterion is called *Akaike's information criterion* (AIC). It is

$$\text{AIC} = N \ln s_p^2 + 2p. \quad (8.30)$$

The term $2p$ is a penalty for higher orders. This criterion also has its shortcomings and tends to overestimate model order. In spite of these shortcomings, both of these criteria are used quite frequently in practical applications. An example will help illustrate the use of the FPE and AIC.

Example 8.4

Consider the signal generated by the fourth-order AR system,

$$y(n) - 2.7607y(n-1) + 3.816y(n-2) - 2.6535y(n-3) + 0.9238y(n-4) = x(n),$$

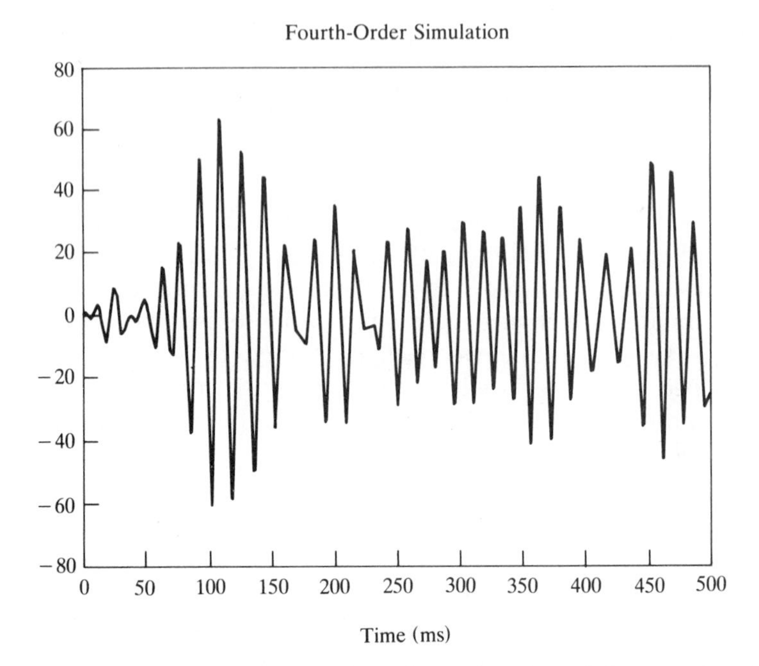

Figure 8.7 A sample function of a realization of the fourth-order process described in Example 8.4 (T = 2 ms).

where $x(n)$ is white noise with unit variance and $T = 2$ ms. A realization is plotted in Figure 8.7. The signal was modeled with orders up to 20 and the FPE is plotted in Figure 8.5. Notice that the first local minima appear for $p = 4$. As the model order increases, a global minimum appears for $p = 6$. Because, ordinarily, one does not know the model order, one would choose either a fourth- or sixth-order model for the signal, depending upon the complexity or simplicity desired. One could also argue that the percentage decrease in criterion values does not warrant the increased model complexity. In any event a judgment is necessary.

The parameter sets for the two models are

$$[1 \quad -1.8046 \quad 1.6128 \quad -0.5079 \quad 0.0897],$$

$$[1 \quad -1.6292 \quad 1.1881 \quad 0.1896 \quad -0.1482 \quad -0.2105 \quad 0.3464].$$

This signal model is used in the signal processing literature and one can find extensive analyses of it. For instance, refer to Robinson and Treital (1980) and Ulrych and Bishop (1975).

Example 8.5

Meteorological data are always of great interest to modelers and scientists, especially with the problems of droughts and hunger. It is then of interest to model data accurately (such as rainfall data) in order to predict future trends. Figure 8.8 is the plot of the deviation of the average annual rainfall in the eastern United States for the years 1817–1922 ($\sigma^2 = 89.9$). The values are listed in Appendix 8.4. It is our task to model this time series. AR models with orders from 1 through 20 were estimated for it. The FPE and AIC criteria are plotted in Figure 8.9. There are local minima for

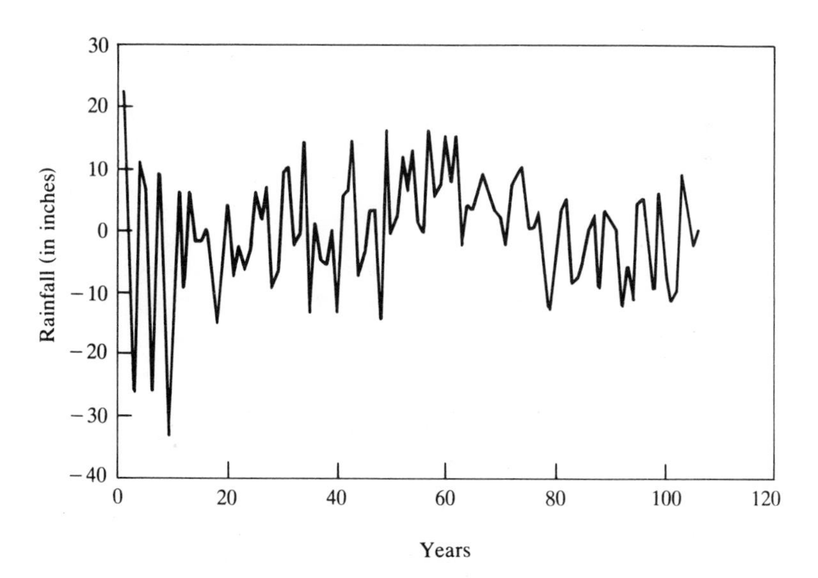

Figure 8.8 Plot of the average annual rainfall in the eastern United States for the years 1817–1922.

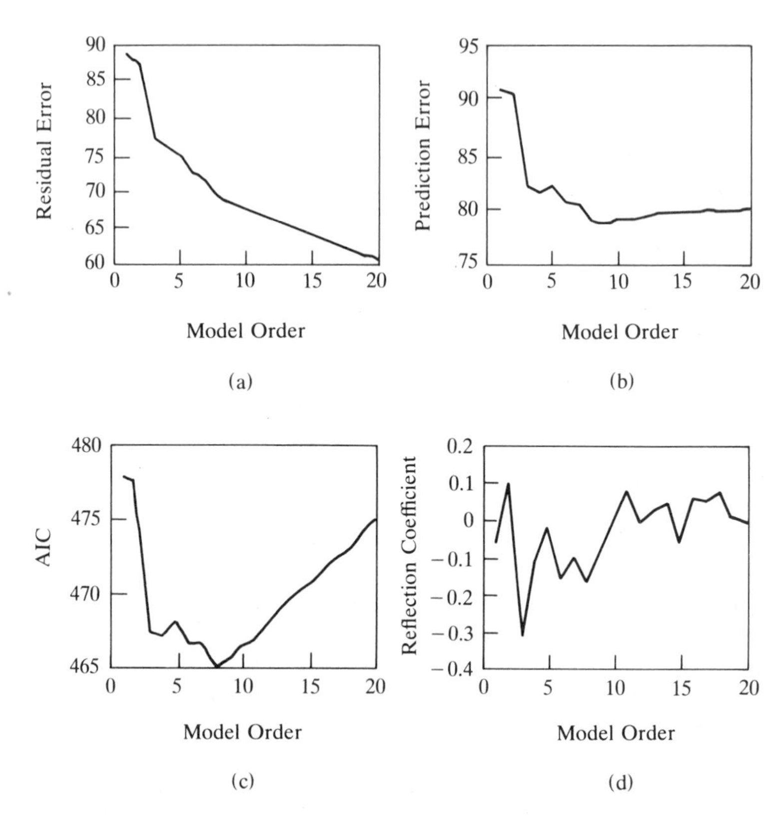

Figure 8.9 The modern selection criteria for the rainfall time series are plotted versus model order: (a) residual error; (b) FPE; (c) AIC; (d) partial correlation coefficient. (The Yule–Walker method was used.)

order 4 and global minima for order 8. For the sake of simplicity, order 4 shall be chosen and the model coefficients are [1, -0.065, 0.122, -0.325, -0.128].

Paralleling this concept is studying the values of the parameters $\hat{a}(i)$. Intuitively, one would guess that, when the values of the parameters become very small, they are insignificant and should be considered zero. The largest value of p associated with a nonzero value of $a(i)$ is probably the model order. Testing whether specific values of $a(i)$ are nonzero, although based on a complicated aspect of correlation, can be easily implemented. Reconsider now the first-order model of the grinding wheel profile in Example 8.2. The estimate of $a(1)$ is actually the value of the sample NACF at lag 1. One can test whether a correlation coefficient is nonzero using the Gaussian approximation studied in Section 5.5.5. Thus, if $\hat{a}(1)$ lies within the range $\pm 1.96/\sqrt{N}$, then it is actually zero at the 95% confidence level and the signal can be considered as white noise. However, if $a(1) \neq 0$, then the signal is at least a first-order process. For the profile, $N = 250$, and the confidence limits are ± 0.124. Because $\hat{a}(1) = -0.627$, the profile is at least a first-order process.

It would be easy if all the other coefficients could be tested as simply. Notice from the Yule–Walker equations that $a(i)$ depends on all the other parameters. Thus, it is not a simple correlation coefficient. Another concept must be introduced: the idea of *partial correlation*. When fitting the signal to a model of order p, designate the last coefficient as $\hat{\pi}_p = \hat{a}_p$. The parameter $\hat{\pi}_p$ is the pth partial correlation coefficient. It is a measure of the amount of correlation at lag p *not accounted for by a* $(p-1)$*-order model*. Thus, it can be tested as an ordinary correlation coefficient. Its derivation will be examined in the next section. A plot of $\hat{\pi}_p$ versus p with the correlation confidence limits superimposed presents another order-testing modality. This plot is shown in Figure 8.9(d) for the process used in the previous example. Notice that the value of $\hat{\pi}_p$ is always within the confidence limits for $p \geq 3$. Thus, this criterion indicates that a third-order model is appropriate.

8.3.4 Levinson – Durbin Algorithm

Autoregressive modeling has become a very powerful approach for signal modeling and spectral estimation. One simply solves the Yule–Walker matrix equation with increasing dimensionality until the best model order is determined. The major shortcoming of this approach is that the solution of this matrix equation 15 or 20 times in a batch mode requires too much computational time. Thus, a recursive algorithm has been developed. The recursive scheme is based upon the concept of estimating the parameters of a model of order p from the parameters of a model of order $p-1$. The scheme is also based upon matrix equation (8.24) and is dependent upon the Toeplitz form of the correlation matrix. At this point it is helpful to rewrite the basic equations of the AR system that were presented in Chapter 6. The equation for a system of order p is

$$y(n) + a(1)y(n-1) + a(2)y(n-2) + \cdots + a(p)y(n-p) = x(n). \tag{8.31}$$

When $x(n)$ is a white noise process, all of the types of correlation function have a recursive relationship among magnitudes at different lag times. The relationship for the autocovariance function is

$$C(k) + a(1)C(k-1) + a(2)C(k-2) + \cdots + a(p)C(k-p) = 0, \tag{8.32}$$

for $k \neq 0$. For $k = 0$,

$$C(0) + a(1)C(1) + a(2)C(2) + \cdots + a(p)C(p) = \sigma_x^2. \tag{8.33}$$

Remember that, in general, the mean values of $x(n)$ and $y(n)$ are zero so that $R(k) = C(k)$. The development of the recursive scheme will begin by deriving the recursive algorithm for a second-order model. The matrix equation is

$$\begin{bmatrix} r(0) & r(1) \\ r(1) & r(0) \end{bmatrix} \begin{bmatrix} \hat{a}_2(1) \\ \hat{a}_2(2) \end{bmatrix} = \begin{bmatrix} -r(1) \\ -r(2) \end{bmatrix}. \tag{8.34}$$

Notice now that the symbols for the estimates of the parameters have a subscript. The value of the subscript indicates the model order. This is because, in general, the value of $a(2)$ for a second-order model is different than that for a third-order model. From Section 8.3.1 it was found that for a measured signal, a first-order model has the solution

$$\hat{a}_1(1) = -\frac{r(1)}{r(0)}, \tag{8.35}$$

with MSE

$$\sigma_{\epsilon,1}^2 = r(0)\left(1 - \hat{a}_1(1)^2\right). \tag{8.36}$$

Now rewrite the first equation of matrix equation (8.34), solving for $\hat{a}_2(1)$, or

$$r(0)\hat{a}_2(1) = -r(1) - r(1)\hat{a}_2(2). \tag{8.37}$$

This can be expressed in terms of parameters as

$$\hat{a}_2(1) = -\frac{r(1)}{r(0)} - \hat{a}_2(2)\frac{r(1)}{r(0)} = \hat{a}_1(1) + \hat{a}_2(2)\hat{a}_1(1). \tag{8.38}$$

The first parameter of the second-order model is expressed in terms of the parameter of the first-order model and $\hat{a}_2(2)$. In order to find the latter parameter, the augmented Yule–Walker equations must be used. The augmented equations use the equation for the noise variance [equation (8.26)]. The augmented matrix equation containing sample autocovariances is

$$\begin{bmatrix} r(0) & r(1) & r(2) \\ r(1) & r(0) & r(1) \\ r(2) & r(1) & r(0) \end{bmatrix} \begin{bmatrix} 1 \\ \hat{a}_2(1) \\ \hat{a}_2(2) \end{bmatrix} = \begin{bmatrix} \sigma_\epsilon^2 \\ 0 \\ 0 \end{bmatrix}. \tag{8.39}$$

The critical point is expressing the vector on the left-hand side of equation (8.39) in terms of $\hat{a}_2(2)$. Using equation (8.38),

$$\begin{bmatrix} 1 \\ \hat{a}_2(1) \\ \hat{a}_2(2) \end{bmatrix} = \begin{bmatrix} 1 \\ \hat{a}_1(1) + \hat{a}_1(1)\hat{a}_2(2) \\ \hat{a}_2(2) \end{bmatrix}$$

$$= \begin{bmatrix} 1 \\ \hat{a}_1(1) \\ 0 \end{bmatrix} + \hat{a}_2(2)\begin{bmatrix} 0 \\ \hat{a}_1(1) \\ 1 \end{bmatrix}. \tag{8.40}$$

Expanding the left-hand side of equation (8.39) using equation (8.40) gives a sum of two vectors. Each will be found in succession. For the first one,

$$\mathbf{T}_1 = \begin{bmatrix} r(0) & r(1) & r(2) \\ r(1) & r(0) & r(1) \\ r(2) & r(1) & r(0) \end{bmatrix} \begin{bmatrix} 1 \\ \hat{a}_1(1) \\ 0 \end{bmatrix} = \begin{bmatrix} r(0) + r(1)\hat{a}_1(1) \\ r(1) + r(0)\hat{a}_1(1) \\ r(2) + r(1)\hat{a}_1(1) \end{bmatrix}.$$

Examine the elements of the vector $\mathbf{T}_1$ with respect to the equations for a first-order model. The first element is equal to the error variance. The second element is given the symbol Δ_2. If order 1 were sufficient, this term would equal zero. The third element is given the symbol Δ_3 and now,

$$\mathbf{T}_1 = \begin{bmatrix} \sigma^2_{\epsilon,1} \\ \Delta_2 \\ \Delta_3 \end{bmatrix}. \tag{8.41}$$

The second term of equation (8.39) is

$$\mathbf{T}_2 = \hat{a}_2(2) \begin{bmatrix} r(0) & r(1) & r(2) \\ r(1) & r(0) & r(1) \\ r(2) & r(1) & r(0) \end{bmatrix} \begin{bmatrix} 0 \\ \hat{a}_1(1) \\ 1 \end{bmatrix}$$

$$= \hat{a}_2(2) \begin{bmatrix} r(1)\hat{a}_1(1) + r(2) \\ r(0)\hat{a}_1(1) + r(1) \\ r(1)\hat{a}_1(1) + r(0) \end{bmatrix}.$$

Examining the elements of the vector on a term-by-term basis shows that the elements are the same as in $\mathbf{T}_1$ except that they are in reverse order. Now

$$\mathbf{T}_2 = \hat{a}_2(2) \begin{bmatrix} \Delta_3 \\ \Delta_2 \\ \sigma^2_{\epsilon,1} \end{bmatrix} \tag{8.42}$$

Summing the two terms together and equating them to equation (8.39) gives

$$\mathbf{T}_1 + \mathbf{T}_2 = \begin{bmatrix} \sigma^2_{\epsilon,1} \\ \Delta_2 \\ \Delta_3 \end{bmatrix} + \hat{a}_2(2) \begin{bmatrix} \Delta_3 \\ \Delta_2 \\ \sigma^2_{\epsilon,1} \end{bmatrix} = \begin{bmatrix} \sigma^2_{\epsilon,2} \\ 0 \\ 0 \end{bmatrix}. \tag{8.43}$$

The second parameter is found using the third element of equation

(8.43) as

$$\hat{a}_2(2) = -\frac{\Delta_3}{\sigma_{\epsilon,1}^2} = -\frac{r(1)\hat{a}_1(1) + r(2)}{\sigma_{\epsilon,1}^2}. \tag{8.44}$$

Similarly, the error variance is found using the first element of equation (8.43) and equation (8.44) as

$$\sigma_{\epsilon,2}^2 = \sigma_{\epsilon,1}^2 + \hat{a}_2(2)\Delta_3 = \left(1 - \hat{a}_2(2)^2\right)\sigma_{\epsilon,1}^2. \tag{8.45}$$

Example 8.6

Consider again the temperature signal. What is the residual error for the first-order model? From equation (8.45) it is known that

$$\sigma_{\epsilon,1}^2 = r(0)\left(1 - \hat{a}_1(1)^2\right) = 21.9(1 - 0.472) = 11.56.$$

Calculate the parameters of a second-order model. From equations (8.45) and (8.38),

$$\hat{a}_2(2) = -\frac{r(1)\hat{a}_1(1) + r(2)}{\sigma_{\epsilon,1}^2}$$

$$= -\frac{15.1(-0.687) + 7.05}{11.56} = 0.285,$$

$$\hat{a}_2(1) = \hat{a}_1(1) + \hat{a}_2(2)\hat{a}_1(1)$$

$$= -0.687 + 0.285(-0.687) = -0.882,$$

$$\sigma_{\epsilon,2}^2 = \left(1 - \hat{a}_2(2)^2\right)\sigma_{\epsilon,1}^2 = (1 - 0.08267)11.56 = 10.6.$$

This process can be continued and general equations can be written that summarize this approach. The algorithm is called

Levinson's algorithm, whose steps are as follows:

1. Start with a zero-order model, $p = 0$, $\hat{a}_0(0) = 1$, $\sigma_\epsilon^2 = r(0)$.
2. Generate the last parameter of the next higher order model and designate it $\hat{\pi}_{p+1}$,

$$\hat{\pi}_{p+1} = \frac{\sum_{i=0}^{p} r(p+1-i)\hat{a}_p(i)}{\sigma_{\epsilon,p}^2}.$$

3. Find the error variance,

$$\sigma_{\epsilon,p+1}^2 = \left(1 - \hat{\pi}_{p+1}^2\right)\sigma_{\epsilon,p}^2.$$

4. Find the remaining parameters $\hat{a}_{p+1}(i)$ for this order model,

$$\hat{a}_{p+1}(0) = 1, \qquad \hat{a}_{p+1}(p+1) = \hat{\pi}_{p+1},$$

and

$$\hat{a}_{p+1}(i) = \hat{a}_p(i) + \hat{\pi}_{p+1}\hat{a}_p(p+1-i), \qquad \textit{for } 1 \le i \le p.$$

5. Set $p = p + 1$ and return to step 2 until parameters for the maximum-order model are calculated.

The additional parameter needed for the next higher order model is given a special name, the *reflection coefficient*. Examining the error equation in step 3, it is desired that

$$0 \le |\hat{\pi}_p| \le 1, \tag{8.46}$$

so that

$$\sigma_{\epsilon,p+1}^2 \le \sigma_{\epsilon,p}^2. \tag{8.47}$$

This means that the successive error variances associated with higher-order models decrease. In Section 8.3.3 it is shown that $\hat{\pi}_p$ is a correlation coefficient; therefore equation (8.46) is true. A close examination of the numerator of the equation for $\hat{\pi}_{p+1}$ shows that it equals zero if the model order is correct, $\hat{a}_p(i) = a(i)$ and $r(k) = R(k)$ for the signal.

All of the equations can be derived in an elegant manner using matrices. One derivation is included in Appendix 8.2. An analysis of the steps in this algorithm shows that not only are all lower-order models estimated when estimating a model of order p but the number of mathematical operations is less. For a pth-order model, a conventional linear equation solution of the Yule-Walker equations, such as the Gaussian elimination, requires approximately p^3 operations. For the Levinson algorithm, each iteration for an mth-order model requires $2m$ operations. Thus a pth-order model requires $\sum_{m=1}^{p} 2m = p(p-1)$ operations, which is much fewer [Kay (1988)].

8.3.5 Burg Method

The Burg method was developed because of the inaccuracies of the parameter estimation that sometimes occurred when the Yule–Walker equations were used directly. This became evident when models were used to simulate data and then the simulated data were used to estimate model parameters. This is evident in Example 8.4. One of the hypothesized sources of error was the bias in autocorrelation function estimate. A possible solution was to change the solution equations so that the data could be used directly. Also, perhaps more signal points could be utilized simultaneously. The Burg method was developed from the Levinson–Durbin algorithm. Notice that the only parameter that is directly a function of $r(k)$ is the reflection coefficient in step 2. Thus, an error criterion needed to be formed that is not only a function of $\hat{\pi}_p$ but also of more signal points. In order to utilize more points, a *backward prediction error* $\epsilon^b(n)$ is defined. The error that has been used is called the *forward prediction error* $\epsilon^f(n)$. The new error criterion is the average of the mean square value of both errors. The backward predictor uses future points to predict values in the past; for a one-step predictor,

$$\hat{y}(n-p) = -\hat{a}_p(1)y(n-p+1) - \hat{a}_p(2)y(n-p+2) - \cdots - \hat{a}_p y(n) \quad (8.48)$$

and

$$\epsilon_p^b(n) = \hat{y}(n-p) + \hat{a}_p(1)y(n-p+1) + \hat{a}_p(2)y(n-p+2) + \cdots + \hat{a}_p y(n). \quad (8.49)$$

The forward prediction error is equation (8.27). It is rewritten explicitly as a function of model order,

$$\epsilon_p^f(n) = \hat{y}(n) + \hat{a}_p(1)y(n-1) + \hat{a}_p(2)y(n-2) + \cdots + \hat{a}_p y(n-p). \tag{8.50}$$

These errors can now be made a function of the reflection coefficient. Substitute the recursive relationship for the AR parameters in step 4 of Levinson's algorithm into equations (8.49) and (8.50). This yields

$$\epsilon_p^b(n) = \epsilon_{p-1}^b(n-1) + \hat{\pi}_p \epsilon_{p-1}^f(n), \tag{8.51}$$

$$\epsilon_p^f(n) = \epsilon_{p-1}^f(n) + \hat{\pi}_p \epsilon_{p-1}^b(n-1), \tag{8.52}$$

where

$$\epsilon_0^b(n) = \epsilon_0^f(n) = y(n). \tag{8.53}$$

The average prediction error is now

$$\sigma_\epsilon^2 = \frac{1}{2}\left(\frac{1}{N-p}\sum_{n=p}^{N-1}|\epsilon_p^f(n)|^2 + \frac{1}{N-p}\sum_{n=p}^{N-1}|\epsilon_p^b(n)|^2\right). \tag{8.54}$$

Substituting equations (8.51) and (8.52) into equation (8.54) produces

$$\sigma_\epsilon^2 = \frac{1}{2(N-p)}\sum_{n=p}^{N-1}\left(|\epsilon_{p-1}^f(n) + \hat{\pi}_p \epsilon_{p-1}^b(n-1)|^2 + |\epsilon_{p-1}^b(n-1) + \hat{\pi}_p \epsilon_{p-1}^f(n)|^2\right). \tag{8.55}$$

Differentiating this equation with respect to $\hat{\pi}_p$ and setting the result to zero will yield the solution for the reflection coefficient.

$$\frac{\partial}{\partial \hat{\pi}_p} = 0 = \frac{1}{N-p} * \sum_{n=p}^{N-1} \Big(\big(\epsilon_{p-1}^f(n) + \hat{\pi}_p \epsilon_{p-1}^b(n-1)\big) \epsilon_{p-1}^b(n-1)$$

$$+ \big(\epsilon_{p-1}^b(n-1) + \hat{\pi}_p \epsilon_{p-1}^f(n)\big) \epsilon_{p-1}^f(n) \Big) \quad (8.56)$$

Rearranging terms produces

$$\hat{\pi}_p = \frac{-2\sum_{n=p}^{N-1} \epsilon_{p-1}^f(n) \epsilon_{p-1}^b(n-1)}{\sum_{n=p}^{N-1} \big(\epsilon_{p-1}^f(n)\big)^2 + \big(\epsilon_{p-1}^b(n-1)\big)^2}. \quad (8.57)$$

Study in detail the terms in equation (8.57). The numerator is a sum of cross products and the denominator contains two sums of squares. This would be the definition of the correlation coefficient between the forward and backward prediction errors if their variances were equal. (Then the factor 2 would cancel.) This reflection coefficient is a partial correlation coefficient and thus,

$$-1 \le \hat{\pi}_p \le 1. \quad (8.58)$$

The steps for implementing Burg's algorithm are as follows:

1. *Initial conditions.*

$$\hat{\pi}_0 = r(0),$$

$$\epsilon_0^f(n) = y(n), \qquad n = 1, 2, \ldots, N-1,$$

$$\epsilon_0^b(n) = y(n), \qquad n = 0, 1, \ldots, N-2.$$

2. *Reflection coefficients.* For $p = 1, \ldots, P$,

$$\hat{\pi}_p = \frac{-2\sum_{n=p}^{N-1} \epsilon_{p-1}^f(n) \epsilon_{p-1}^b(n-1)}{\sum_{n=p}^{N-1} \big(\epsilon_{p-1}^f(n)\big)^2 + \big(\epsilon_{p-1}^b(n-1)\big)^2},$$

$$s_p^2 = \big(1 - |\hat{\pi}_p|^2\big) s_{p-1}^2.$$

For $p = 1$,

$$\hat{a}_1(1) = \hat{\pi}_1.$$

For $p > 1$,

$$a_p(i) = \begin{cases} a_{p-1}(i) + \hat{\pi}_p a_{p-1}(p-i) & \text{for } i = 1, 2, \ldots, p-1, \\ \hat{\pi}_p & \text{for } i = p. \end{cases}$$

3. *Prediction errors for next order.*

$$\epsilon_p^f(n) = \epsilon_{p-1}^f(n) + \hat{\pi}_p \epsilon_{p-1}^b(n-1), \qquad n = k+1, \ldots, N-1,$$

$$\epsilon_p^b(n) = \epsilon_{p-1}^b(n-1) + \hat{\pi}_p \epsilon_{p-1}^f(n), \qquad n = k+1, \ldots, N-2.$$

Example 8.7

The parameters of the fourth-order signal model were estimated with the Burg method. The estimates of $a(1)$ through $a(4)$ are

$$[1 \quad -2.7222 \quad 3.7086 \quad -2.5423 \quad 0.8756].$$

These are almost exactly the theoretical values.

Example 8.8

Because the order criteria were contradictory for the rainfall data, the criteria and model parameters were recalculated using the Burg method. The residual error and the criteria are plotted in Figure 8.10. If the residual error in Figure 8.10(a) is compared with that in Figure 8.9(a), it is seen that the Burg method produces a slightly larger error at all orders. Examining Figures 8.10(b)–8.10(d)

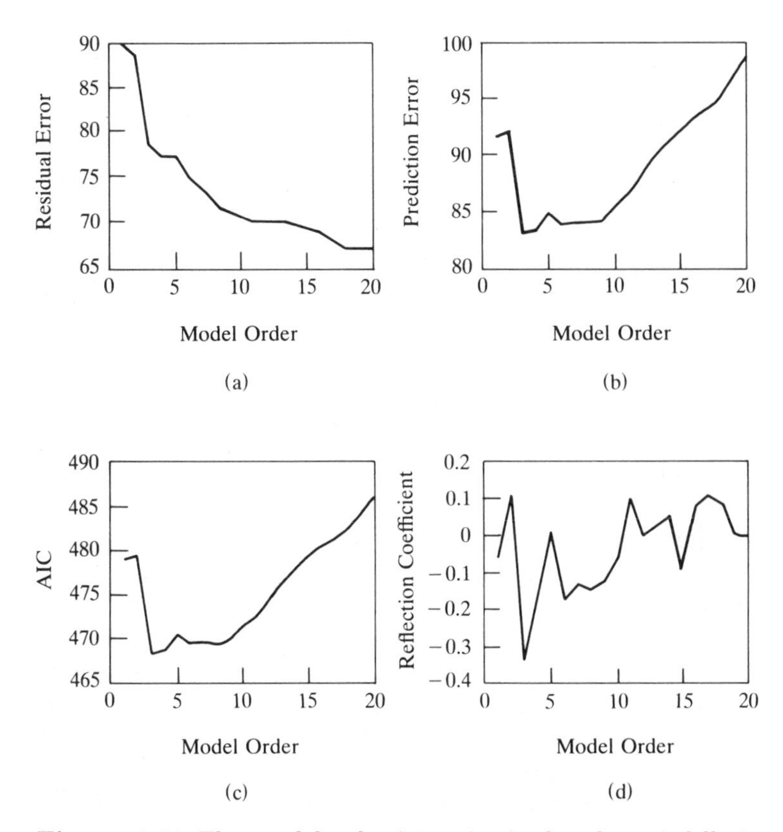

Figure 8.10 The model selection criteria for the rainfall time series are plotted versus model order: (a) residual error; (b) FPE; (c) AIC; (d) partial correlation coefficient. (The Burg method was used.)

reveals that all criteria suggest the same model order, that is, 3. The parameters are

$$[1 \quad -0.1088 \quad 0.14 \quad -0.3392], \quad \text{with } s_3^2 = 78.47.$$

8.3.6 Summary of Signal Modeling

The minimization of the error of linear prediction is the basic modern approach to signal modeling. Many books treat this topic, for instance, refer to Cadzow (1987), Kay (1988), or Marple

(1987). The direct application of this principle leads to the Yule–Walker equations. The use of estimates of the autocovariance or autocorrelation function in the YW matrix equation comprises the autocorrelation method for estimating the parameters of the signal model. It is a direct approach for which there are well-known techniques for solving the matrix equations. The need for faster solution techniques led to the development of the Levinson algorithm for solving the YW equations recursively. It has the benefit of being computationally faster as well as producing all the lower-order models.

Because it is easy to produce models of any order, it is necessary to have rational criteria for estimating model order. Several criteria were described: the final prediction error, Akaike's information criterion, and the partial correlation coefficient. It seems that they all have their shortcomings, but if they are used with these shortcomings in mind, they are good criteria.

In examining the models of simulated signals it was discovered that the autocorrelation method is not always accurate. It was hypothesized that this is because the estimates of the autocorrelation function are used. Thus, another method was developed by Burg. It uses the same signal values but minimizes the forward and backward prediction errors simultaneously. In general it has proven to be a much more accurate method.

Notice that, because of the random nature of signals, not all variations can be modeled. That is, some signals have an inherently large white noise component. This is why the rainfall signal could not be very well modeled compared with the other signals. This is ascertained by the fact that the residual error for the appropriate model has only decreased by 13% compared to decreases on the order of 35% for the temperature and grinding wheel signals.

8.4 Spectral Density Function Estimation

8.4.1 Definition and Properties

The power spectral density function of the signal can be estimated after a model has been found. Once the appropriate order has been decided, then it is assumed that the optimal prediction error sequence is a white-noise sequence and $s_p^2 = \hat{\sigma}_x^2$. Using the system input–output relationship, it is known that

$$S_y(f) = |H(f)|^2 S_x(f) = |H(f)|^2 \sigma_x^2 T. \tag{8.59}$$

Also, it is known that for an AR system the transfer function is

$$H(f) = \frac{1}{\sum_{i=0}^{p} a(i) e^{-2\pi f i T}}. \tag{8.60}$$

The PSD for a signal is estimated using the two previous equations and substituting estimates for the noise variance and system parameters, that is,

$$\hat{S}(f) = \frac{s_p^2 T}{\left|\sum_{i=0}^{p} \hat{a}(i) e^{-2\pi f i T}\right|^2}. \tag{8.61}$$

This estimator is the foundation for what is called *modern spectral estimation*. There are some advantages of this approach over the classical approach. One of them is in the frequency spacing. Examination of equation (8.61) shows that frequency is still a continuous parameter. Thus, the picket fence effect is eliminated and one is not concerned with zero padding and the number of signal points for this purpose. Another advantage is that only p parameters are being estimated directly instead of all the spectral values. This enables a reliable estimate of a spectrum with small values of N (naturally, $p < N$). Several differences exist between the two techniques because of some underlying assumptions. In the periodogram approach the Fourier transform induces a periodic repetition upon the signal outside the time span of its measurement. In the modern approach, one predicts nonmeasured future values using the linear predictor. (However, the prediction is not of concern to us now.) Another difference comes about in the assumptions about the ACF. In the BT approach, it is assumed that the ACF is zero for lags greater than a maximum lag M, whereas the AR model has a recursion relationship for the ACF and it has values for lags greater than M. As with all techniques, there are also some inaccuracies in modern spectral estimation. These will be explained in subsequent paragraphs. Many of the numerical inaccuracies have been reduced by using the Burg method for the parameter estimation.

Example 8.9

The signal described by the fourth-order AR system in a previous example has the PSD

$$S(f) = \frac{T}{\left|\sum_{i=0}^{4} a(i) e^{-2\pi f i T}\right|^2},$$

with $T = 2$ ms and $a(1) = -2.76$, $a(2) = 3.82$, $a(3) = -2.65$, and $a(4) = 0.92$. As we know, the range for frequency is $0 \leq f \leq 250$ Hz. For a fine resolution for plotting, let f equal an integer. A plot of the PSD is shown in Figure 8.11.

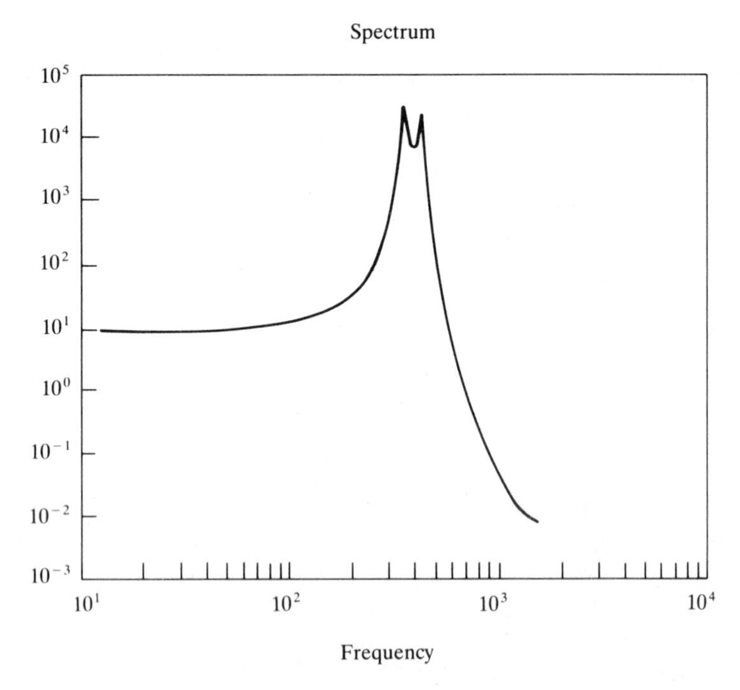

Figure 8.11 The PSD for theoretical fourth-order AR model.

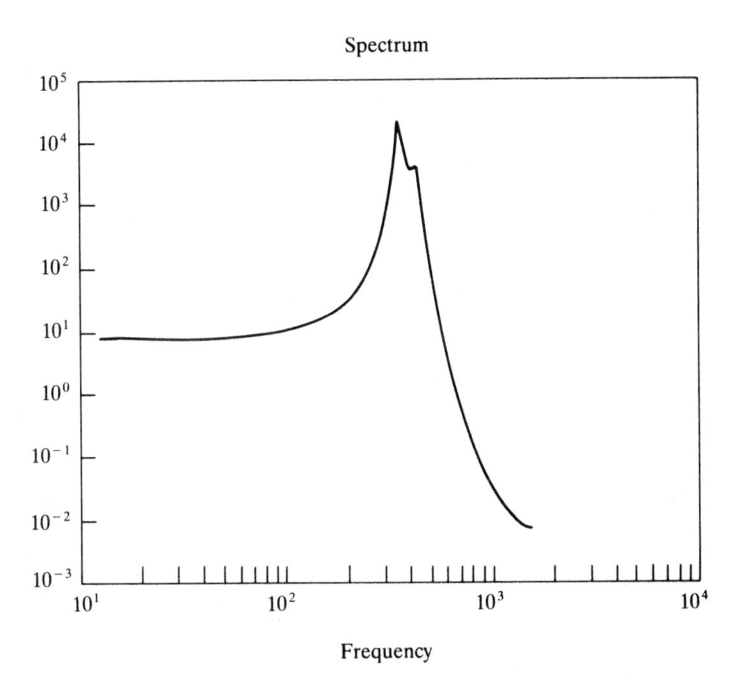

Figure 8.12 Estimate of a PSD from a simulated random signal ($p = 4$, $N = 250$).

Example 8.10

A random signal is generated using the model in the previous example. Its time series and estimated parameters, using the Burg method, are described in Examples 8.4 and 8.7. The spectral estimate from this model is plotted in Figure 8.12. Notice that it agrees quite well with the theoretical spectrum.

At this point it is advantageous to demonstrate the need for a second methodology for spectral estimation. Because the AR model is the result of an all-pole system, narrow-band spectral peaks can be estimated with more accuracy and closely spaced spectral peaks can be distinguished. Consider target reflections in radar systems. The doppler frequencies are used to estimate target velocities. Figure 8.13 shows the output spectrum estimated with both the AR method and the BT method. Two spectral peaks are

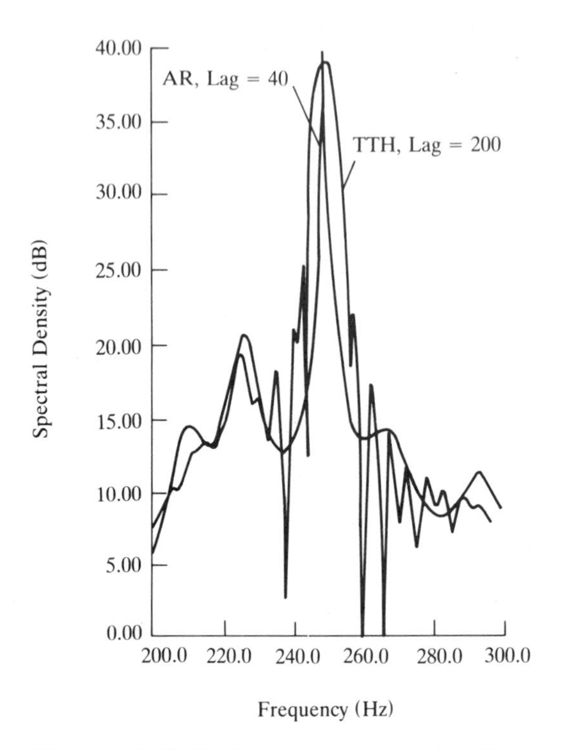

Figure 8.13 Radar output spectrum for a 2000-point doppler signal with a sampling interval of 1 ms. TTH indicates the BK estimate with a Hanning window and maximum ACF lag of 200; AR indicates the AR estimate using 40 ACF lags. [From Kaveh and Cooper (1976, IEEE), Figure 10, with permission.]

evident at approximately 225 and 245 Hz. The stronger energy peak is from a discrete target. The AR spectrum possesses a more narrow peak at 245 Hz, which is consistent with a narrow frequency band of power, whereas the BT spectrum tends to blend the two peaks together [Kaveh and Cooper (1976)].

This same good characteristic can also produce artifactual peaks in the estimated spectra. If the model order is too high, then false peaks appear; this has been proved using simulations (Figure 8.14 is an example). The theoretical spectrum comes from the second-order AR system

$$y(n) = 0.75y(n-1) - 0.5y(n-2) + x(n), \qquad \sigma_x^2 = 1, T = 1. \tag{8.62}$$

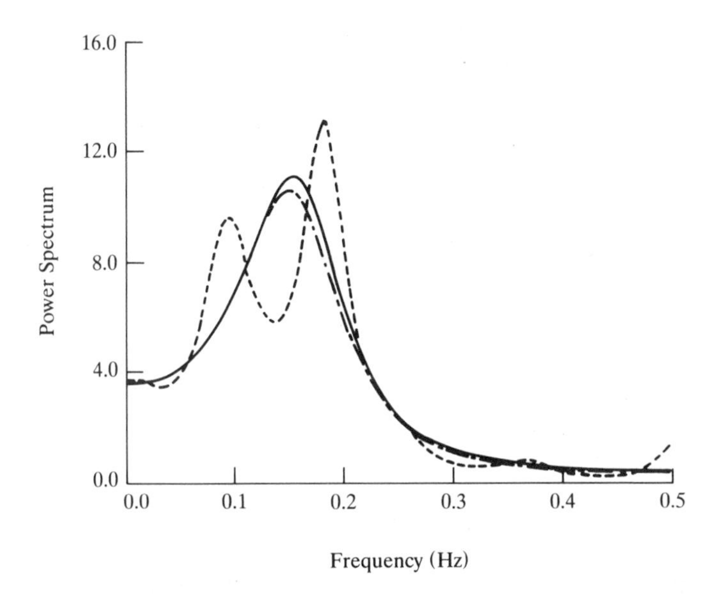

Figure 8.14 Illustration of line splitting caused by overestimating model order: (——) actual PSD of second-order signal; PSD estimates using 2d-order (—·—) and 11th-order (---) models. [From Ulrych and Bishop (1975, IEEE), Figure 1b, with permission.]

A realization containing 50 points was generated and its PSD estimated with a 2d- and an 11th- order model. It can be seen quite clearly in the figure that the spectrum of the second-order model estimates the actual spectrum very closely. However, in the other model, the broad spectral peak has been split into two peaks. This phenomenon is called *line splitting* [Ulrych and Bishop (1975)], thus emphasizing even more the importance of model order determination.

An application that illustrates the possible occurrence of line splitting is the study of geomagnetic micropulsations. These micropulsations are disturbances in the earth's geomagnetic field that have magnitudes on the order of several hundred gamma (1 gamma is 10^{-9} tesla), whereas the earth's field strength is on the order of 50,000 gammas. The disturbances arise from the magnetosphere and their spectra are calculated so the models of this disturbance can be developed. Figure 8.15(a) shows the Y component of a 15-min record that was sampled every 3.5 s ($f_s =$

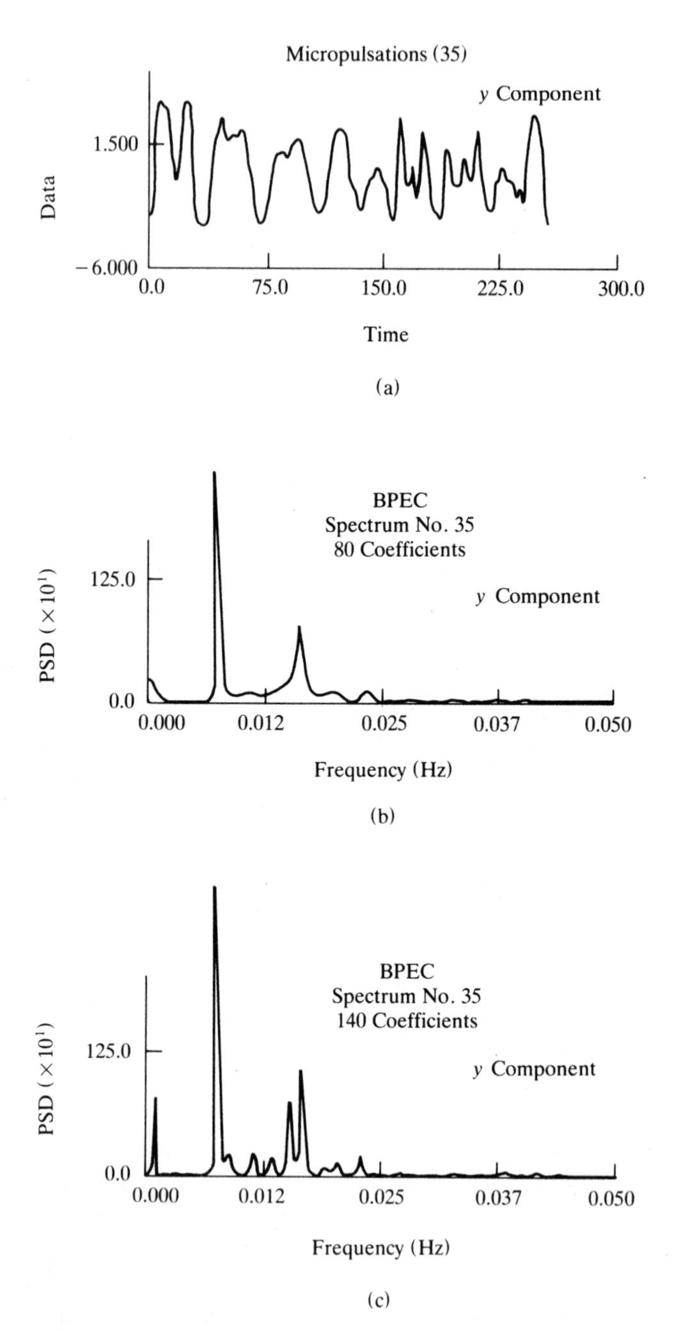

Figure 8.15 Micropulsations and spectra: (a) 15-min record of micropulsations; (b) AR PSD using 80 parameters; (c) AR PSD using 140 parameters. [From Radowski, Zawalick, and Fougere (1976), Figures 2, 5, and 7, with permission.]

0.283 Hz), giving $N = 257$. The AR spectrum was estimated using 80 parameters and is shown in Figure 8.15(b). Notice that this is quite high compared to the modeling, emphasizing the time-domain properties. Two narrow-band spectral peaks occur at 7 and 17 mHz (millihertz). In order to determine if any of these peaks are composed of multiple peaks closely spaced in frequency, the spectrum was reestimated using 140 parameters [Figure 8.15(c)]. All the peaks are sharper and the one at 17 mHz appears as two peaks. One must now be careful that this is not an artifact of line splitting. One must resort to the theory of the phenomenon or to other spectral estimation methods, such as Pisarenko harmonic decomposition, to verify these closely spaced peaks [Kay (1988)].

8.4.2 Statistical Properties

The exact results for the statistics of the AR spectral estimator are not available. Approximations based upon large samples show that for stationary processes, $\hat{S}(f)$ is distributed according to a Gaussian pdf and is an asymptotically unbiased and consistent estimator of the PSD. Its variance depends upon the model order. In particular,

$$E[\hat{S}(f)] = S(f)$$

and

$$\mathrm{Var}[\hat{S}(f)] = \begin{cases} \dfrac{4p}{N} S^2(f) & \text{for } f = 0 \text{ and } \pm\dfrac{1}{2T}, \\ \dfrac{2p}{N} S^2(f) & \text{otherwise.} \end{cases} \tag{8.63}$$

As with the periodogram method, the magnitudes are uncorrelated, that is

$$\mathrm{Cov}[\hat{S}(f_1), \hat{S}(f_2)] = 0 \quad \text{for } f_1 \neq f_2. \tag{8.64}$$

The upper and lower limits for the $100(1 - \alpha)\%$ confidence interval are

$$\hat{S}(f)\left[1 \pm \sqrt{\frac{2p}{N}}\, z\left(1-\frac{\alpha}{2}\right)\right], \tag{8.65}$$

where $z(\beta)$ represents the β percentage point of a zero-mean, unit-variance Gaussian pdf [Kay (1988)].

Example 8.11

The PSD estimate for the rainfall data using the AR(4) model in Example 8.5 is

$$\hat{S}(f) = \frac{s_p^2 T}{\left|\sum_{i=0}^{p} \hat{a}(i) e^{-j2\pi f i T}\right|^2} = \frac{77.19}{\left|\sum_{i=0}^{4} \hat{a}(i) e^{-j2\pi f i}\right|^2},$$

with the coefficients being $[-0.065,\ 0.122,\ -0.325,\ -0.128]$. The 95% confidence limits for this estimate are

$$\hat{S}(f)\left[1 \pm \sqrt{\frac{2p}{N}}\, z\left(1-\frac{\alpha}{2}\right)\right] = \hat{S}(f)\left[1 \pm \sqrt{\frac{2\cdot 4}{106}}\, 1.96\right]$$

$$= \hat{S}(f)[1 \pm 0.54].$$

The estimate and its limits are plotted in Figure 8.16.

Example 8.12

Speech is a short-term stationary signal whose stationary epochs are approximately 120 ms long. Speech signals are sampled at the rate of 10 kHz and 30-ms epochs are studied for their frequency content. Each epoch overlaps 20 ms with the preceding epoch. The LPC method is the preferred approach because of the

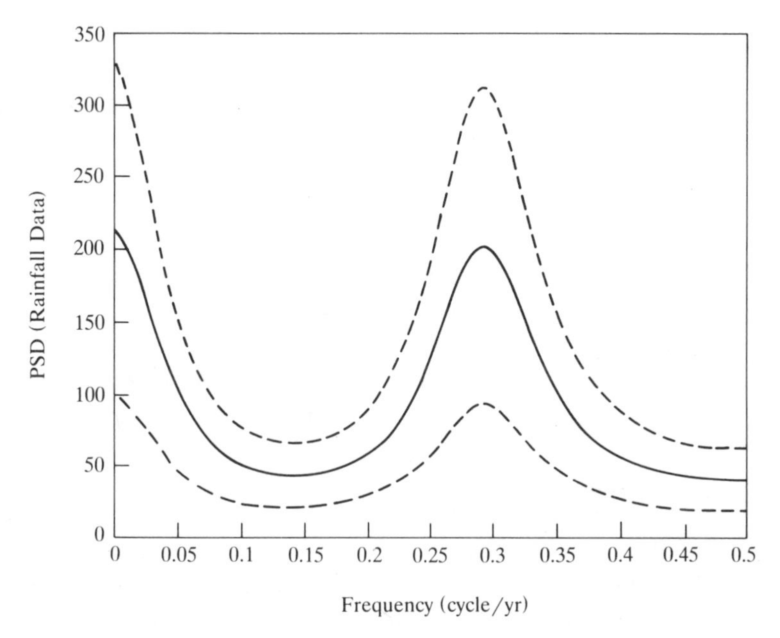

Figure 8.16 The LPC PSD estimate of the rainfall data (—) and the 95% confidence limits (---).

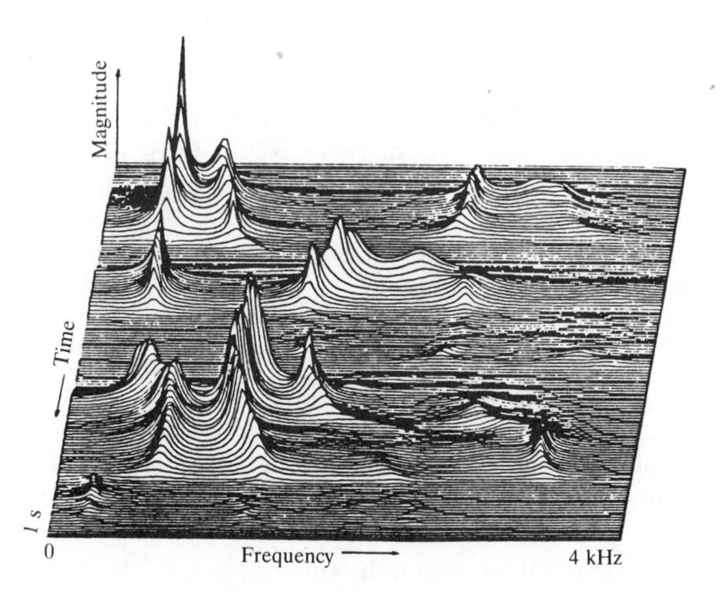

Figure 8.17 Digital spectrogram using AR(14) models of the utterance "oak is strong." [From Veeneman (1988), Figure 17.23, with permission.]

small values of N (approximately 300) available and the speed and low variance required. AR(14) models are accurate. Figure 8.17 shows the spectrogram produced for performing a spectral analysis of the phrase "oak is strong."

8.4.3 Other Spectral Estimation Methods

There are many other spectral estimation methods. Most of them are designed for special purposes, such as accurately estimating the frequencies of sinusoids whose frequencies are close together and estimating the PSD for signals with a small number of points. The study of these topics is undertaken in advanced treatments. The one method that must be mentioned explicitly here is the *maximum entropy method* (MEM). It is also an AR modeling method that places a constraint on the data that are not available. The principle is that, given the known or estimated autocorrelation function, one can predict future or unknown signal values such that they are not only consistent with the known values but also are the most random set of signal points possible. The MEM produces the Yule–Walker equations and hence is the same as the LPC method.

There are alternative algorithms for solving the Yule–Walker equations and most of them depend upon the Levinson–Durbin recursion. They are more accurate and are necessary for short-duration signals. One of these is the Burg algorithm. This algorithm uses not only the forward prediction error but also the backward prediction error and minimizes the sum of both of them. For long-duration signals, it yields the same results as the autocorrelation method studied here.

8.4.4 Comparison of Modern and Classical Methods

At this point it is worthwhile to summarize the comparison of the applicability of classical and modern AR methods for estimating power spectra.

1. The AR method is definitely superior when the signal being analyzed is appropriately modeled by an AR system. In addition, one is not concerned about frequency resolution.

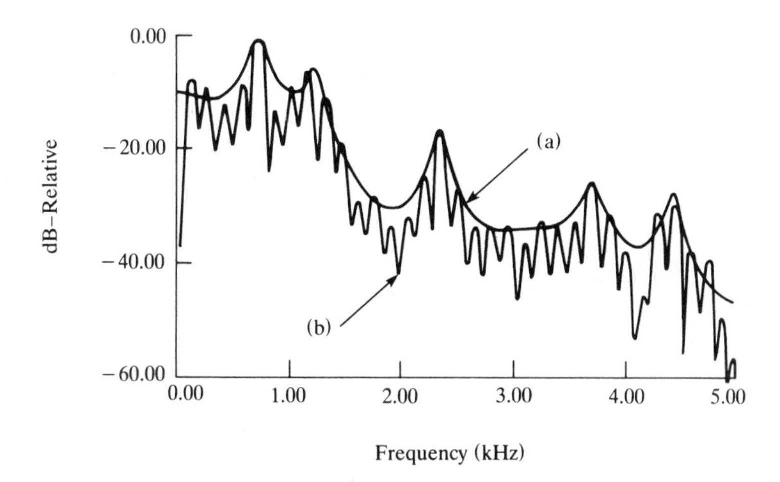

Figure 8.18 The periodogram and AR spectra ($p = 16$) of a speech segment ($N = 256$, $T = 0.1$ ms). [From Childers (1982), Figure 7, with permission.]

2. The AR approach is superior for estimating narrow-band spectra. If the spectra are smooth, the classical approach is adequate and more reasonable.
3. When additive white noise components dominate the signal, the classical approach is more accurate. AR spectral estimates are very sensitive to high noise levels.
4. The variances of both methods are comparable when the maximum autocorrelation lag in the BT method or twice the segment length in the periodogram averaging method is approximately equal to the model order.

There is a major qualitative difference between the two spectral estimation methods, which arises because of the difference in the number of parameters being estimated. In the classical method the magnitude of the PSD is being calculated at every harmonic frequency. However, in the modern methodology, only several parameters are being calculated and the spectrum from the model is used. This results in the estimates from the latter approach being much smoother. Figure 8.18 illustrates this quite readily.

There are other methods of spectral estimation based upon MA and ARMA signal models. Their use is not as widespread, and the theory and implementation are more complex and hence are left as topics of advanced study. One important generality is that MA-based methods are good for estimating broadband and narrowband reject spectra.

References

Ayres, F. (1962). *Theory and Problems of Matrices*. Schaum Publishing Co., New York.

Burg, J. (1968). A new analysis technique for time series data. NATO Advanced Study Institute on Signal Processing with Emphasis on Underwater Acoustics, August, 1968. [Reprinted in *Modern Spectral Analysis* (D. Childers, ed.), IEEE Press, New York (1978).]

Cadzow, J. (1987). *Foundations of Digital Signal Processing and Data Analysis*. Macmillan Publishing Co., New York.

Childers, D. (1982). Digital spectral analysis. In *Digital Waveform Processing and Recognition* (C. Chen, ed.). CRC Press, Boca Raton, FL.

Fante, R. (1988). *Signal Analysis and Estimation—An Introduction*. John Wiley and Sons, New York.

Graupe, D. (1984). *Time Series Analysis, Identification and Adaptive Filtering*. Robert E. Krieger Publishing Co., Malabar, FL.

Hefftner, G., W. Zucchini, and G. Jaros (1988). The electromyogram (EMG) as a control signal for functional neuromuscular stimulation. Part I: Autoregressive modeling as a means of EMG signature discrimination. *IEEE Trans. Biomed. Engrg.* 35:230–237.

Jansen, B. (1985). Analysis of biomedical signals by means of linear modeling. *CRC Critical Reviews in Biomed. Engrg.* 12(4):343–392.

Jenkins, G. and D. Watts (1968). *Spectral Analysis and Its Applications*. Holden-Day, San Francisco.

Kaveh, M. and G. Cooper (1976). An empirical investigation of the properties of the autoregressive spectral estimator. *IEEE Trans. Inform. Theory* 22:{313–323. [Reprinted in *Modern Spectral Analysis* (D. Childers, ed.), IEEE Press, New York (1978).]

Kay, S. (1988). *Modern Spectral Estimation, Theory and Applications*. Prentice-Hall, Englewood Cliffs, NJ.

Kay, S. and S. Marple (1981). Spectrum analysis—A modern perspective. *Proc. IEEE* 69:1380–1419.

Lim, J. and A. Oppenheim (1988). *Advanced Topics in Signal Processing*. Prentice-Hall, Englewood Cliffs, NJ.

Marple, S. (1987). *Digital Spectral Analysis with Applications*. Prentice-Hall, Englewood Cliffs, NJ.

Priestley, M. (1981). *Spectral Analysis and Time Series*, Vol. 1: *Univariate Series*. Academic Press, New York.

Radowski, H., E. Zawalick, and P. Fougere (1976). The superiority of maximum entropy power spectrum techniques applied to geomagnetic disturbances. *Phys. Earth Planetary Interiors* 12:208–216. [Reprinted in *Modern Spectral Analysis* (D. Childers, ed.), IEEE Press, New York (1978).]

Robinson, E. and S. Treital (1980). *Geophysical Analysis*. Prentice-Hall, New York.

Triolo, R., D. Nash, and G. Moskowitz (1988). The identification of time series models of lower extremity EMG for the control of prostheses using Box–Jenkins criteria. *IEEE Trans. Biomed. Engrg.* 35:584–594.

Ulrych, T. and T. Bishop (1975). Maximum entropy spectral analysis and autoregressive decomposition. *Rev. Geophysics and Space Phys*. 13:183–200. [Reprinted in *Modern Spectral Analysis* (D. Childers, ed.), IEEE Press, New York (1978).]

Veeneman, D. (1988). Speech Signal Analysis. In *Signal Processing Handbook* (C. Chen, ed.), Marcel Dekker, New York.

Exercises

8.1 Prove that minimizing the MSE in equation (8.5) with respect to $h(2)$ yields equation (8.9).

8.2 Prove that the expression for the total squared error of a first-order AR signal model is that given in Example 8.1.

8.3 Starting with equation (8.19), show that the estimate of the minimum error variance is equation (8.25).

8.4 Write the Yule–Walker equations for a fourth-order signal model.

8.5 Modeling the temperature signal in Example 8.3 produced parameter estimates of $a(1) = -0.833$ and $a(2) = 0.285$. Use the unbiased covariance estimator and show that the ACF and parameter estimates change.

8.6 Derive the solutions for the parameters of a second-order signal model. Prove that they are

$$\hat{a}_1 = \frac{r(1)(r(2) - r(0))}{r^2(0) - r^2(1)},$$

$$\hat{a}_2 = \frac{(r^2(1) - r(0)r(2))}{r^2(0) - r^2(1)}.$$

8.7 The NACF of an EMG signal is plotted in Figure 8.2(a).
a. Find the AR parameters for model orders of 1, 2, and 3.
b. Assume that $\hat{R}(0) = 0.04\ V^2$; find the TSE for each model. Does the third-order model seem better than the first- or second-order one?

8.8 Write the augmented Yule–Walker equations for a first-order AR signal model. Solve for the parameter and squared error in the form of the recursive equations.

8.9 The Yule–Walker equations can be derived directly from the definition of an all-pole system with a white-noise input. Show this by starting with the system definition and making the autocorrelations

$$E[y(n)y(n-k)] \quad \text{for } 0 \leq k \leq p.$$

8.10 Do the following for the second-order model of the temperature signal in Example 8.3:

a. Calculate the residual error sequence (signal).

b. Estimate $\hat{\rho}_\epsilon(1)$ and $\hat{\rho}_\epsilon(2)$.

c. Do these correlation values test as zero?

8.11 For the grinding wheel signal in Examaple 8.2, generate the error sequence. Is it white noise?

8.12 Write the Levinson recursion equations for a fourth-order signal model. Expand the summation in step 2 and write all the equations in step 4 (refer to Section 8.3.4).

8.13 Redo Exercise 8.7a using the Levinson–Durbin algorithm.

8.14 In Appendix A8.2.2, the recursion relationships for the reflection coefficient and squared error are derived. Perform this derivation for $p = 4$ without partitioning the matrices. That is, keep all of the elements explicit in the manipulations.

8.15 Derive the recursive relationship for the forward and backward prediction errors in equations (8.51) and (8.52).

8.16 Prove that the denominator $D(p-1)$ in the reflection coefficient as defined by step 2 in Burg's algorithm can be derived in a recursive manner by

$$D(p) = \left(1 - |\hat{\pi}_{p-1}|^2\right)D(p-1)$$
$$- \epsilon_{p-1}^f(p-1)^2 - \epsilon_{p-1}^b(N-1)^2.$$

8.17 What is the PSD of the temperature signal? Plot it.

8.18 Generate a first-order AR signal with $a(1) = 0.7$, unit-variance white noise input, a sampling interval of 1, and $N = 100$.

a. What is the theoretical power spectrum?

b. What are $a(1)$, the error variance, and the estimated PSD?

c. What are any differences between the theoretical and estimated spectra?

8.19 For the signal in Exercise 8.18, the model order is estimated to be 3. Estimate the PSD using the Burg algorithm. How does it differ from what is expected?

8.20 Consider the spectral analysis of the speech signal in Example 8.12.
a. What are the degrees of freedom in the LPC analysis?
b. Consider using a BT approach with $M = 10$ and 20. What are the resulting degrees of freedom? What are the frequency spacings?
c. State which method is preferable and give reasons.

8.21 A discrete-time signal with $T = 1$ has an ACF

$$R(k) = \delta(k) + 5.33\cos(0.3\pi k) + 10.66\cos(0.4\pi k).$$

a. Plot $R(k)$ for $0 \le k \le 20$.
b. Find the coefficients and white-noise variance for an AR(3) model. Use the Levinson–Durbin algorithm and perform the calculations manually or write your own software to perform them.

8.22 An AR(4) model of a biological signal was found to have the parameters $[1, -0.361, 0.349, 0.212, -0.005]$, $\sigma_y^2 = 0.059$, and $T = 2$ ms. What is its PSD?

Computer Exercises

8.23 Generate two white-noise sequences with zero mean and variance 6.55. Use them to drive the first-order model of the grinding wheel signal in Example 8.2. How does each of them compare visually to the original signal?

8.24 Vibration data from an operating mechanical system are listed in Appendix 8.5.
a. Model these data for $1 \le p \le 20$.
b. Plot the FPE and σ_ϵ^2 versus model order.
c. What is the best model for this signal?

8.25 The hospital census data in Appendix 7.4 are modeled well with an ARMA(7, 6) system. The data have a mean of 461.5 and $\sigma_\epsilon^2 = 119.4$. The parameters are

$$a(i) = [1, 0.27, 0.24, 0.25, 0.28, 0.28, 0.20, -0.72],$$

$$b(i) = [1, 0.91, 0.98, 1.24, 1.30, 1.53, 1.26].$$

Find a good AR model for these data. Use any model order selection criteria studied.

8.26 A discrete time signal with $T = 1$ has an ACF

$$R(k) = \delta(k) + 5.33\cos(0.3\pi k) + 10.66\cos(0.4\pi k).$$

a. Calculate its PSD using an AR(10) model.
b. Calculate its PSD using a BT approach having a triangular lag window and maximum lag of 10.
c. What are the differences between these PSDs?

8.27 Estimate the PSD of the hospital census data using the YW and Burg algorithms.
a. What are the orders and residual errors of the chosen AR spectra?
b. Does either of them make more sense when compared to what is expected from the time series?

Appendices

Appendix 8.1 Mean Squared Error for Two-Term Linear Prediction

For the two-term linear prediction the MSE, or the error variance, is

$$\begin{aligned} \text{MSE} &= \sigma_\epsilon^2 \\ &= E\left[(y(n) - h(1)y(n-1) - h(2)y(n-2))^2\right]. \quad \text{(A8.1)} \end{aligned}$$

This expression is expanded and simplified using the recursive forms for the ACF,

$$R_y(1) - h(1)R_y(0) - h(2)R_y(1) = 0, \qquad \text{(A8.2)}$$

$$R_y(2) - h(1)R_y(1) - h(2)R_y(0) = 0. \qquad \text{(A8.3)}$$

The term within the square brackets of equation (A8.1) is expanded as

$$
\begin{aligned}
y(n)y(n) - h(1)y(n-1)y(n) - h(2)y(n-2)y(n) \\
- h(1)(y(n)y(n-1) - h(1)y(n-1)y(n-1) \\
-h(2)y(n-2)y(n-1)) \\
- h(2)(y(n)y(n-2) - h(1)y(n-1)y(n-2) \\
-h(2)y(n-2)y(n-2)). \quad \text{(A8.4)}
\end{aligned}
$$

Examine the sum of terms within the parentheses of the second and third lines in equation (A8.4). If expectations are made, they become identical to equations (A8.2) and (A8.3) and equal zero. Thus the error variance is the expectation of the first line,

$$
\sigma_\epsilon^2 = R_y(0) - h(1)R_y(1) - h(2)R_y(2). \qquad \text{(A8.5)}
$$

Appendix 8.2 Matrix Form of Levinson – Durbin Recursion

A8.2.1 General Coefficients

The recursion form can be developed from the matrix representation of the Yule–Walker equations. The general form is,

$$
\begin{bmatrix}
c(0) & c(1) & \cdots & c(p-1) \\
c(1) & c(0) & \cdots & c(p-2) \\
\vdots & \vdots & & \vdots \\
c(p-1) & c(p-2) & \cdots & c(0)
\end{bmatrix}
\begin{bmatrix}
\hat{a}(1) \\ \hat{a}(2) \\ \vdots \\ \hat{a}(p)
\end{bmatrix}
=
\begin{bmatrix}
-c(1) \\ -c(2) \\ \vdots \\ -c(p)
\end{bmatrix}, \qquad \text{(A8.6)}
$$

where $c(k)$ is the sample autocovariance function. Using bold letters to represent matrices and vectors,

$$\mathbf{C}_p \mathbf{a}_p(\mathbf{p}) = -\mathbf{c}_p, \tag{A8.7}$$

where the subscript denotes the order of the square matrix or of the vector, and

$$\mathbf{a}_p(\mathbf{m}) = \left[\hat{a}_p(1), \hat{a}_p(2), \ldots, \hat{a}_i(m)\right]'. \tag{A8.8}$$

Begin with the solution to the first-order model from Section 8.3.1,

$$\hat{a}_1(1) = -\frac{c(1)}{c(0)}, \qquad \sigma_{\epsilon,1}^2 = c(0)\left(1 - |\hat{a}_1(1)|^2\right). \tag{A8.9}$$

For a second-order system

$$\begin{bmatrix} c(0) & c(1) \\ c(1) & c(0) \end{bmatrix} \begin{bmatrix} \hat{a}_2(1) \\ \hat{a}_2(2) \end{bmatrix} = \begin{bmatrix} -c(1) \\ -c(2) \end{bmatrix}. \tag{A8.10}$$

Using the first equation of equation (A8.10), and equation (A8.9) yields

$$\begin{aligned} \hat{a}_2(1) &= \hat{a}_1(1) - \hat{a}_2(2)\frac{c(1)}{c(0)} \\ &= \hat{a}_1(1) + \hat{a}_2(2)\hat{a}_1(1). \end{aligned} \tag{A8.11}$$

For third-order and larger models, the same general procedure is utilized and implemented through matrix partitioning. The goal of the partitioning is to isolate the (p, p) element of $\mathbf{C}_p$. Thus, for a third-order model,

$$\begin{bmatrix} c(0) & c(1) & c(2) \\ c(1) & c(0) & c(1) \\ c(2) & c(1) & c(0) \end{bmatrix} \begin{bmatrix} \hat{a}_3(1) \\ \hat{a}_3(2) \\ \hat{a}_3(3) \end{bmatrix} = \begin{bmatrix} -c(1) \\ -c(2) \\ -c(3) \end{bmatrix} \tag{A8.12}$$

becomes

$$\begin{bmatrix} \mathbf{C}_2 & \boldsymbol{\psi}_2 \\ \boldsymbol{\psi}_2' & c(0) \end{bmatrix} \begin{bmatrix} \mathbf{a}_3(2) \\ \hat{a}_3(3) \end{bmatrix} = \begin{bmatrix} -\mathbf{c}_2 \\ -c(3) \end{bmatrix}, \tag{A8.13}$$

where

$$\boldsymbol{\psi}_p = [c(p), c(p-1), \ldots, c(1)]'. \tag{A8.14}$$

Now, solving for $\mathbf{a}_3(2)$ yields

$$\mathbf{a}_3(2) = -\mathbf{C}_2^{-1}\boldsymbol{\psi}_2\hat{a}_3(3) - \mathbf{C}_2^{-1}\mathbf{c}_2. \tag{A8.15}$$

From the second-order model it is known that $\mathbf{C}_2\mathbf{a}_2(2) = \mathbf{c}_2$. Thus,

$$\mathbf{a}_2(2) = -\mathbf{C}_2^{-1}\mathbf{c}_2. \tag{A8.16}$$

Because of the property of Toeplitz matrices, $\mathbf{C}_p\boldsymbol{\alpha}_p(p) = -\boldsymbol{\psi}_p$ and

$$\boldsymbol{\alpha}_2(2) = -\mathbf{C}_2^{-1}\boldsymbol{\psi}_2. \tag{A8.17}$$

Now equation (A8.15) can be written in recursive form as

$$\mathbf{a}_3(2) = \mathbf{a}_2(2) + \hat{a}_3(3)\boldsymbol{\psi}_2(2). \tag{A8.18}$$

Notice that this procedure yields all but the reflection coefficient. The recursive relationship for this parameter and the error variance will be developed in the next section.

The general recursive rule for the reflection coefficient is also found by first partitioning the $p \times p$ covariance matrix in order to isolate the $\hat{a}_p(p)$ parameter. Equation (A8.6) is written as

$$\begin{bmatrix} \mathbf{C}_{p-1} & \boldsymbol{\psi}_{p-1} \\ \boldsymbol{\psi}'_{p-1} & c(0) \end{bmatrix} \begin{bmatrix} \mathbf{a}_p(p-1) \\ \hat{a}_p(p) \end{bmatrix} = \begin{bmatrix} -\mathbf{c}_{p-1} \\ -c(p) \end{bmatrix}. \tag{A8.19}$$

The first equation is solved for $\mathbf{a}_p(p-1)$,

$$\mathbf{a}_p(p-1) = -\mathbf{C}_{p-1}^{-1}\boldsymbol{\psi}_{p-1}\hat{a}_p(p) - \mathbf{C}_{p-1}^{-1}\mathbf{c}_{p-1}. \tag{A8.20}$$

Because

$$\boldsymbol{\alpha}_{p-1}(p-1) = -\mathbf{C}_{p-1}^{-1}\boldsymbol{\psi}_{p-1}$$

and

$$\mathbf{a}_{p-1}(p-1) = -\mathbf{C}_{p-1}^{-1}\mathbf{c}_{p-1}, \tag{A8.21}$$

substituting the equations (A8.21) into equation (A8.20) yields

$$\mathbf{a}_p(p-1) = \mathbf{a}_{p-1}(p-1) + \hat{a}_p(p)\boldsymbol{\alpha}_{p-1}(p-1). \quad \text{(A8.22)}$$

Thus, the first $p-1$ parameters of the model for order p are found.

A8.2.2 Reflection Coefficient and Variance

The reflection coefficient and the squared error are found using the augmented Yule–Walker equations,

$$\begin{bmatrix} c(0) & c(1) & \cdots & c(p) \\ c(1) & c(0) & \cdots & c(p-1) \\ \vdots & \vdots & & \vdots \\ c(p) & c(p-1) & \cdots & c(0) \end{bmatrix} \begin{bmatrix} 1 \\ \hat{a}_p(1) \\ \vdots \\ \hat{a}_p(p) \end{bmatrix}$$

$$= \begin{bmatrix} \boldsymbol{\sigma}^2_{\epsilon,p} \\ 0 \\ \vdots \\ 0 \end{bmatrix}. \quad \text{(A8.23)}$$

The $\mathbf{a}_p(p+1)$ vector is expanded using equation (A8.22) and becomes

$$\mathbf{a}_p(p+1) = \begin{bmatrix} 1 \\ \hat{a}_p(1) \\ \vdots \\ \hat{a}_p(p) \end{bmatrix}$$

$$= \begin{bmatrix} 1 \\ \hat{a}_{p-1}(1) \\ \vdots \\ 0 \end{bmatrix} + \hat{a}_p(p) \begin{bmatrix} 0 \\ \hat{a}_{p-1}(p-1) \\ \vdots \\ 1 \end{bmatrix},$$

or

$$\mathbf{a}_p(p+1) = \begin{bmatrix} 1 \\ \mathbf{a}_{p-1}(p-1) \\ 0 \end{bmatrix} + \hat{a}_p(p) \begin{bmatrix} 0 \\ \boldsymbol{\alpha}_{p-1}(p-1) \\ 1 \end{bmatrix}. \tag{A8.24}$$

The $\mathbf{C}_{p+1}$ covariance matrix is partitioned such that the multiplication by equation (A8.24) can be accomplished. For the first vector,

$$\mathbf{T}_1 = \begin{bmatrix} c(0) & \mathbf{c}'_{p-1} & c(p) \\ \mathbf{c}_{p-1} & \mathbf{C}_{p-1} & \boldsymbol{\psi}_{p-1} \\ c(p) & \boldsymbol{\psi}'_{p-1} & c(0) \end{bmatrix} \begin{bmatrix} 1 \\ \mathbf{a}_{p-1}(p-1) \\ 0 \end{bmatrix} = \begin{bmatrix} \sigma^2_{\epsilon,p-1} \\ \boldsymbol{\Delta}_2 \\ \Delta_3 \end{bmatrix} \tag{A8.25}$$

For the second vector,

$$\mathbf{T}_2 = \begin{bmatrix} c(0) & \mathbf{c}'_{p-1} & c(p) \\ \mathbf{c}_{p-1} & \mathbf{C}_{p-1} & \boldsymbol{\psi}_{p-1} \\ c(p) & \boldsymbol{\psi}'_{p-1} & c(0) \end{bmatrix} \begin{bmatrix} 0 \\ \boldsymbol{\alpha}_{p-1}(p-1) \\ 1 \end{bmatrix} = \begin{bmatrix} \Delta_3 \\ \boldsymbol{\Delta}_2 \\ \sigma^2_{\epsilon,p-1} \end{bmatrix} \tag{A8.26}$$

Combining the last three equations with equation (A8.23) produces

$$\begin{bmatrix} \sigma^2_{\epsilon,p} \\ \mathbf{0}_{p-1} \\ 0 \end{bmatrix} = \begin{bmatrix} \sigma^2_{\epsilon,p-1} \\ \boldsymbol{\Delta}_2 \\ \Delta_3 \end{bmatrix} + \hat{a}_p(p) \begin{bmatrix} \Delta_3 \\ \boldsymbol{\Delta}_2 \\ \sigma^2_{\epsilon,p-1} \end{bmatrix}. \tag{A8.27}$$

The reflection coefficient is found using the $p+1$ equation from matrix equation (A8.27),

$$\hat{a}_p(p) = -\frac{\Delta_3}{\sigma^2_{\epsilon,p-1}}$$
$$= -\frac{c(p) + \mathbf{c}_{p-1} \cdot \boldsymbol{\alpha}_{p-1}(p-1)}{\sigma^2_{\epsilon,p-1}}$$
$$= -\frac{1}{\sigma^2_{\epsilon,p-1}} \sum_{i=0}^{p-1} c(p-i-1)\hat{a}_{p-1}(i). \qquad \text{(A8.28)}$$

The recursion equation for the squared error is found using the first equation of matrix equation (A8.27) and equation (A8.28).

$$\sigma^2_{\epsilon,p} = \sigma^2_{\epsilon,p-1} + \hat{a}_p(p)\Delta_3 = \sigma^2_{\epsilon-1}\left(1 - |\hat{a}_p|^2\right) \qquad \text{(A8.29)}$$

Appendix 8.3 Surface Roughness for a Grinding Wheel

Surface Roughness Height × 10^3 in. (Sampling Interval = 0.002 in.)*

13.5*	4.0	4.0	4.5	3.0	3.0	10.0	10.2	9.0	10.0
8.5	7.0	10.5	7.5	7.0	10.5	9.5	7.0	12.0	13.5
12.5	15.0	13.0	11.0	9.0	10.5	10.5	11.0	10.5	9.0
8.2	8.5	9.2	8.5	10.0	14.5	13.0	2.0	6.0	6.0
11.0	9.5	12.5	13.8	12.0	12.0	12.0	13.0	12.0	14.0
14.5	13.5	12.3	7.0	7.0	7.0	6.5	12.5	15.0	12.5
11.6	11.0	10.0	8.5	3.0	11.5	11.5	11.5	11.0	9.0
2.5	7.0	6.0	6.6	14.0	11.0	9.0	6.5	4.0	6.0
12.0	11.0	12.0	12.5	12.5	13.6	13.0	8.0	6.5	6.8
6.0	7.2	10.2	8.0	7.5	11.0	11.8	11.8	6.5	8.0
9.0	8.0	8.0	9.0	9.5	10.0	9.0	12.0	13.5	13.8
15.0	12.5	11.0	11.5	14.5	11.5	11.8	13.0	15.0	14.5
13.0	9.0	11.0	9.0	10.0	14.0	13.5	3.0	2.2	6.0
8.0	9.0	9.0	9.0	7.0	6.0	6.5	7.0	7.5	8.5
9.0	9.5	10.0	11.5	11.2	12.5	11.6	8.0	7.0	6.0
6.0	6.0	9.0	12.0	13.5	13.0	3.5	1.8	1.6	7.5
8.0	7.9	11.6	12.5	10.5	8.0	9.0	11.6	11.8	12.6
10.2	10.0	5.0	7.0	−1.0	0.0	0.0	3.0	11.0	12.0
12.2	11.0	8.0	7.0	5.5	10.0	11.5	7.0	4.0	7.0
7.0	10.0	9.0	8.0	10.0	13.0	10.0	6.5	11.0	13.0
13.0	14.0	13.0	12.5	12.0	9.0	8.5	7.0	8.5	10.0
8.0	4.0	3.0	10.0	13.0	13.0	13.0	12.5	11.0	11.0
11.0	14.5	14.0	14.0	13.5	10.0	9.5	10.0	12.5	10.0
9.0	9.0	4.0	3.0	6.0	5.0	7.0	6.0	5.0	8.5
10.5	11.1	11.0	10.0	11.2	8.0	2.5	5.0	13.2	14.0

*Total 250 observations (read row-wise).

Appendix 8.4 Annual Rainfall in the Eastern United States

22.54*	−17.46	−26.46	11.54	6.54	−26.46	9.54	−1.46	−33.46	−14.46
6.54	−9.46	6.54	−1.46	−1.46	0.54	−6.46	−15.46	−5.46	4.54
−7.46	−2.46	−6.46	−2.46	6.54	1.54	7.54	−9.46	−6.46	9.54
10.54	−2.46	−0.46	14.54	−13.46	1.54	−4.46	−5.46	0.54	−13.46
5.54	6.54	14.54	−7.46	−3.46	3.54	3.54	−14.46	16.54	−0.46
2.54	12.54	6.54	13.54	1.54	−0.46	16.54	5.54	7.54	15.54
7.54	15.54	−2.46	4.54	3.54	6.54	9.54	6.54	3.54	2.54
−2.46	7.54	8.54	10.54	0.54	0.54	3.54	−6.46	−12.46	−5.46
3.54	5.54	−8.46	−7.46	−4.46	0.54	3.54	−9.46	3.54	1.54
−0.46	−12.46	−5.46	−11.46	4.54	5.54	−3.46	−9.46	6.54	−6.46
−11.46	−9.46	9.54	3.54	−2.46	0.54				

*Total 106 observations (read row-wise).

Appendix 8.5 Vibration in a Mechanical System

30.0*	28.0	25.0	24.0	23.0	21.0	20.0	22.0	24.0	27.0
30.0	31.0	34.0	37.0	33.0	28.0	25.0	23.0	21.0	19.0
18.0	17.0	16.0	17.0	18.5	22.0	29.0	32.0	32.0	30.0
25.0	20.0	17.0	14.0	13.0	17.0	22.0	27.0	33.0	30.0
21.0	15.0	12.0	10.0	9.0	6.0	6.0	8.0	10.0	12.0
15.0	16.0	16.5	18.0	21.0	15.0	7.0	4.0	3.0	7.0
15.5	22.0	30.0	40.0	40.0	39.0	38.0	35.0	30.0	25.0
20.0	18.0	20.0	22.5	27.0	32.0	32.5	33.0	32.0	30.0
25.5	23.3	23.3	24.0	27.0	31.5	35.0	36.0	34.0	30.5
29.0	25.0	20.0	19.0	21.0	23.5	28.0	33.5	36.0	37.5
38.0	36.0	33.0	29.5	28.0	28.0	30.0	30.5	30.0	30.0
28.0	25.0	23.0	24.5	27.0	31.0	34.0	33.0	25.0	16.0
13.0	14.0	17.0	22.0	29.0	32.0	30.0	26.0	24.0	24.0

*Total 130 observations (read row-wise), sampling interval = 0.02 s. *Source*: Data from H. J. Stedudel (1971), "A Time Series Approach to Modeling Second Order Mechanical Systems," M.S. thesis, Univ. of Wisconsin-Milwaukee.

Chapter 9 Theory and Application of Cross Correlation and Coherence

9.1 Introduction

The concept of cross correlation was introduced in Chapter 6. It was defined in the context of linear systems and described a relationship between the input and output of a system. In general, a relationship can exist between any two signals whether or not they are intrinsically related through a system. For instance, we can understand that two signals, such as the temperatures in two cities or the noise in electromagnetic transmissions and atmospheric disturbances, can be related; however, their relationship is not well defined. An example is demonstrated in Figure 9.1, in which the time sequence of speed measurements made at different locations on a highway is plotted. It is apparent that these signals have similar shapes and that some, such as v_1 and v_2, are almost identical except for a time shift. How identical these signals are and the time shift that produces the greatest similarity can be ascertained quantitatively with the cross correlation functions. This same information can be ascertained in the frequency domain with cross spectral density and coherence functions. In addition, the latter functions can be used to ascertain synchronism in frequency components within two signals. They will be described in the second half of this chapter. There are several comprehensive refer-

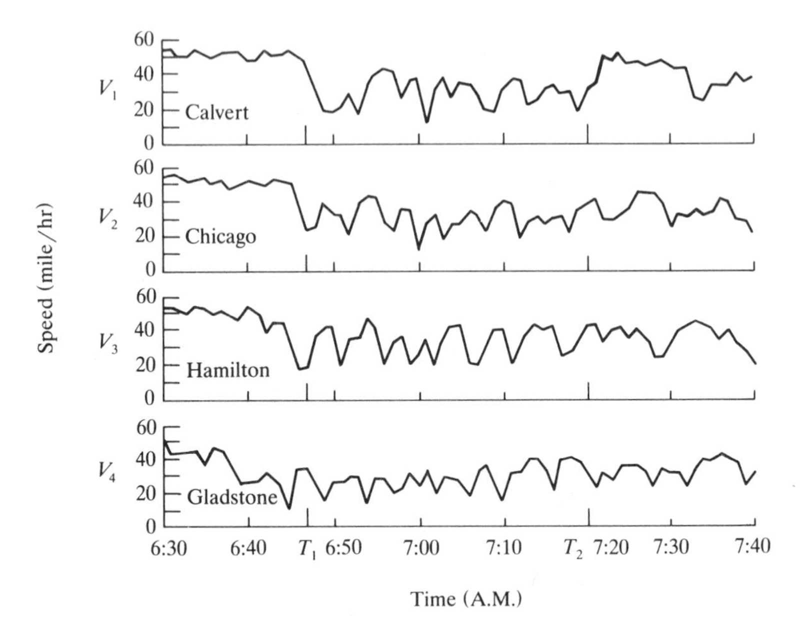

Figure 9.1 Records of average speeds at four different locations on a highway. Averages are computed over 1-min intervals. [Adapted from Schwartz and Shaw (1975), Figure 4.7, with permission.]

ences treating cross correlation and coherence. Good treatments with applications are Bendat and Piersol (1980) and Jenkins and Watts (1968). References that focus on recent developments and give detailed procedures are Carter (1988), Silvia (1987), and the special issue of the *IEEE Transactions on Acoustics, Speech and Signal Processing* in 1981 [IEEE (1981)].

The *cross correlation function* (CCF) is defined as

$$R_{yx}(k) = E[y(n)x(n+k)], \tag{9.1}$$

where $\tau_d = kT$ is the amount of time that the signal $x(n)$ is delayed with respect to $y(n)$. As with the definition of the autocorrelation functions, there are several forms. The *cross covariance function* (CCVF) is defined as

$$C_{yx}(k) = E\big[(y(n) - m_y)(x(n+k) - m_x)\big] = R_{yx}(k) - m_y m_x, \tag{9.2}$$

and the *normalized cross covariance function* (NCCF) is

$$\rho_{yx}(k) = \frac{C_{yx}(k)}{\sigma_y \sigma_x}. \tag{9.3}$$

The information in these functions is explained in the context of the highway speed signals. Figure 9.2 shows the NCCF between each pair of locations. Examination of $\rho_{21}(k)$ reveals that, at a lag time of 1 min, it achieves a maximum magnitude of 0.55. Thus, the signals are most similar at this time difference and the correlation

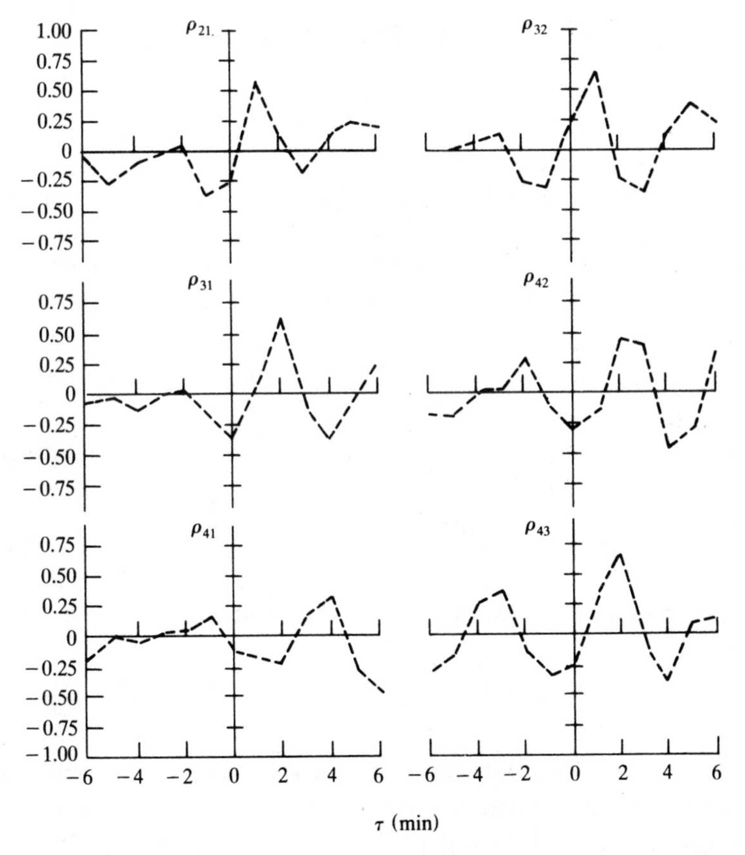

Figure 9.2 Cross correlations of speed records in Figure 9.1. [Adapted from Schwartz and Shaw (1975), Figure 4.8, with permission.]

coefficient at that delay is 0.55. A realistic interpretation is that this time delay equals the time necessary to travel between locations 2 and 1. The other NCCF give the same information between the other locations.

This same type of information is needed in many other applications. Some other major applications of this measure include the following: the estimation of time delays in order to estimate ranges and bearings in radar and sonar systems; path length determination in multipath environments such as sound studios and auditoriums; and identification of systems.

9.2 Properties of Cross Correlation Functions

9.2.1 Theoretical Function

The cross correlation function has symmetry properties that depend on the ordering of the two signals. Our intuition informs us that signals can be aligned by either shifting $y(n)$ in one direction or shifting $x(n)$ in the opposite direction. Then $R_{yx}(k) = E[y(n)x(n+k)] = E[x(n+k)y(n)]$, which by the notation is simply $R_{xy}(-k)$. Thus,

$$R_{yx}(k) = R_{xy}(-k). \tag{9.4}$$

The CCF also has magnitude boundaries. Start with the inequality

$$E\left[(ay(n) + x(n+k))^2\right] \geq 0. \tag{9.5}$$

Expanding the square and taking expectations yields

$$a^2 R_{yy}(0) + 2aR_{yx}(k) + R_{xx}(0) \geq 0. \tag{9.6}$$

Solving equation (9.6) for the unknown variable a will produce complex roots because the equation is nonnegative. The discriminant then is negative and

$$4R_{yx}^2(k) - 4R_{yy}(0)R_{xx}(0) \leq 0,$$

or

$$R_{yx}^2(k) \le R_{yy}(0)\, R_{xx}(0). \tag{9.7}$$

Similarly, one could use the cross covariances and show that

$$C_{yx}^2(k) \le C_{yy}(0) C_{xx}(0). \tag{9.8}$$

Another useful inequality, whose proof is based on a modification of equation (9.5), is

$$|R_{yx}(k)| \le \tfrac{1}{2}\big(R_{yy}(0) + R_{xx}(0)\big). \tag{9.9}$$

Its derivation is left as an exercise. Because the NCCF is itself a correlation coefficient

$$-1 \le \rho_{yx}(k) \le 1. \tag{9.10}$$

9.2.2 Estimators

The set of cross correlation functions has estimators that are analogous to those for the set of autocorrelation functions. For two signals containing N points and indexed over $0 \le n \le N-1$, the sample CCVF is

$$\hat{C}_{yx}(k) = \begin{cases} \dfrac{1}{N} \displaystyle\sum_{n=0}^{N-k-1} \big(y(n) - \hat{m}_y\big)\big(x(n+k) - \hat{m}_x\big) & \text{for } k \ge 0, \\ \dfrac{1}{N} \displaystyle\sum_{n=0}^{N-k-1} \big(y(n+k) - \hat{m}_y\big)\big(x(n) - \hat{m}_x\big) & \text{for } k \le 0. \end{cases} \tag{9.11}$$

The sample means are defined conventionally. In fact, because the data must always be detrended first, the sample cross covariance function is equal to the sample cross correlation function, or $\hat{R}_{yx}(k) = \hat{C}_{yx}(k)$. Another symbol for $\hat{C}_{yx}(k)$ is $c_{yx}(k)$. The sample NCCF is

$$\hat{\rho}_{yx}(k) = \frac{c_{yx}(k)}{s_y s_x}. \tag{9.12}$$

Using equation (9.11) and assuming the signals have been detrended, the mean of the estimator is

$$E[c_{yx}(k)] = \frac{1}{N} \sum_{n=0}^{N-k-1} E[y(n)x(n+k)] = \frac{N-k}{N} C_{yx}(k)$$

for $k \geq 0$.

The same result exists for $k \leq 0$; thus,

$$E[c_{yx}(k)] = \left(1 - \frac{|k|}{N}\right) C_{yx}(k) \tag{9.13}$$

and the estimator is biased. The derivation of the covariance for the CCVF is quite involved and similar to that for the autocorrelation function. Assuming that $x(n)$ and $y(n)$ have Gaussian distributions, the covariance is

$$\operatorname{Cov}[c_{yx}(k)c_{yx}(l)] \approx \frac{1}{N} \sum_{r=-\infty}^{\infty} C_{yy}(r)C_{xx}(r+l-k) + C_{yx}(r+l)C_{xy}(r-k) \tag{9.14}$$

and indicates that, in general, the magnitudes of CCVF estimates at different lags are correlated themselves. The exact manner is highly dependent upon the inherent correlational properties of the signals themselves and thus is difficult to determine without resorting to some modeling. Letting $k = l$, the variance of the estimator is

$$\operatorname{Var}[c_{yx}(1)] \approx \frac{1}{N} \sum_{r=-\infty}^{\infty} C_{yy}(r)C_{xx}(r) + C_{yx}(r+k)C_{xy}(r-k) \tag{9.15}$$

and is consistent. The variance and covariance expressions for the NCCF are much more complex. Refer to Box and Jenkins (1976) for details. Another significant aspect of equation (9.15) is that the variance is dependent not only upon the CCVF between both processes but also upon the individual ACVFs. If both processes are uncorrelated and are white noise, then obviously,

$$\operatorname{Var}\left[c_{yx}(k)\right] = \frac{\sigma_y^2 \sigma_x^2}{N}. \tag{9.16}$$

Given the procedure for testing for significant correlations in an NCCF, it is tempting to conjecture that this measure could provide a basis for testing the amount of correlation of two signals. However, because the ACVFs affect this variance, caution must be used. Consider two first-order AR processes $x(n)$ and $y(n)$ with parameters α and β, respectively. Then equation (9.15) yields

$$\operatorname{Var}\left[c_{yx}(k)\right] = \frac{\sigma_y^2 \sigma_x^2}{N} \frac{1 + \alpha\beta}{1 - \alpha\beta}. \tag{9.17}$$

This means that the individual signals must be modeled first and then the estimated parameters must be used in the variance estimation.

Example 9.1

The effect of signal structure upon the variance of the estimate of the CCVF is illustrated by generating two first-order AR processes with $a_y(1) = 0.9$, $a_x(1) = 0.7$, $\sigma_\varepsilon^2 = T = 1$, and $N = 100$. The magnitudes of $C_{yx}(k)$ and $\rho_{yx}(k)$ are estimated and $\hat{\rho}_{yx}(k)$ is plotted in Figure 9.3. The values should be close to zero and should be within the bounds $\pm 1.96/\sqrt{N}$; however, they are not. Because of the large estimation variance, knowledge of the signal parameters is not helpful. Another approach is to remove the structure within these signals by modeling them and generating their error processes $\varepsilon_x(n)$ and $\varepsilon_y(n)$. The amount of information about signal $x(n)$ within signal $y(n)$ has not changed. Now the NCCF of the error signals is estimated and also plotted in Figure 9.3. The magnitudes are much less and are closer to what is expected.

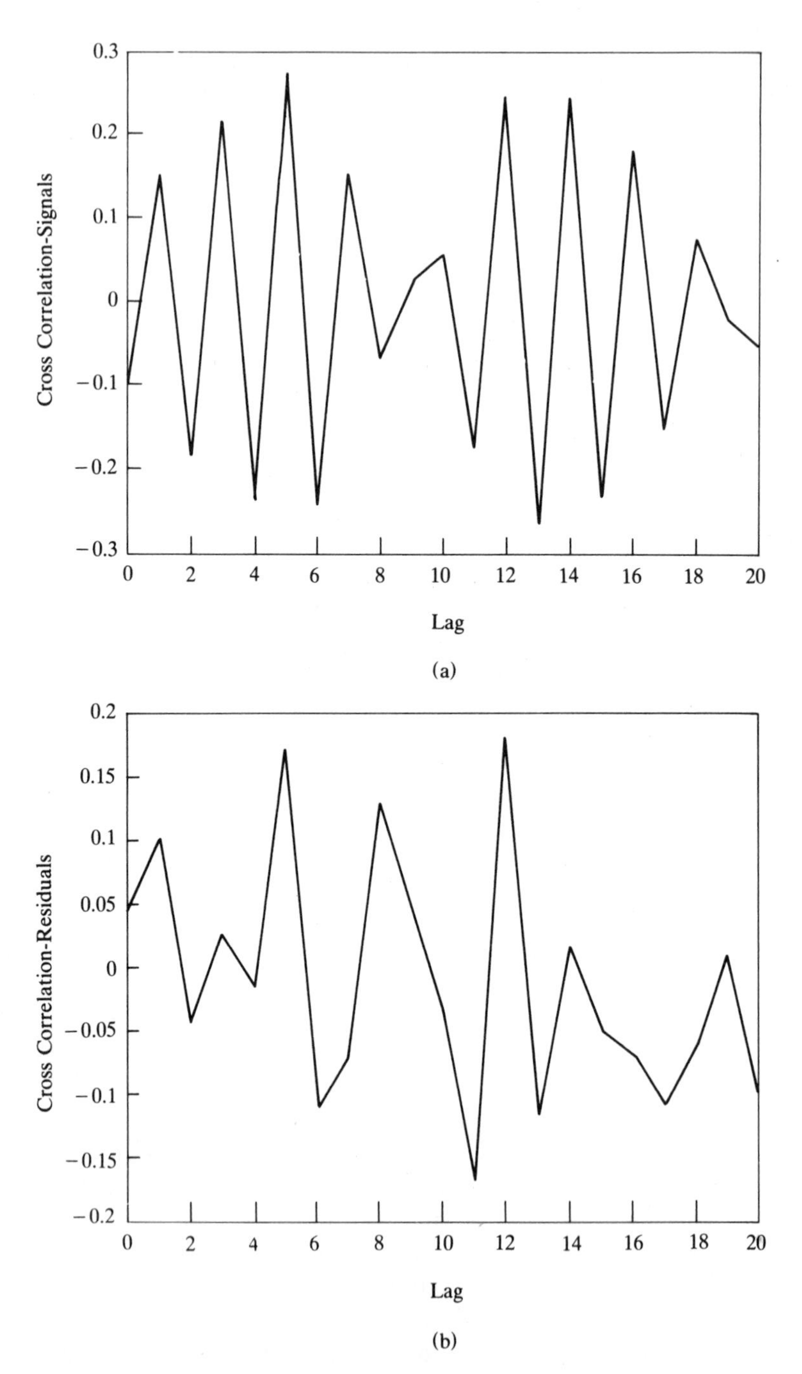

Figure 9.3 The estimate of the NCCF for (a) two independent first-order AR signals and (b) their error sequences.

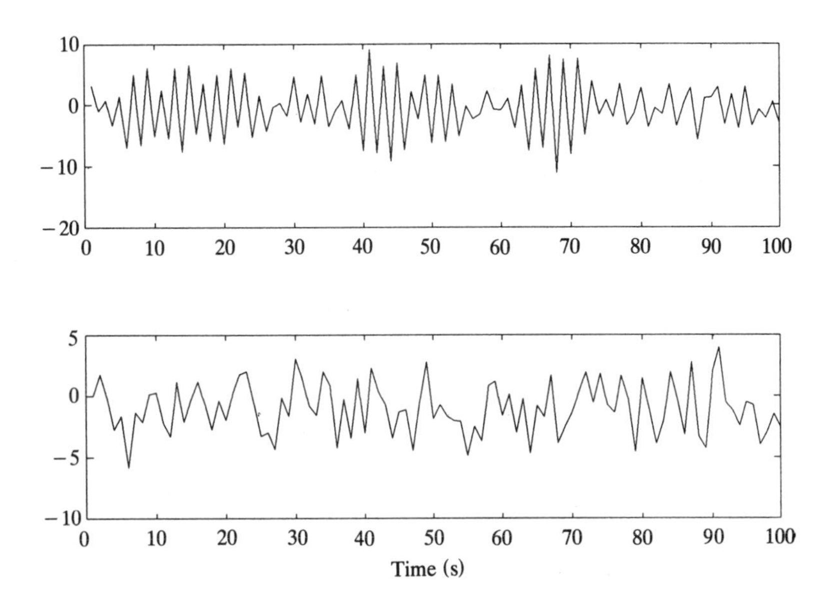

Figure 9.4 The time series for an AR(1) process, with $a(1) = 0.9$, from Example 9.1: (a) sample function, $N = 100$; (b) error process.

To show this effect, $y(n)$ and $\varepsilon_y(n)$ are plotted in Figure 9.4. Notice that the negative correlation present in the process is absent in the error process.

Based on the theory of sampling, for properly testing the significance of the cross correlation between signals, each component signal must be modeled and the variance expression in equation (9.14) must be derived. This expression not only can be quite complex and difficult to derive but also results in a uselessly large sample variance. A simpler approach is to duplicate the general procedure used in Example 9.1. First, model the signals and estimate the CCVF of the error sequences that were generated. Then test this estimate to determine the presence of any correlation between the signals. Referring to Chapter 8, this generation of the error sequence is essentially filtering the signals with an MA system and the process is called *prewhitening*.

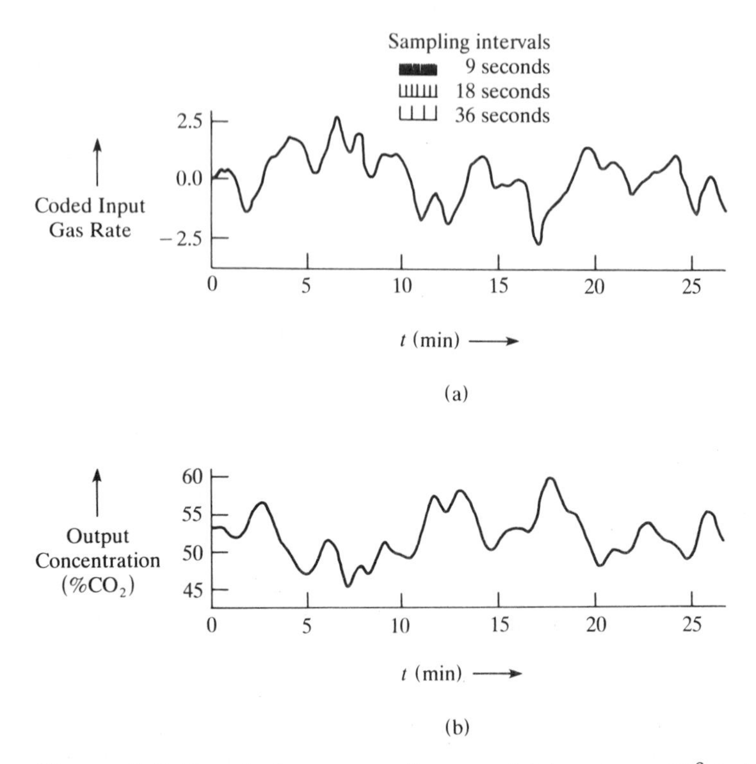

Figure 9.5 Signals from a gas furnace: (a) input gas (ft^3/min); (b) output CO_2 (percentage concentration). [Adapted from Box and Jenkins (1976), Figure 11.1, with permission.]

Example 9.2

A classical correlated signal set that is analyzed in many texts is the gas furnace data that are plotted in Figure 9.5 and tabulated in Appendix 9.1 with $T = 9.0$ s. The normalized correlation function is directly estimated and plotted in Figure 9.6(a). The NCCF peaks at a lag of 5 time units, 45 s, with a magnitude near -1. Thus, we conclude that the output strongly resembles the negative of the input and is delayed by 45 s. Examination of the signals would strongly suggest this. In order to test the significance, the prewhitening procedure must be used. The input and output are prewhitened and modeled as sixth- and fourth-order AR

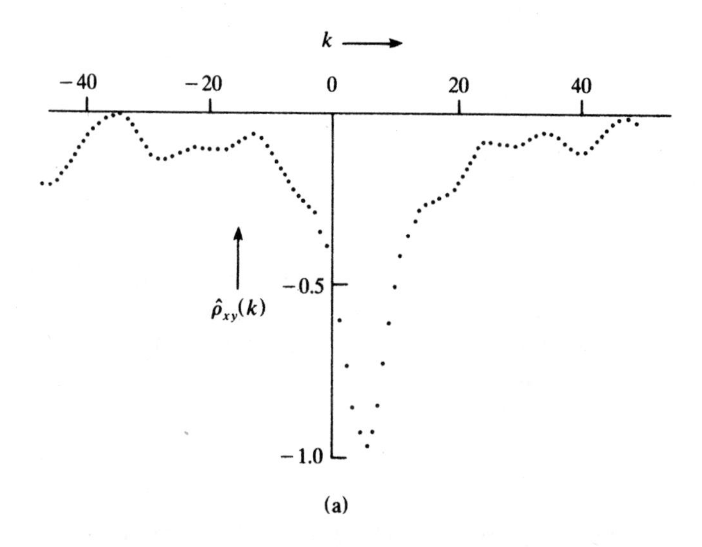

(a)

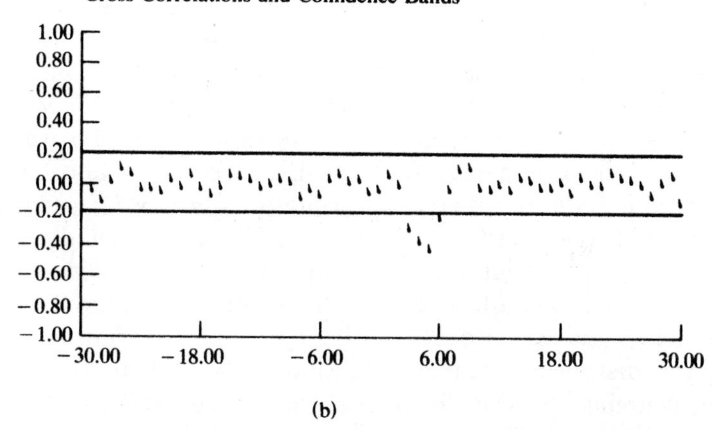

(b)

Figure 9.6 Estimates of the NCCF for the gas furnace signals: (a) direct estimate; (b) estimate after prewhitening. [Adapted from Box and Jenkins (1976), Figure 11.4, and Newton (1988), Figure 4.4, with permission.]

signals with parameter sets

$$\alpha_x = [1 \quad -1.93 \quad 1.20 \quad -0.19 \quad 0.13 \quad -0.27 \quad 0.11],$$

$$\alpha_y = [1 \quad -1.85 \quad 0.84 \quad 0.31 \quad -0.26].$$

The NCCF of the error sequences is plotted in Figure 9.6(b) along with the 95% confidence limits. It is observed that the signals are significantly correlated at lags 3–6, the maximum occurring at lag 5. Thus our initial judgment is verified.

9.3 Detection of Time-Limited Signals

A major implementation of cross correlation is for detecting the occurrence of time-limited signals. These signals can have a random or deterministic nature. One set of applications is those that seek to detect the repeated occurrences of a specific waveform. For instance, it is often necessary to know when the waveforms indicating epileptic seizures in the EEG occur. In another set of applications, it is necessary to estimate the time delay between a transmitted and a received waveform. These are usually in man-made systems so that the shape of the transmitted waveform is designed to be least affected by the transmitting medium. From the time delay and knowledge of the speed of transmission in the medium, distance of the propagation path is then calculated. In radar systems, the waveforms are transmitted electromagnetically through the atmosphere; in sonar and ultrasonic imaging systems, the waveforms are transmitted with sonic energy through a fluid medium. In many instances, there are multiple pathways for transmission that need to be known because they contribute to confounding or distorting the desired received signal. In audio systems, the multiple paths of reflections need to be known in order to avoid echoes. Figure 9.7 shows a schematic of an experiment to determine multiple pathways. Obviously, it is desired to have only the direct path in an auditorium or sound studio. Before expounding in detail it is necessary to formulate the implementation of the cross correlation.

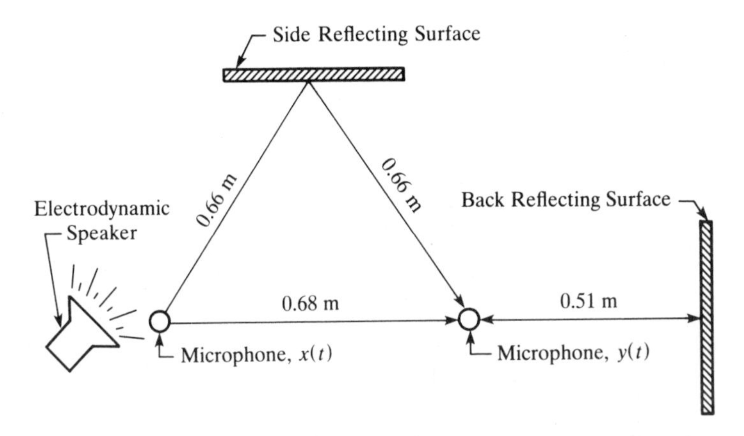

Figure 9.7 Schematic of setup for multiple-path acoustic experiment. [Adapted from Bendat and Piersol (1980), Figure 6.2, with permission.]

9.3.1 Basic Concepts

Assume a ranging system is designed to generate and transmit a square pulse $x(t)$ of amplitude and duration shown in Figure 9.8(a). For now, assume the environment is ideal, that is, the received pulse has exactly the same shape and amplitude as the transmitted pulse; it is only shifted in time, $x(t-\gamma)$. The exact time of arrival $\gamma = dT$ is obtained by cross correlating a template of the transmitted signal with the received signal $y(t) = x(t-\tau_d)$, shown in Figure 9.8(b). Notice that in this situation both the transmitted and received signals are deterministic. Writing this in continuous time gives

$$R_{xy}(\tau) = \frac{1}{W}\int_0^W x(t)y(t+\tau)\,dt$$

$$= \frac{1}{W}\int_0^W x(t)x(t-\tau_d+\tau)\,dt = R_{xx}(\tau-\tau_d) \quad (9.18)$$

and is plotted in Figure 9.8(c). Note that the integration is confined to the duration of the pulse. This is necessary to standardize the magnitude of $R_{xx}(\tau)$. The discrete-time counterpart for $W = MT$

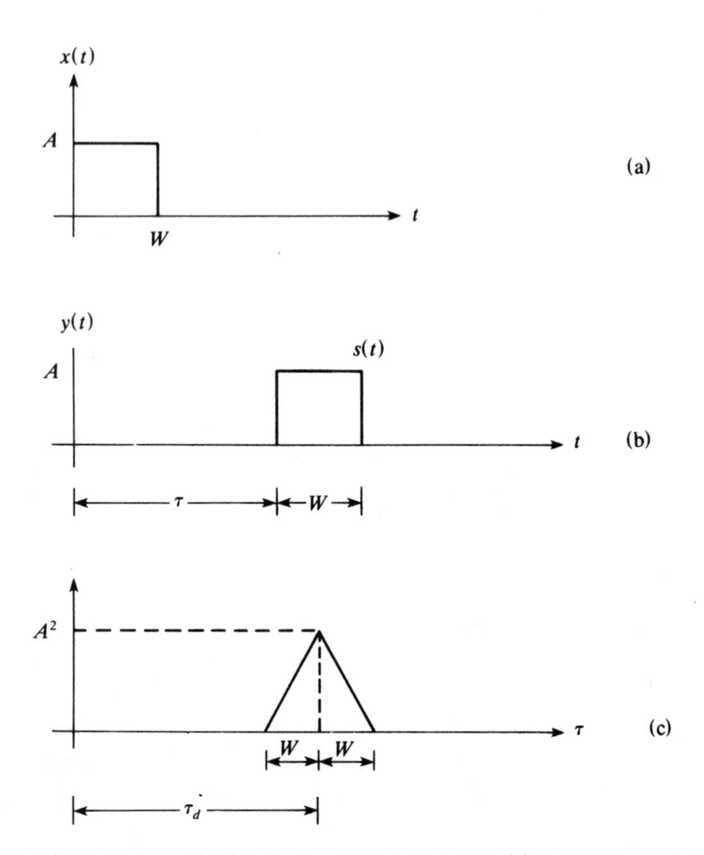

Figure 9.8 Ideal detection situation: (a) transmitted pulse; (b) received pulse; (c) cross correlation between (a) and (b).

and $\tau_d = dT$ is

$$\begin{aligned} R_{xy}(k) &= \frac{1}{M} \sum_{n=0}^{M-1} x(n)y(n+k) \\ &= \frac{1}{M} \sum_{n=0}^{M-1} x(n)x(n+k-d) \\ &= R_{xx}(k-d), \qquad 0 \le k \le N-M. \end{aligned} \tag{9.19}$$

The cross correlation peaks at the delay time and its shape is a triangular pulse. The CCF of the transmitted and received signals is the ACF of the pulse shifted by the delay time. Now proceed to a more realistic situation with a *lossy* medium and a *noisy* environment. The received signal has a reduction in amplitude (the loss coefficient is g) and is contaminated by additive white noise $\eta(n)$, that is,

$$y(n) = gx(n-d) + \eta(n), \qquad |g| \leq 1, \mathrm{Var}[\eta(n)] = \sigma_\eta^2. \quad (9.20)$$

A random signal is involved and statistical moments must be used,

$$\begin{aligned} R_{xy}(k) &= E[x(n)y(n+k)] \\ &= gE[x(n)x(n-d+k)] + E[x(n)\eta(n+k)]. \quad (9.21) \end{aligned}$$

Assuming that the signal and noise are uncorrelated,

$$E[x(n)\eta(n+k)] = R_{x\eta}(k) = m_x m_\eta = 0,$$

and the result is the same as the ideal situation, equation (9.19), except for an attenuation factor. For a multiple-path environment with q paths, the received signal is

$$y(n) = \sum_{i=1}^{q} g_i x(n-d_i) + \eta(n). \quad (9.22)$$

Its cross correlation with the transmitted signal is

$$\begin{aligned} R_{xy}(k) &= E[x(n)y(n+k)] \\ &= \sum_{i=1}^{q} g_i E[x(n)x(n-d_i+k)] + E[x(n)\eta(n+k)] \\ &= \sum_{i=1}^{q} g_i R_{xx}(k-d_i) \quad (9.23) \end{aligned}$$

and contains multiple peaks of different heights $A^2 g_i$.

9.3.2 Application of Pulse Detection

The measurement of the fetal electrocardiogram (FECG) is important for monitoring the status of the fetus during parturition. The electrodes are placed on the abdomen of the mother and the

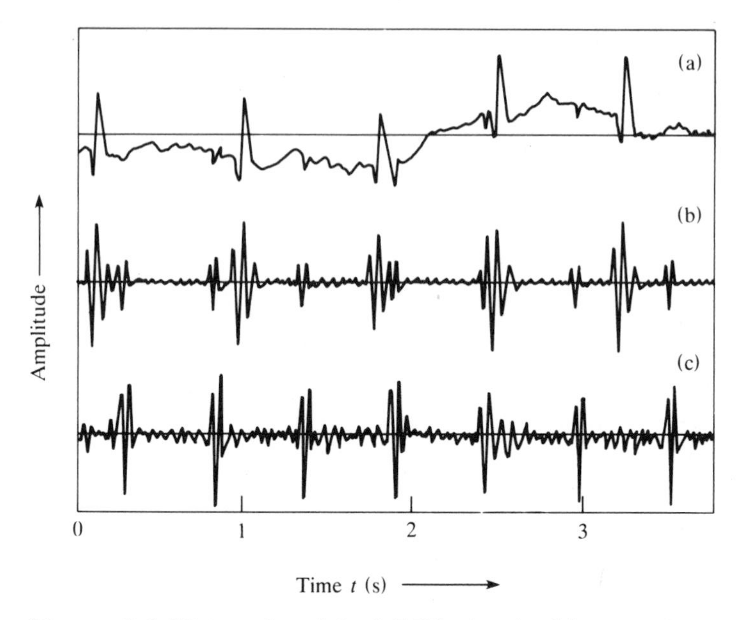

Figure 9.9 Maternal and fetal ECG signals: (a) original abdominal ECG; (b) filtered ECG; (c) FECG after MECG is subtracted from plot (b). [From Nagel (1984), Figure 1, with permission.]

resulting measurement is a summation of the FECG and the maternal electrocardiogram (MECG) with very little noise, as shown in Figure 9.9(a). Often both ECG waveforms are superimposed and it is impossible to ascertain the nature of the fetus's heart activity. Cross correlation is used to obtain a recording of the isolated FECG. First, the signal is low-pass filtered to stabilize the baseline and produce a zero-mean signal [Figure 9.9(b)], producing $y(n)$. Examining this trace shows that sometimes the FECG and MECG are separate and distinct. Through an interactive program one obtains a template of the MECG. This is used as the reference signal $x(n) = m(n)$. Then an estimate of $R_{xy}(k)$ is found,

$$\hat{R}_{xy}(k) = \sum_{i=1}^{q} g_i \hat{R}_{mm}(k - d_i) + \sum_{j=1}^{r} g_j \hat{R}_{mf}(k - d_j), \quad (9.24)$$

where $\hat{R}_{mm}(k)$ is the sample ACF of the MECG waveform and $\hat{R}_{mf}(k)$ is the sample cross correlation between the maternal and fetal ECGs. Because no loss is involved, $g_i = g_j = 1$. Examination of Figure 9.9(b) reveals that the maximum absolute value of $m(n)$ is greater than that of $f(n)$; therefore, $R_{mm}(0) > \max R_{mf}(k)$. $\hat{R}_{xy}(k)$ is searched for peak values and the time of the peaks, $d_1 \ldots d_q$, correspond to the times of occurrence of the MECG. At these times the template $m(n)$ is subtracted from $y(n)$ and the resulting FECG is produced [Figure 9.9(c)].

9.3.3 Random Signals

In the situation where the goal is to determine the relationship between two random signals, the correlation functions do not involve deterministic signals and are interpreted slightly differently. However, because only stationary signals are being considered, the mathematical manipulations are identical. The only practical difference is the number of points in the summation of equation (9.19). Now, for the time delay in a lossy and noisy environment,

$$y(n) = gx(n - d) + \eta(n), \tag{9.25}$$

$$\begin{aligned} R_{xy}(k) &= E[x(n)y(n+k)] \\ &= gE[x(n)x(n+k-d)] + E[x(n)\eta(n+k)] \\ &= gR_{xx}(k-d) + R_{x\eta}(k). \end{aligned} \tag{9.26}$$

The term $R_{x\eta}(k)$ is the cross correlation between the signal and noise. The usual assumption is that $\eta(n)$ is a zero-mean white noise that is uncorrelated with $x(n)$. It comprises the effect of measurement and transmission noise. Again, $R_{x\eta}(k) = m_x m_\eta = 0$, and the cross correlation is simply the autocorrelation function of the reference signal shifted by the time delay and multiplied by the attenuation factor. The peak value is $R_{xy}(d) = gR_x(0) = g\sigma_x^2$.

Consider the multiple path acoustic environment shown in Figure 9.7. The reference signal $x(n)$ is a band-limited noise process with bandwidth B of 8 kHz and is sketched in Figure 9.10(a). With only the direct path equation (9.25) represents the measurement situation, Figure 9.10(b) the received signal, and Figure 9.11(a) shows $R_{xy}(k)$. The cross correlation function peaks at 2 ms, which is consistent with the fact that the path length is 0.68 m and the speed of sound in air is 340 m/s. With one reflecting sur-

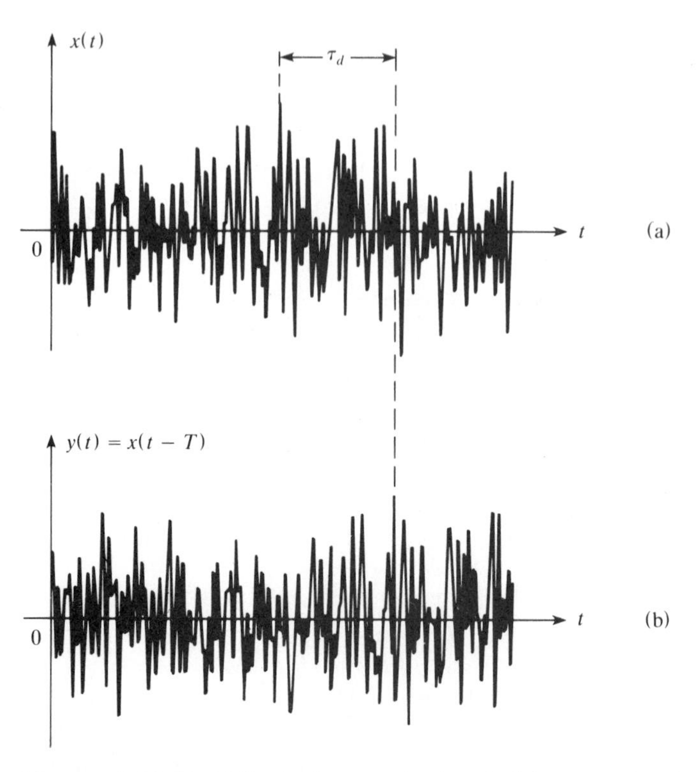

Figure 9.10 Band-limited acoustic signal: (a) transmitted signal; (b) received signal. [From de Coulon (1986), Figure 13.23, with permission.]

face the CCF appears as that in Figure 9.11(b). Another term $g_2 R_{xx}(k - d_2)$ is present and peaks at $\tau_2 = d_2 T = 3.9$ ms. This means that an additional pathway with a length of 1.32 m exists. Study of the geometry of the situation indicates that this reflection comes from a sidewall.

9.3.4 Time Difference of Arrival

In ranging systems, arrays of sensors are used to detect the waveform sent from a transmitter (emitter source) or the reflection from a target. The basic hypothesis of the measurement situation is that the waveform is a plane wave and the sensors are close together so that the two received signals are composed of the same waveform with added noise. The difference in the time of arrival is

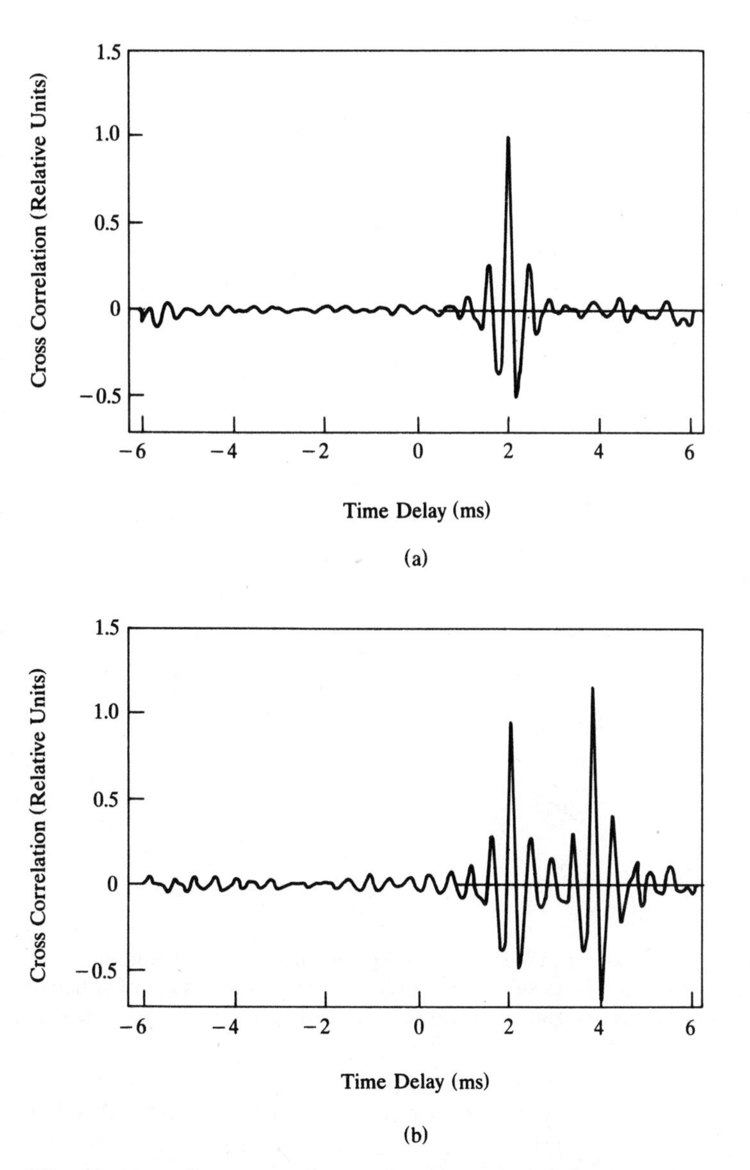

Figure 9.11 Cross correlation functions for multiple-path acoustic experiment with $T = 0.012$ ms: (a) direct path only; (b) direct and side reflection paths present. [Adapted from Bendat and Piersol (1980), Figure 6.3, with permission.]

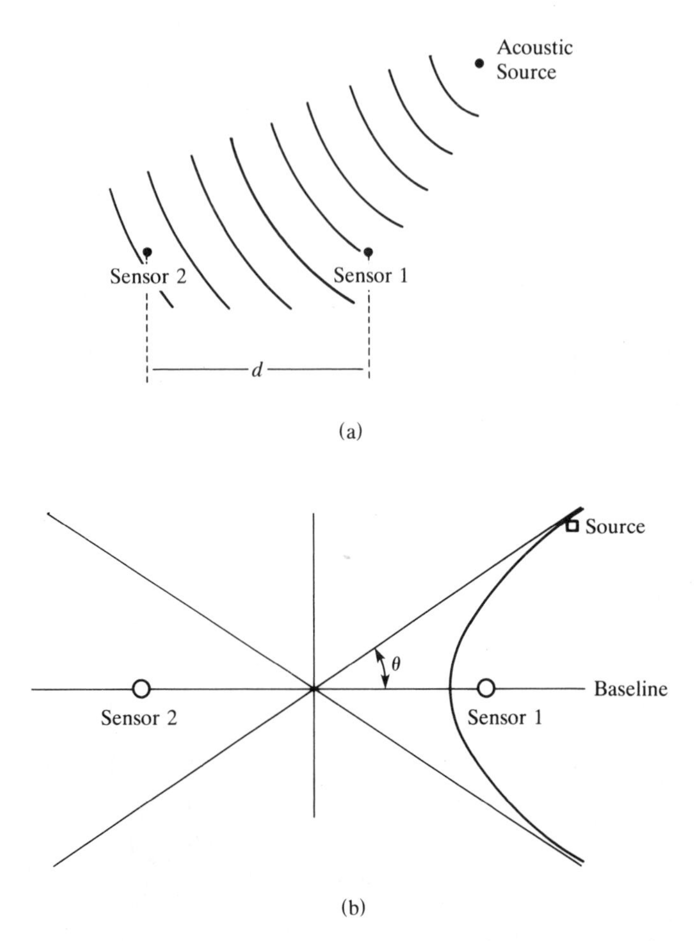

Figure 9.12 Determining bearing angle: geometrical arrangement of (a) acoustic source and two sensors and (b) bearing angle reference point. [From Chen (1982), Figure 3, with permission.]

used to estimate the angular position of the source. A schematic is shown in Figure 9.12 for an acoustic source. The two measured signals are modeled as

$$\begin{aligned} y(n) &= x(n-n_y) + \eta_y(n), \quad \text{sensor 1}, \\ z(n) &= gx(n-n_z) + \eta_z(n), \quad \text{sensor 2}, \end{aligned} \tag{9.27}$$

where g represents a relative attenuation factor. The white noise processes are uncorrelated with each other and the signal waveform $x(n)$. The cross correlation function for these signals is

$$R_{yz}(k) = gR_{xx}(k + n_y - n_z). \tag{9.28}$$

Thus, the CCF will have the shape of the ACF of $x(n)$ and peak at the *time difference of arrival* (TDOA).

9.3.5 Marine Seismic Signal Analysis

Marine seismic explorations are undertaken to determine the structure of the layered media under the water. Identifying the material attributes and the geometry of the layers are important goals not only in searching for hydrocarbon formations but also for geodetic study. A boat tows the acoustical signal source and an array of hydrophones. The energy source is either an explosion or a high-pressure, short-duration air pulse. A typical acoustic source waveform $x(t)$ is plotted in Figure 9.13(a). The underground medium is modeled as a layered medium, as shown in Figure 9.13(b). The source signal undergoes reflections at the layer boundaries and the received signal has the form

$$y(t) = \sum_{i=1}^{\infty} g_i x(t - \tau_i) + \eta(t). \tag{9.29}$$

The amplitude coefficients g_i are related to the reflection coefficients at each layer and the delay times τ_i are related to the travel distances to each layer and the speed of sound propagation through each layer c_i. The information in the signals is usually within the frequency range from 10 Hz to 10 kHz. The delay times are found by digitizing $x(t)$ and $y(t)$ and estimating their cross correlation function. Once they are found, the amplitude coefficients can be found by dividing the amplitude of the peaks of the CCF by the peak value of the ACF of $x(t)$ [El-Hawary (1988)].

9.3.6 Procedure for Estimation

In summary, these steps must be performed in order to implement the concepts and procedures for estimating cross correlation functions.

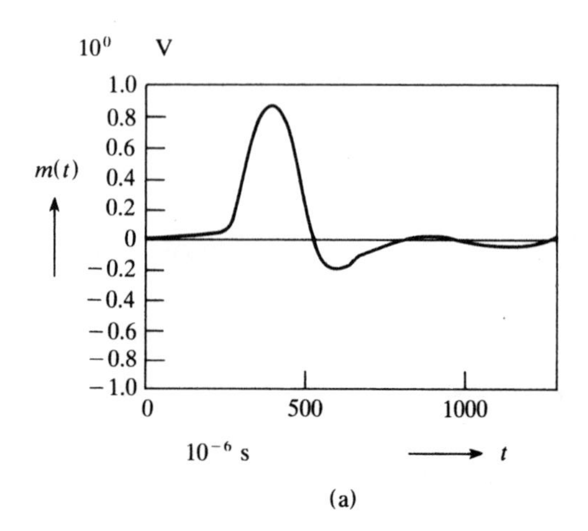

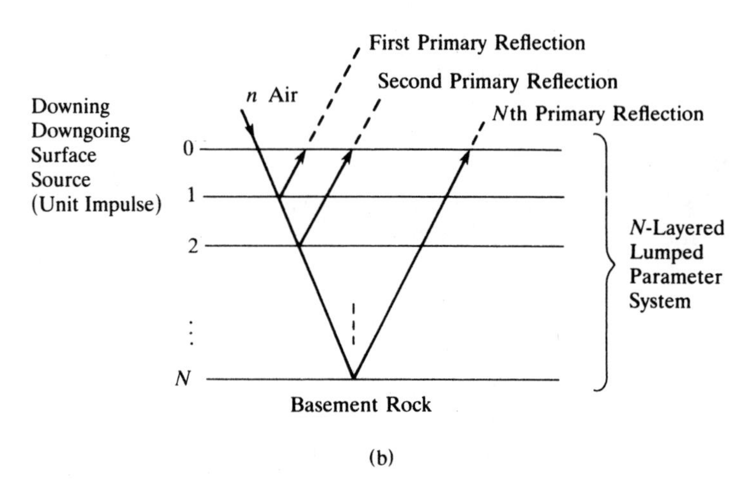

Figure 9.13 Marine seismology: (a) acoustic source waveform: (b) schematic of layers of reflection. [Adapted from El-Hawary (1988), Figure 20.1, and Chen (1982), Figure 3, with permission.]

Pulse Detection

1. *Detrending.* Both signals must be examined for trends and detrended (cf. Section 3.4.6).
2. *Correlation detection.* Select the reference signal or waveform and compute estimates of the cross correlation functions.

Correlation between Signals

1. *Detrending*. Both signals must be examined for trends and detrended.
2. *Alignment and prewhitening*. Compute the cross correlation function estimates. Look for a dominant lag and, if one exists, align the two signals and store the lag time for next step. Develop an AR model for the two signals and generate their residual sequences.
3. *Correlation functions*. Compute the autocorrelation and cross correlation function estimates using the residual sequences and shift them using the lag time from the previous step.
4. *Significance*. Test the magnitudes of the correlation function to determine which, if any, are different from zero.

9.4 Cross Spectral Density Functions

9.4.1 Definition and Properties

In Chapter 6 the cross spectral density function was defined in the context of linear discrete-time systems with input $x(n)$ and output $y(n)$. In review, the correlation function relationship is

$$R_{yx}(k) = R_x(k) * h(k),$$
$$\text{DTFT}\left[R_{yx}(k)\right] = S_{yx}(f) = S_x(f)H(f), \tag{9.30}$$

where $S_{yx}(f)$ is the *cross spectral density function* (CSD). It can be seen from equation (9.30) that the CSD can be used to determine a system's transfer function if the PSD of the input signal is also known. It is also used directly as another methodology for determining signal relationships. The CSD has a magnitude relationship with the PSDs of the component signals,

$$|S_{yx}(f)|^2 \le S_x(f)S_y(f). \tag{9.31}$$

Proof of this is left as an exercise. A function called the

magnitude-squared coherence function (MSC) is defined as

$$K_{yx}^2(f) = \frac{|S_{yx}(f)|^2}{S_y(f)S_x(f)} = \frac{S_{yx}(f)S_{yx}(f)^*}{S_y(f)S_x(f)}. \tag{9.32}$$

With the inequality of equation (9.31), the bound on the MSC is

$$0 \leq K_{yx}^2(f) \leq 1. \tag{9.33}$$

An important interpretation of the coherence function arises if it is assumed that $y(n)$ is a system output; then $Y(f) = H(f)X(f)$. Using equation (9.19) and system relationships, the coherence is

$$K_{yx}^2(f) = \frac{H(f)S_x(f)H(f)^*S_x(f)^*}{H(f)H(f)^*S_x(f)S_x(f)} = 1. \tag{9.34}$$

Thus, if two signals are linearly related, their coherence is unity, and coherence becomes a good basis for determining the linear relationship of frequency components.

Because the phase angle is an important component of the CSD, the *complex coherence function* is also defined,

$$K_{yx}(f) = +\sqrt{K_{yx}^2(f)}\,\angle S_{yx}(f). \tag{9.35}$$

The importance of the phase angle is that it reflects the time shift between waveforms in the signals $y(n)$ and $x(n)$. The time shift comes forth in a direct manner from the situations concerning ranging in Section 9.3. It was shown that, even in a noisy and lossy environment, the correlation functions have the relationship

$$R_{xy}(k) = gR_x(k-d). \tag{9.26}$$

Taking the DTFT, then

$$S_{xy}(f) = gS_x(f)e^{-j2\pi fdT} \tag{9.36}$$

and the slope of the phase angle curve is proportional to the time delay $\tau = dT$. The attenuation factor is easily found by

$$g = \frac{|S_{xy}(f)|}{S_x(f)}. \tag{9.37}$$

As with the ordinary DTFT, the CSD has its real part that is an even function and its imaginary part that is an odd function. This can be shown by writing the CCF as a summation of even and odd functions,

$$\begin{aligned} \lambda_{yx}(k) &= \tfrac{1}{2}\big(R_{yx}(k) + R_{yx}(-k)\big), \\ \psi_{yx}(k) &= \tfrac{1}{2}\big(R_{yx}(k) - R_{yx}(-k)\big), \end{aligned} \tag{9.38}$$

and

$$R_{yx}(k) = \lambda_{yx}(k) + \psi_{yx}(k). \tag{9.39}$$

The CSD can then be expressed as

$$\begin{aligned} S_{yx}(f) &= \sum_{k=-\infty}^{\infty} \big(\lambda_{yx}(k) + \psi_{yx}(k)\big)e^{-j2\pi fkT} \\ &= \sum_{k=-\infty}^{\infty} \lambda_{yx}(k)e^{-j2\pi fkT} + \sum_{k=-\infty}^{\infty} \psi_{yx}(k)e^{-j2\pi fkT} \\ &= \Lambda_{yx}(f) + j\Psi_{yx}(f), \end{aligned} \tag{9.40}$$

where

$$\Lambda_{yx}(f) = \Re\big[S_{yx}(f)\big] \quad \text{and} \quad \Psi_{yx}(f) = \Im\big[S_{yx}(f)\big].$$

$\Lambda_{yx}(f)$ is called the *cospectrum* and $\Psi_{yx}(f)$ is called the *quadrature spectrum*. As can be anticipated, these form the *cross magnitude* $|S_{yx}(f)|$ and *cross phase* $\phi_{yx}(f)$ spectra, where

$$\begin{aligned} |S_{yx}(f)| &= \sqrt{\Lambda_{yx}^2(f) + \Psi_{yx}^2(f)}, \\ \phi_{yx}(f) &= \tan^{-1}\frac{\Psi_{yx}(f)}{\Lambda_{yx}(f)}. \end{aligned} \tag{9.41}$$

9.4.2 Properties of Cross-Spectral Estimators

9.4.2.1 Definition

There are many different estimators and properties of cross correlation and cross spectral density functions that can be studied. The ones being emphasized are those that are needed to test the independence between two time series and which lay a foundation for further study of system identification. The properties of the estimators for the CSD are derived exactly as the estimators for the PSD in Chapter 7. Simply substitute the signal $y(n)$ for the first of the two $x(n)$ signals in the equations. The estimator for the CSD is

$$\hat{S}_{yx}(f) = T \sum_{k=-M}^{M} \hat{R}_{yx}(k) e^{-j2\pi fkT}, \tag{9.42}$$

where the CSD is evaluated at frequencies $f = \pm m/2MT$, with $-M \leq m \leq M$. In terms of the frequency number this is

$$\hat{S}_{yx}(m) = T \sum_{k=-M}^{M} \hat{R}_{yx}(k) e^{-j2\pi mk/N}. \tag{9.43}$$

The analog to the periodogram is

$$\hat{S}_{yx}(m) = \frac{1}{NT} Y^*(m) X(m). \tag{9.44}$$

9.4.2.2 Mean and Variance for Uncorrelated Signals

The variance of the estimate of the CSD is a function of the ACFs of the component signals. The variance and distributional properties of the CSD of independent signals will be presented because they provide the basis for developing statistical tests for assessing the correlation between the signals in the frequency domain. Using the same spectral approach as in Chapter 7 for deriving the statistical properties of the estimators, the definition of the DFT is

$$\frac{X(m)}{\sqrt{NT}} = A(m) - jB(m),$$

with

$$A(m) = \sqrt{\frac{T}{N}} \sum_{n=0}^{N-1} x(n)\cos\left(\frac{2\pi mn}{N}\right),$$
$$B(m) = \sqrt{\frac{T}{N}} \sum_{n=0}^{N-1} x(n)\sin\left(\frac{2\pi mn}{N}\right). \tag{9.45}$$

With the parameters T and N being part of the real and imaginary components, then

$$\hat{S}_{yx}(m) = \frac{T}{N} Y^*(m) X(m)$$
$$= \left(A_y(m) + jB_y(m)\right)\left(A_x(m) - jB_x(m)\right). \tag{9.46}$$

Dropping the frequency number for simplicity in this explanation, the sample cospectrum and quadrature spectrum are

$$\hat{\Lambda}_{yx}(m) = (A_y A_x + B_y B_x),$$
$$\hat{\Psi}_{yx}(m) = (A_x B_y - A_y B_x). \tag{9.47}$$

It is known from Chapter 7 that, for Gaussian random processes, the real and imaginary components of the sample spectra are Gaussian random variables with a zero mean and variance equal to $S_y(m)/2$ or $S_x(m)/2$. If the two processes are uncorrelated, then

$$E\left[\hat{\Lambda}_{yx}(m)\right] = E\left[\hat{\Psi}_{yx}(m)\right] = 0. \tag{9.48}$$

The variance for the sample cospectrum is

$$\mathrm{Var}\left[\hat{\Lambda}_{yx}(m)\right] = E\left[A_y^2 A_x^2 + B_y^2 B_x^2 + 2A_y A_x B_y B_x\right]$$
$$= \left(E\left[A_y^2\right]E\left[A_x^2\right] + E\left[B_y^2\right]E\left[B_x^2\right]\right)$$
$$= \left(\frac{S_y(m)}{2}\frac{S_x(m)}{2} + \frac{S_y(m)}{2}\frac{S_x(m)}{2}\right)$$
$$= \frac{S_y(m) S_x(m)}{2}. \tag{9.49}$$

The variance for the sample quadrature spectrum is the same and the covariance between $\hat{\Lambda}_{yx}(m)$ and $\hat{\Psi}_{yx}(m)$ is zero.

The distribution of the magnitude of the sample CSD estimator is derived through its square,

$$|\hat{S}_{yx}(m)|^2 = \frac{T^2}{N^2} Y^*(m) X(m) Y(m) X^*(m)$$

$$= \hat{S}_y(m) \hat{S}_x(m). \tag{9.50}$$

Now introduce the random variable

$$\Gamma^2(m) = \frac{4|\hat{S}_{yx}(m)|^2}{S_y(m) S_x(m)} = \frac{2\hat{S}_y(m)}{S_y(m)} \frac{2\hat{S}_x(m)}{S_x(m)} = UV. \tag{9.51}$$

Knowing that each sample PSD has a chi-square distribution and that they are independent of each other, then

$$E[\Gamma^2(m)] = E[U]E[V] = 2 \cdot 2 = 4,$$

$$E[\Gamma^4(m)] = E[U^2]E[V^2] = 8 \cdot 8 = 64,$$

and

$$\mathrm{Var}[\Gamma^2(m)] = 48. \tag{9.52}$$

Using equations (9.51) and (9.52), the mean and variance of the squared sample CSD are found to be

$$E\left[|\hat{S}_{yx}(m)|^2\right] = S_y(m) S_x(m),$$

$$\mathrm{Var}\left[|\hat{S}_{yx}(m)|^2\right] = 3S_y^2(m) S_x^2(m). \tag{9.53}$$

The sample phase spectrum is

$$\hat{\phi}_{yx}(m) = \tan^{-1}\left(\frac{\hat{\Psi}_{yx}(m)}{\hat{\Lambda}_{yx}(m)}\right) = \tan^{-1}\left(\frac{A_y B_x - A_x B_y}{A_y A_x + B_y B_x}\right). \tag{9.54}$$

Because the terms A_i and B_i are independent Gaussian random variables ranging from $-\infty$ to ∞, the numerator and denominator terms are approximately Gaussian, independent, and possess the same variance. Thus, it can be stated that $\hat{\phi}_{yx}(m)$ has an approximately uniform distribution ranging between $-\pi/2$ and $\pi/2$ [Jenkins and Watts (1968)].

9.4.2.3 Adjustments for Correlated Signals

It can be seen that for two uncorrelated signals $\hat{S}_{yx}(m)$ is unbiased and inconsistent. Remember that this development is based on a large value of N. As is known from the study of PSD estimators, these CSD estimators also have a bias because the CCF is biased as shown in equation (9.13). The exception is when $y(n)$ and $x(n)$ are uncorrelated. Thus,

$$E\left[\hat{S}_{yx}(m)\right] = S_{yx}(m) * W(m), \qquad (9.55)$$

where $W(m)$ again represents the lag spectral window. The same procedures are used to reduce this inherent bias:

a. A lag window must be applied to the sample ACF when using equation (9.43).
b. Data windows must be applied to the acquired signals when using equation (9.44) and the power corrected for process loss.

The CSD is not used in general for the same reason that the CCVF is not used much in the time domain; it is unit sensitive. The normalized version of the CSD, the MSC, and the phase angle of the CSD are used in practice. Thus concentration will be upon estimating the coherence and phase spectra. Before developing the techniques, several applications will be briefly summarized in the next section.

9.5 Applications

Noise and vibration problems cause much concern in manufacturing. Sound intensity measurements are used to quantify the radiation of noise from vibrating structures and to help locate the noise generation sources. At one point on a vibrating structure

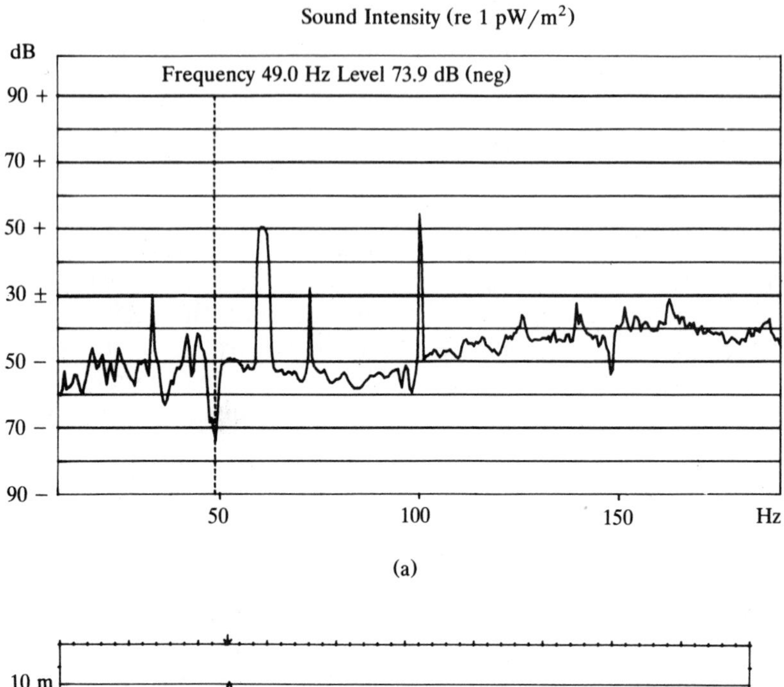

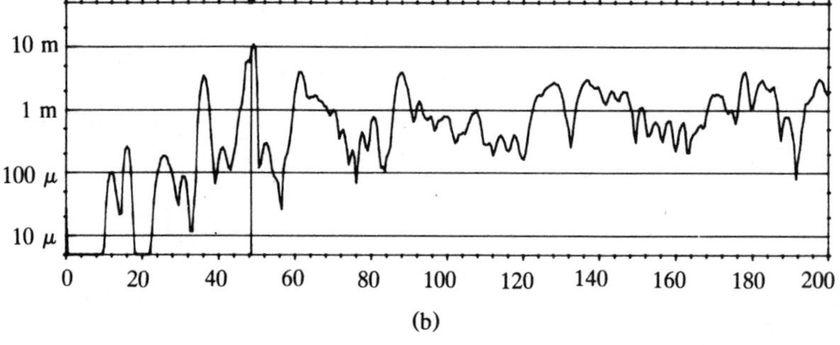

Figure 9.14 Noise measurements from a vibrating structure: (a) PSD of sound intensity; (b) CSD of vibration and sound intensity. [From Thrane and Gade (1988), Figures 1 and 2, with permission.]

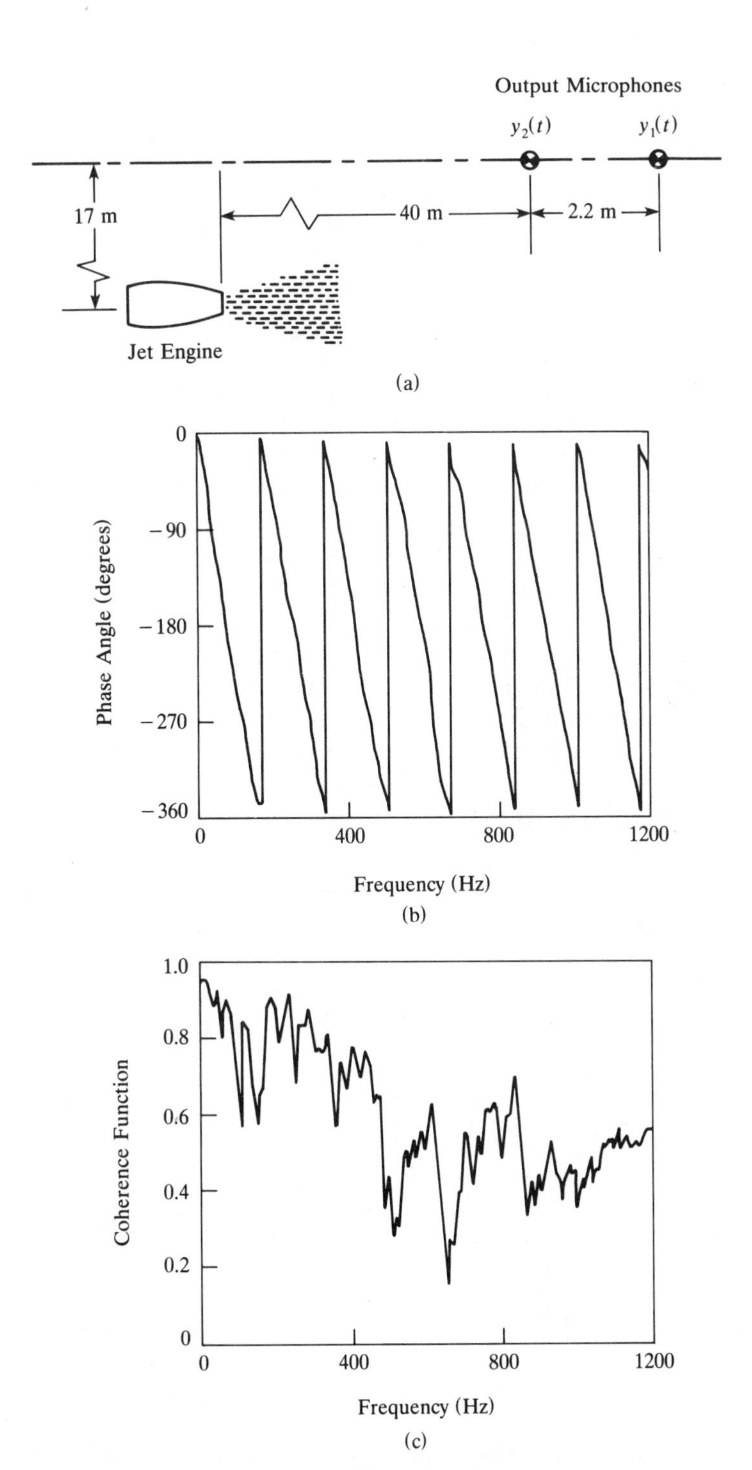

Figure 9.15 Jet exhaust sound measurements at two locations: (a) measurement configuration; (b) phase spectra; (c) coherence spectra. [From Bendat and Piersol (1980), Figures 7.5 and 7.6, with permission.]

the deflections were measured using accelerometers. The sound intensity was also measured at a place in space close to the point of vibration measurement. The PSD for the sound intensity is plotted in Figure 9.14(a). There are definite peaks at 49, 61, and 100 Hz. The negative peak at 49 Hz indicates a 180-degree phase shift with respect to the other frequencies. The CSD estimate is plotted in Figure 9.14(b). There are distinct peaks at 49 and 61 Hz but not at 100 Hz. This means that the surface is contributing to the noise at the lower frequencies but not at 100 Hz. Another part of the structure is creating this latter noise frequency [Thrane and Gade (1988)].

The study of the noise generated by a jet engine requires knowing if the noise is coherent or diffuse, and at what distances from the engine does any noise become diffuse. The coherence function will quantify the strength of synchronous noise between two positions. Figure 9.15(a) shows the configuration for measuring the noise from the exhaust of a jet engine. Measurements were made from both locations in order to average 256 estimates of the needed spectra. Figures 9.15(b) and 9.15(c) show the phase and coherence spectra (the bandwidth is 4 Hz). The shape of the phase spectrum is consistent with the distance between the two locations. The MSC shows that the noise is coherent from 0 to 500 Hz and

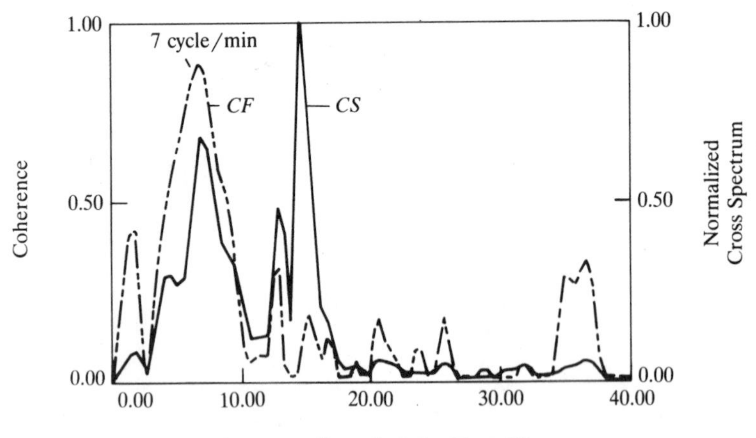

Figure 9.16 Coherence function (CF) and normalized cross spectral density function (CS) for the ECG in the colon. [From Reddy, Collins, and Daniel (1987), Figure 10b, with permission.]

higher-frequency components are diffuse [Bendat and Piersol (1980)].

Investigating how the colon moves food along the digestive tract involves coherence analysis. The muscle activity of the colon is measured at two sites that are 3 cm apart. The signals are sampled at 5 Hz and 30 segments of data lasting 51.2 s are acquired. The CSD and MSC are estimated and plotted in Figure 9.16. The CSD shows peaks at 7 and 15 cycle/min (cycles per minute, c/m in the figure), whereas the MSC has a major peak only at 7 cycle/min. The interpretation of the MSC is that the 7 cycle/min oscillation is propagated down the colon. The existence of the 15 cycle/min oscillation in the CSD means that it exists strongly at one of the sites but is not propagated [Reddy, Collins, and Daniel (1987)].

9.6 Tests for Correlation between Time Series

9.6.1 Coherence Estimators

The sample CSD provides the basis for a set of tests complementary to the CCF for testing the amount of correlation between two time series. The coherency, being a normalized magnitude, is a logical test criterion. However, if the sample squared coherency is considered, it provides no information because

$$\hat{K}_y^2(m) = \frac{Y^*(m)X(m)Y(m)X^*(m)}{Y^*(m)Y(m)X(m)X^*(m)} = 1. \tag{9.56}$$

Thus, let us examine the effect that spectral smoothing will have on the properties of this estimator. It was seen before that

$$E\left[\hat{R}_{yx}(k)\right] = \left(1 - \frac{|k|}{N}\right) R_{yx}(k), \tag{9.57}$$

which resulted in a biased CSD estimator. As in autospectral estimation, a lag window is applied to the CCF estimate and a smoothed estimate of the CSD is produced, with the mean being

$$E\left[\tilde{S}_{yx}(m)\right] = T \sum_{k=-M}^{M} w(k)\left(1 - \frac{|k|}{N}\right) R_{yx}(k) e^{-j2\pi mk/N}. \quad (9.58)$$

Remember the characteristics of the lag window dominate the effects of the data window, which results in a smoothed version of the theoretical CSD with

$$E\left[\tilde{S}_{yx}(m)\right] \approx S_{yx}(m) * W(m)$$

$$= \bar{S}_{yx}(m) = \bar{\Lambda}_{yx}(m) + j\bar{\Psi}_{yx}(m). \quad (9.59)$$

Extreme care must be exercised when using this approach. *If the value of M is not large enough* to encompass any significant CCF values, then equation (9.59) is highly and erroneously biased. To remove this effect, one must find the major significant time shift, using equation (9.58), and then *align the signals $y(n)$ and $x(n)$ and then perform the CSD estimation*. For PSD estimation, it has been found in general that the quality of the estimate was the same whether the ensemble averaging, correlational smoothing, or spectral smoothing was used to reduce the variance of the estimate. The same principle is true concerning CSD estimation, the important parameters being the spectral resolution and the number of degrees of freedom. From this point, a tilde is used to represent estimates obtained using any of the approaches to reduce the variance. It seems that the most-used smoothing procedure in CSD estimation is the Welch method (WOSA) with 50% overlap.

9.6.2 Statistical Properties of Estimators

The coherence spectrum, squared coherence spectrum, and phase spectrum are now estimated using these averaged, or smoothed, estimates; that is,

$$\tilde{K}_{yx}^2(m) = \frac{\tilde{\Lambda}_{yx}^2(m) + \tilde{\Psi}_{yx}^2(m)}{\tilde{S}_y(m)\tilde{S}_x(m)}, \quad (9.60)$$

$$\tilde{\phi}_{yx}(m) = \tan^{-1}\left(-\frac{\tilde{\Psi}_{yx}(m)}{\tilde{\Lambda}_{yx}(m)}\right). \quad (9.61)$$

As with the autospectral estimation, the window properties dictate the amount of variance reduction in the smoothing process. Recall that in the BT smoothing,

$$\operatorname{Var}\left[\tilde{S}(m)\right] = \frac{S^2(m)}{N} \sum_{k=-M}^{M} w^2(k) = \frac{S^2(m)}{\mathrm{VR}}, \tag{9.62}$$

where VR is the variance reduction factor. Also recall that, for ensemble averaging and spectral smoothing, VR equals the number of spectral values being averaged together. Similar results occur for estimating the cross spectral density functions. However, the derivation of their sampling distributions is very complex and voluminous and is reserved for advanced study. Detailed derivations can be found in Jenkins and Watts (1968) and Fuller (1976). Thankfully, however, the estimation techniques are similarly implementable and will be summarized and used. The variance for the smoothed coherency and squared coherency estimators are

$$\begin{aligned} \operatorname{Var}\left[|\tilde{K}_{yx}|\right] &= \frac{1}{2\,\mathrm{VR}}\left(1 - K_{yx}^2\right)^2, \\ \operatorname{Var}\left[\tilde{K}_{yx}^2\right] &= \frac{1}{2\,\mathrm{VR}} 4K_{yx}^2\left(1 - K_{yx}^2\right)^2. \end{aligned} \tag{9.63}$$

The variance of the smoothed phase estimator (in radians) is

$$\operatorname{Var}\left[\tilde{\phi}_{yx}(m)\right] = \frac{1}{2\,\mathrm{VR}}\left(\frac{1}{K_{yx}^2} - 1\right). \tag{9.64}$$

The covariance between the smoothed coherence and phase spectral estimators is approximately zero. It is important to notice that these variance expressions are dominated by two terms, VR and MSC. The important practical reality is that *the variance reduction factor is controllable whereas the coherence spectrum is not, which can defeat any averaging or smoothing efforts*.

Because the CSD is biased by the inherent spectral windows, likewise is the estimate of the MSC. Another source of bias exists because the signal component of one measurement is delayed with respect to the other measurement. It has been shown that

$$E\left[\tilde{K}_{yx}^2(m)\right] \approx \left(1 - \frac{\tau_d}{NT}\right) K_{yx}^2(m), \tag{9.65}$$

where τ_d is the time shift [Carter (1988)]. Simulation studies have demonstrated that this bias can be appreciable. Fortunately, this source of bias can be controlled by *aligning* both signals as the first step in the estimation procedure.

As with any estimators that are intricate, the sampling distributions are quite complicated. This is certainly true for $\tilde{K}_{yx}^2(m)$ and $\tilde{\phi}_{yx}^2(m)$ [Carter (1988)]. The distributions depend upon whether $K_{yx}^2(m)$ is zero. If $K_{yx}^2(m) = 0$, then $\tilde{\phi}_{yx}(m)$ is uniformly distributed on the interval $(-\pi/2, \pi/2)$ and $\tilde{K}_{yx}^2(m)$ has an F distribution. If $K_{yx}^2(m) \neq 0$, then $\tilde{\phi}_{yx}(m)$ converges to a normally distributed random variable and $\tilde{K}_{yx}^2(m)$ converges to 1 for a multiple correlation coefficient. Several variable transformations have been developed so that only one test is needed for each estimator. These will be reviewed in Section 9.6.3. If the number of degrees of freedom is large the sample coherence, complex coherence, and phase become unbiased estimators with normal distributions and the variances stated above.

9.6.3 Confidence Limits

It has been recognized that the variance of the smoothed coherence estimator is identical to the variance of an ordinary correlation coefficient. Hence, the Fisher z transformation can be applied and the estimator becomes

$$\tilde{Z}_{yx}(m) = \tanh^{-1}\left(|\tilde{K}_{yx}(m)|\right) = \frac{1}{2}\ln\left(\frac{1 + |\tilde{K}_{yx}(m)|}{1 - |\tilde{K}_{yx}(m)|}\right). \tag{9.66}$$

The function $\tilde{Z}_{yx}(m)$ is a biased and consistent estimator with

$$E\left[\tilde{Z}_{yx}(m)\right] = \tanh^{-1}\left(|K_{yx}(m)|\right) + \frac{1}{\nu - 2},$$

$$\text{bias} = b = \frac{1}{\nu - 2}, \tag{9.67}$$

$$\text{Var}\left[\tilde{Z}_{yx}(m)\right] = \frac{1}{\nu - 2},$$

where ν is the number of degrees of freedom of the variance reduction process used, $\nu = 2B_e NT$ for a smoothing procedure. This transformation is valid for $\nu \geq 20$ and $0.3 \leq K_{yx}^2(m) \leq 0.98$. Empirical improvements have been developed that enable statistical testing for the entire range of actual coherence and lower degrees of freedom. A correction factor for bias in the coherence domain is

$$B = \frac{1}{2\nu}\left(1 - \tilde{K}_{yx}^2(m)\right), \tag{9.68}$$

so that in equation (9.66) the estimate for the coherence function should be replaced by

$$\tilde{K}_{yx}^2(m) \Rightarrow \tilde{K}_{yx}^2(m) - \frac{1}{2\nu}\left(1 - \tilde{K}_{yx}^2(m)\right). \tag{9.69}$$

The actual bias of the transformed variable does not change. A variance correction (VC) is used in the z domain,

$$\text{VC} = 1 - 0.004^{1.6\tilde{K}^2 + 0.22}, \tag{9.70}$$

so that the new variance of $\tilde{Z}_{yx}(m)$ is

$$\text{Var}\left[\tilde{Z}_{yx}(m)\right] = \text{VC}\frac{1}{\nu - 2}. \tag{9.71}$$

The confidence intervals are established by assuming that the transformed estimator has a Gaussian distribution. Then, for an estimated transformed spectrum $\tilde{Z}_{yx}(m)$, the confidence limits are

$$\tilde{Z}_{yx}(m) - b \pm \Xi\left(\frac{1 - \alpha}{2}\right)\sqrt{\frac{\text{VC}}{\nu - 2}}, \tag{9.72}$$

where $\Xi(1 - \alpha/2)$ indicates the value of the $N(0, 1)$ random variable for the probability of $1 - \alpha/2$. The confidence limits for the coherence function are established by making the hyperbolic tangent transformation of the limits in equation (9.72) [Otnes and Enochson (1972)].

A usable sample distribution of the phase spectrum estimator is difficult to derive because an accurate approximation is unwieldy. A transformation is used to make the resulting variable approximately Gaussian. A tangent transformation is used and

$$\tilde{\theta}_{yx}(m) = \tan\left(\tilde{\phi}_{yx}(m)\right), \tag{9.73}$$

which produces a variance

$$\operatorname{Var}\left[\tilde{\theta}_{yx}(m)\right] = \sigma_{\theta}^{2} \approx \sec^{4}\left(\phi_{yx}(m)\frac{1}{2\,\mathrm{VR}}\left(\frac{1}{K_{yx}^{2}(m)} - 1\right)\right). \tag{9.74}$$

The confidence limits for a $(1 - \alpha)$ confidence level are

$$\tilde{\theta}_{yx}(m) \pm \Xi(1 - \alpha/2)\sigma_{\theta}. \tag{9.75}$$

Because the components of the actual complex coherence spectrum needed in equation (9.74) are unknown, they must be replaced by their estimates. However, because $\tilde{\phi}_{yx}(m)$ and $\phi_{yx}(m)$ are independent, it is expected that when the limits in equation (9.75) are transformed back into angle units that they will be independent of the actual phase angle [Jenkins and Watts (1968)].

Example 9.3

The coherence spectrum in Figure 9.16 shows the synchronous oscillation at 7 cycle/min of two locations in a colon. The confidence limits for this peak magnitude will be calculated. The signals were acquired at 5 Hz and the spectra were estimated using segment averaging with $K = 30$ and segment lengths being 51.2 s long. The frequency number of the peak is $m = 53$ and $\tilde{K}_{yx}^{2}(m) = 0.9$. The z transformation is

$$\tilde{Z}_{yx}(m) = \frac{1}{2}\ln\left(\frac{1 + 0.95}{1 - 0.95}\right) = 1.818,$$

with $b = 1/(\nu - 2) = 1/(60 - 2) = 0.0172$. The bias correction is

$$B = \frac{1}{2\nu}\left(1 - \tilde{\phi}_{yx}^2(m)\right) = \frac{1}{120}(1 - 0.9) = 0.00083.$$

The corrected z value is

$$\tilde{Z}_{yx}(m) = \frac{1}{2}\ln\left(\frac{1 + 0.944}{1 - 0.944}\right) = 1.776.$$

The variance correction factor is

$$\text{VC} = 1 - 0.004^{1.6k^2 + 0.22} = 0.99993$$

and the corrected variance is

$$\text{Var}\left[\tilde{Z}_{yx}(m)\right] = \frac{0.99993}{60} = 0.01667.$$

The 95% confidence limits are

$$\tilde{Z}_{yx}(m) - b \pm \Xi(0.975)\sqrt{0.01667}$$

$$= 1.75876 \pm 0.12588 = 1.88464, 1.63287.$$

In the coherence domain the limits become

$$K_{\text{UL}} = \tanh(1.88464) = 0.95490,$$

$$K_{\text{LL}} = \tanh(1.63287) = 0.92647.$$

The magnitude estimated was 0.94868. Thus, the estimation procedure is quite good as the limits are close to $\tilde{K}_{yx}(53)$.

9.6.4 Procedure for Estimation

In summary, these following steps must be performed to implement the concepts and procedures for estimating cross magnitude and phase spectra.

Correlation Approach

1. *Detrending.* Both signals must be examined for trends and detrended. These trends will contribute to artifactual low-frequency components in all the magnitude spectra.
2. *Correlation functions.* Compute the autocorrelation and cross correlation function estimates. Look for the minimum number of lags needed for smoothing the autocorrelation estimates and conspicuous lags between the two signals. If a dominant lag exists, align the two signals and store the lag time for a later step. Recalculate the cross correlation estimate.
3. *Spectra.* Estimate the autospectrum, cross-spectrum, and the coherence spectrum for several values of bandwidths (maximum correlation lags). Perform the window closing, that is, look for the minimum lag that causes convergence of the estimates.
4. *Estimation.* Compensate the phase spectra for alignment and transform the coherence and phase spectra into their testing variables. Calculate the confidence limits and inverse-transform the spectra and their associated confidence limits to the coherence and phase spectra. Interpret these spectra according to the hypotheses necessary.

Direct Spectral Approach

1. *Detrending.* Perform as previously described.
2. *Spectra.* Estimate the autospectrum and cross-spectrum using the periodogram approach. Smooth or average as necessary to reduce the variance. If segment averaging is used, one must be aware that the segments must be long enough to encompass any delayed phenomena. Perhaps a signal alignment will also be necessary.
3. *Estimation.* Perform as previously described.

9.6.5 Application

The study of human balance has become very important because of the effect of drugs and aging on the control of balance. An effective and valid analytic procedure is an important compo-

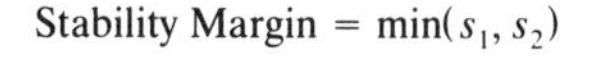

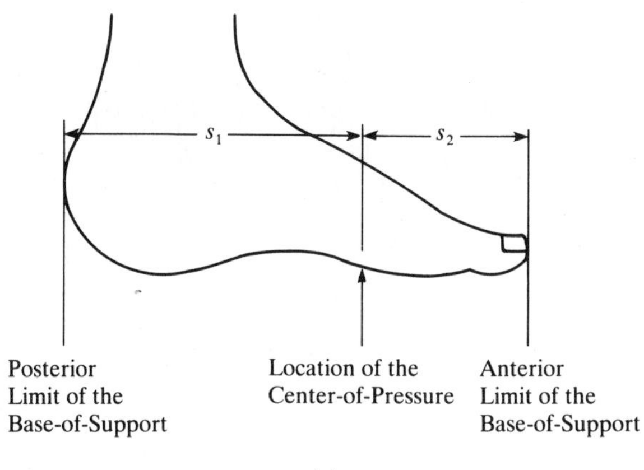

(a)

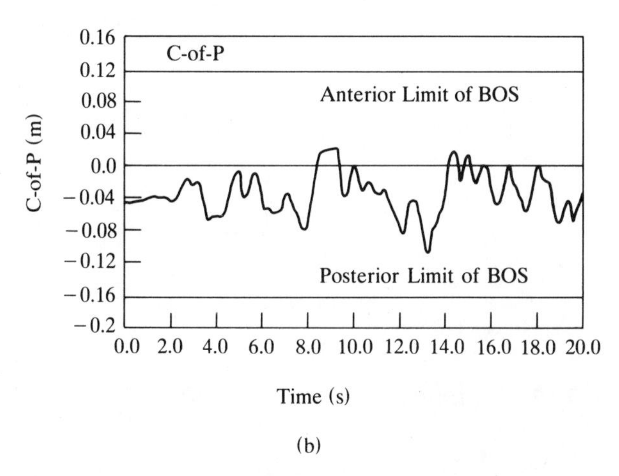

(b)

Figure 9.17 Investigation of balance and stability: (a) schematic of foot and position of center of pressure: (b) sample function of the movement of center of pressure; (c) sample function of the acceleration of the floor; (d) estimate of the coherence function. [From Maki, Holliday, and Fermie (1987, IEEE), Figures 1, 4, and 5, with permission.]

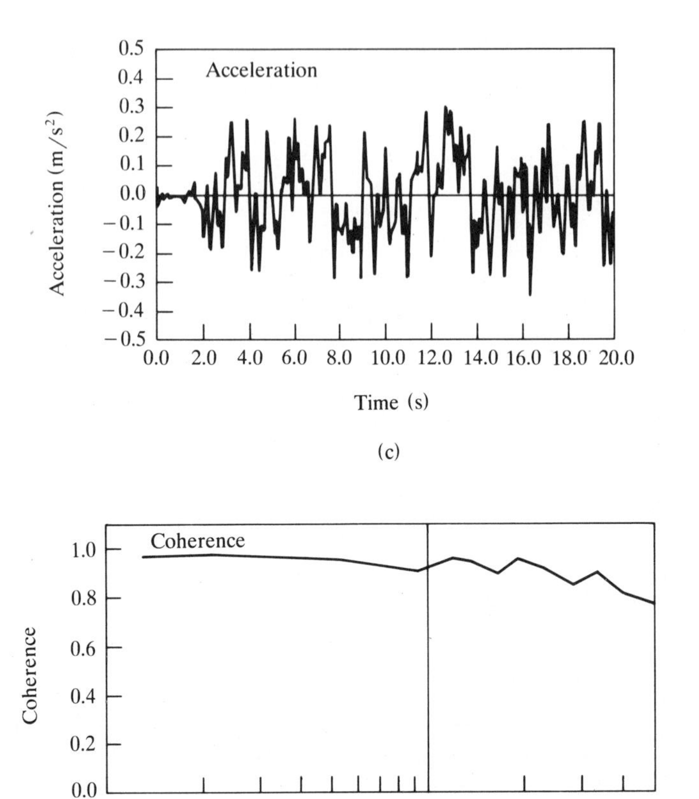

Figure 9.17 (*Continued*).

nent of this study. The coherence function was used to assess if the results of a certain experimental procedure could be analyzed using linear system relationships.

The rationale is explained using Figure 9.17(a). As a person stands, the resulting force on the floor is located within the boundary of the foot and is called the center of pressure. It will move as the body is perturbed. The body is perturbed by moving the floor slightly and recording the acceleration of the floor and the location of the center of pressure. The floor only moves the foot forward and backward. A sample of the recordings of the movement of the center of pressure and the acceleration are plotted in Figures 9.17(b) and 9.17(c). The signals were recorded for 185 s with $T = 0.06$ s. The MSC is estimated with the WOSA method. Each segment contained 256 points and a Hamming window with 50% overlap was used. The resulting $\tilde{K}_{yx}^2(m)$ is plotted in Figure 9.17(d). Appreciate that almost all magnitudes of the MSC have a value of 1. Thus, a linear systems approach for signal analysis seems valid.

References

Bendat, J. and A. Piersol (1980). *Engineering Applications of Correlation and Spectral Analysis*. John Wiley and Sons, New York.

Bendat, J. and A. Piersol (1986). *Random Data, Analysis and Measurement Procedures*. John Wiley and Sons, New York.

Box, G. and G. Jenkins (1976). *Time Series Analysis, Forecasting and Control*. Holden-Day, Oakland, CA.

Carter, G. (1988). Coherence and time delay estimation. In *Signal Processing Handbook* (C. Chen, ed.). Marcel Dekker, New York.

Chen, C. (1982). Sonar signal processing. In *Digital Waveform Processing and Recognition* (C. Chen, ed.). CRC Press, Boca Raton, FL.

de Coulon, F. (1986). *Signal Theory and Processing*. Artech House, Dedham, MA.

El-Hawary, F. (1988). Marine geophysical signal processing. In *Signal Processing Handbook* (C. Chen, ed.). Marcel Dekker, New York.

Fuller, W. (1976). *Introduction to Statistical Time Series*. John Wiley and Sons, New York.

IEEE (1981). Special issue on time delay estimation. *IEEE Trans. Acoust., Speech, Signal Process.* Vol. 29, June.

Jenkins, G. and D. Watts (1968). *Spectral Analysis and Its Applications.* Holden-Day, San Francisco.

Maki, B., P. Holliday, and G. Fermie (1987). A posture control model and balance test for the prediction of relative postural stability. *IEEE Trans. Biomed. Engrg.* 34:797–810.

Nagel, J. (1984). Progresses in fetal monitoring by improved data acquisition. *IEEE Engineering in Medicine and Biology Magazine* 3(3):9–13.

Newton, H. (1988). *TIMESLAB: A Time Series Analysis Laboratory.* Wadsworth and Brooks/Cole, Pacific Grove, CA.

Otnes, R. and L. Enochson (1972). *Digital Time Series Analysis.* John Wiley and Sons, New York.

Reddy, S., S. Collins, and E. Daniel (1987). Frequency analysis of gut EMG. *CRC Critical Reviews in Biomed. Engrg* 15(2):95–116.

Schwartz, M. and L. Shaw (1975). *Signal Processing: Discrete Spectral Analysis, Detection, and Estimation.* McGraw-Hill, New York.

Silvia, M. (1987). Time delay estimation. In *Handbook of Digital Signal Processing—Engineering Applications* (D. Elliott, ed.). Academic Press, New York.

Thrane, N. and S. Gade (1988). Use of operational deflection shapes for noise control of discrete tones. *Brüel & Kjaer Technical Review* No. 1.

Exercises

9.1 Prove that $C_{yx}^2(k) \leq C_{yy}(0)C_{xx}(0)$.

9.2 Derive the inequality $|R_{yx}(k)| \leq \frac{1}{2}(R_{yy}(0) + R_{xx}(0))$. [*Hint*: Start with equation (9.5) and make one minor change and let $a = 1$.]

9.3 Equation (9.17) states the variance of the cross covariance estimator for two independent first-order AR processes. Derive the expression.

9.4 What happens to the bias in the estimate of the CCVF after the signals are prewhitened?

9.5 Derive the cross correlation function [equation (9.23)] for the time difference of arrival situation. What is the TDOA? Knowing the distance between the sensors and the speed of sound in water (1,500 m/s), estimate the radial position of the source θ as shown in Figure 9.12(b).

9.6 A single receiver is used to track a source using passive sonar and an autocorrelation technique. The acoustic signal received is a multipath one whose paths are sketched in Figure E9.6. The multiple-path reflections can be used to estimate the depth h_s of the source. Assume that the source produces pulses of sound with a rectangular wave-shape of duration 2 ms. What are the constraints on the

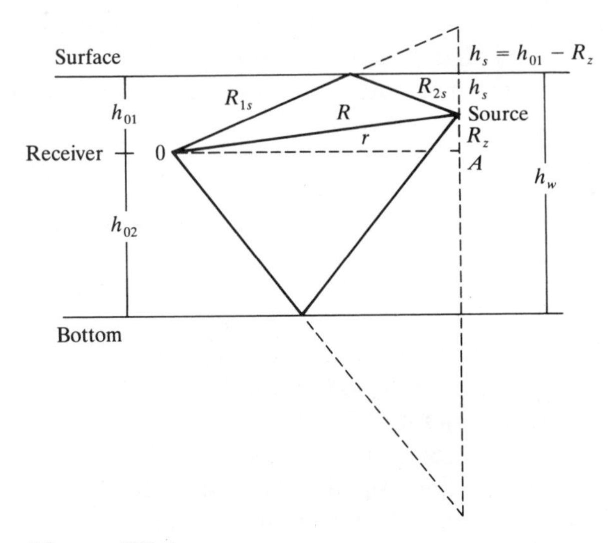

Figure E9.6

path lengths so that the multiple reflections produce distinct pulses in the ACF of the received signal?

9.7 Prove the relationship $|S_{yx}(m)|^2 \le S_{yy}(m)S_{xx}(m)$. Start with the relationship

$$\left(\frac{Y^*(m)}{\sqrt{S_{yy}(m)}} - \frac{X(m)}{\sqrt{S_{xx}(m)}} \right)^2 \ge 0$$

and take expectations.

9.8 Prove that the variance of the sample quadrature spectrum for two independent white noise signals is $\sigma_y^2\sigma_x^2/2$.

9.9 Prove that the covariance between the sample cospectrum and quadrature spectrum for any uncorrelated signals is zero. One can use the same information used in the derivation of the variance of the sample cospectrum.

9.10 For the coherence function in Figure 9.17, find the confidence limits at 1.0 Hz.

Computer Exercises

9.11 The bivariate signals relating average temperatures and birthrate are listed in Appendix 9.2.

a. Estimate the normalized cross correlation function between these two signals.
b. Determine AR models for these signals. What are their parameters?
c. Prewhiten these signals and estimate the NCCF between the noise signals.
d. What does $\hat{\rho}_{yx}(k)$ indicate? Is there any difference between this function and that in part a?

9.12 This is an exercise to practice detecting the delay of a random signal in a noisy environment.

a. Generate 100 points of a Gaussian AR(1) process with a variance of 10; $a(1) = 0.5$. This is $x(n)$.
b. Form three other signals by doing the following: (1) delaying $x(n)$ by 10 time units and attenuating it by a factor of 2; (2) adding uniform white noise with variances

of 1, 2 and 5 to the process in step 1, that is,

$$y_i(n) = 0.5x(n-10) + \eta_i(n), \qquad i = 1,2,3.$$

c. Estimate the cross correlation functions between $x(n)$ and each of the $y_i(n)$. How do they differ and why?

Appendices

Appendix 9.1 Bivariate Gas Furnace Data

t	X_t*	Y_t†	t	X_t*	Y_t†
1	−0.109	53.8	32	1.767	47.9
2	0.000	53.6	33	1.608	47.6
3	0.178	53.5	34	1.265	47.5
4	0.339	53.5	35	0.790	47.5
5	0.373	53.4	36	0.360	47.6
6	0.441	53.1	37	0.115	48.1
7	0.461	52.7	38	0.088	49.0
8	0.348	52.4	39	0.331	50.0
9	0.127	52.2	40	0.645	51.1
10	−0.180	52.0	41	0.960	51.8
11	−0.588	52.0	42	1.409	51.9
12	−1.055	52.4	43	2.670	51.7
13	−1.421	53.0	44	2.834	51.2
14	−1.520	54.0	45	2.812	50.0
15	−1.302	54.9	46	2.483	48.3
16	−0.814	56.0	47	1.929	47.0
17	−0.475	56.8	48	1.485	45.8
18	−0.193	56.8	49	1.214	45.6
19	0.088	56.4	50	1.239	46.0
20	0.435	55.7	51	1.608	46.9
21	0.771	55.0	52	1.905	47.8
22	0.866	54.3	53	2.023	48.2
23	0.875	53.2	54	1.815	48.3
24	0.891	52.3	55	0.535	47.9
25	0.987	51.6	56	0.122	47.2
26	1.263	51.2	57	0.009	47.2
27	1.775	50.8	58	0.164	48.1
28	1.976	50.5	59	0.671	49.4
29	1.934	50.0	60	1.019	50.6
30	1.866	49.2	61	1.146	51.5
31	1.832	48.4	62	1.155	51.6

Appendix 9.1 Bivariate Gas Furnace Data (cont'd)

t	X_l^*	$Y_i^\dagger$	t	X_l^*	$Y_i^\dagger$
63	1.112	51.2	105	−0.255	53.4
64	1.121	50.5	106	−0.229	53.6
65	1.223	50.1	107	−0.007	53.7
66	1.257	49.8	108	0.254	53.8
67	1.157	49.6	109	0.330	53.8
68	0.913	49.4	110	0.102	53.8
69	0.620	49.3	111	−0.423	53.3
70	0.255	49.2	112	−1.139	53.0
71	−0.280	49.3	113	−2.275	52.9
72	−1.080	49.7	114	−2.594	53.4
73	−1.551	50.3	115	−2.716	54.6
74	−1.799	51.3	116	−2.510	56.4
75	−1.825	52.8	117	−1.790	58.0
76	−1.456	54.4	118	−1.346	59.4
77	−0.944	56.0	119	−1.081	60.2
78	−0.570	56.9	120	−0.910	60.0
79	−0.431	57.5	121	−0.876	59.4
80	−0.577	57.3	122	−0.885	58.4
81	−0.960	56.6	123	−0.800	57.6
82	−1.616	56.0	124	−0.544	56.9
83	−1.875	55.4	125	−0.416	56.4
84	−1.891	55.4	126	−0.271	56.0
85	−1.746	56.4	127	0.000	55.7
86	−1.474	57.2	128	0.403	55.3
87	−1.201	58.0	129	0.841	55.0
88	−0.927	58.4	130	1.285	54.4
89	−0.524	58.4	131	1.607	53.7
90	0.040	58.1	132	1.746	52.8
91	0.788	57.7	133	1.683	51.6
92	0.943	57.0	134	1.485	50.6
93	0.930	56.0	135	0.993	49.4
94	1.006	54.7	136	0.648	48.8
95	1.137	53.2	137	0.577	48.5
96	1.198	52.1	138	0.577	48.7
97	1.054	51.6	139	0.632	49.2
98	0.595	51.0	140	0.747	49.8
99	−0.080	50.5	141	0.900	50.4
100	−0.314	50.4	142	0.993	50.7
101	−0.288	51.0	143	0.968	50.9
102	−0.153	51.8	144	0.790	50.7
103	−0.109	52.4	145	0.399	50.5
104	−0.187	53.0	146	−0.161	50.4

Appendix 9.1 Bivariate Gas Furnace Data (cont'd)

t	X_l^*	$Y_i^\dagger$	t	X_l^*	$Y_i^\dagger$
147	−0.553	50.2	191	0.782	53.6
148	−0.603	50.4	192	0.858	53.6
149	−0.424	51.2	193	0.918	53.2
150	−0.194	52.3	194	0.862	52.5
151	−0.049	53.2	195	0.416	52.0
152	0.060	53.9	196	−0.336	51.4
153	0.161	54.1	197	−0.959	51.0
154	0.301	54.0	198	−1.813	50.9
155	0.517	53.6	199	−2.378	52.4
156	0.566	53.2	200	−2.499	53.5
157	0.560	53.0	201	−2.473	55.6
158	0.573	52.8	202	−2.330	58.0
159	0.592	52.3	203	−2.053	59.5
160	0.671	51.9	204	−1.739	60.0
161	0.933	51.6	205	−1.261	60.4
162	1.337	51.6	206	−0.569	60.5
163	1.460	51.4	207	−0.137	60.2
164	1.353	51.2	208	−0.024	59.7
165	0.772	50.7	209	−0.050	59.0
166	0.218	50.0	210	−0.135	57.6
167	−0.237	49.4	211	−0.276	56.4
168	−0.714	49.3	212	−0.534	55.2
169	−1.099	49.7	213	−0.871	54.5
170	−1.269	50.6	214	−1.243	54.1
171	−1.175	51.8	215	−1.439	54.1
172	−0.676	53.0	216	−1.422	54.4
173	0.033	54.0	217	−1.175	55.5
174	0.556	55.3	218	−0.813	56.2
175	0.643	55.9	219	−0.634	57.0
176	0.484	55.9	220	−0.582	57.3
177	0.109	54.6	221	−0.625	57.4
178	−0.310	53.5	222	−0.713	57.0
179	−0.697	52.4	223	−0.848	56.4
180	−1.047	52.1	224	−1.039	55.9
181	−1.218	52.3	225	−1.346	55.5
182	−1.183	53.0	226	−1.628	55.3
183	−0.873	53.8	227	−1.619	55.2
184	−0.336	54.6	228	−1.149	55.4
185	0.063	55.4	229	−0.488	56.0
186	0.084	55.9	230	−0.160	56.5
187	0.000	55.9	231	−0.007	57.1
188	0.001	55.2	232	−0.092	57.3
189	0.209	54.4	233	−0.620	56.8

Appendix 9.1 Bivariate Gas Furnace Data (cont'd)

t	X_l*	Y_i†	t	X_l*	Y_i†
234	−1.086	55.6	266	0.866	54.0
235	−1.525	55.0	267	0.527	55.1
236	−1.858	54.1	268	0.093	54.5
237	−2.029	54.3	269	−0.458	52.8
238	−2.024	55.3	270	−0.748	51.4
239	−1.961	56.4	271	−0.947	50.8
240	−1.952	57.2	272	−1.029	51.2
241	−1.794	57.8	273	−0.928	52.0
242	−1.302	58.3	274	−0.645	52.8
243	−1.030	58.6	275	−0.424	53.8
244	−0.918	58.8	276	−0.276	54.5
245	−0.798	58.8	277	−0.158	54.9
246	−0.867	58.6	278	−0.033	54.9
247	−1.047	58.0	279	0.102	54.8
248	−1.123	57.4	280	0.251	54.4
249	−0.876	57.0	281	0.280	53.7
250	−0.395	56.4	282	0.000	53.3
251	0.185	56.3	283	−0.493	52.8
252	0.662	56.4	284	−0.759	52.6
253	0.709	56.4	285	−0.824	52.6
254	0.605	56.0	286	−0.740	53.0
255	0.501	55.2	287	−0.528	54.3
256	0.603	54.0	288	−0.204	56.0
257	0.943	53.0	289	0.034	57.0
258	1.223	52.0	290	0.204	58.0
259	1.249	51.6	291	0.253	58.6
260	0.824	51.6	292	0.195	58.5
261	0.102	51.1	293	0.131	58.3
262	0.025	50.4	294	0.017	57.8
263	0.382	50.0	295	−0.182	57.3
264	0.922	50.0	296	−0.262	57.0
265	1.032	52.0			

*X, 0.60–0.04 (input gas rate in ft^3 min).†Y, %CO_2 in outlet gas. Sampling interval 9 s and $N = 296$ pairs of data points.

Appendix 9.2 Bivariate Temperature and Birth Data

Temperatures and Births in New York City*

NYC Monthly Temperature 1946–59

11.506	11.022	14.405	14.442	16.524	17.918	18.959	18.309
18.160	16.691	14.480	17.862	12.082	10.558	12.138	14.442
16.152	17.714	19.015	19.164	17.900	16.933	13.364	11.468
10.000	10.985	12.993	14.480	16.134	17.862	19.238	19.089
18.067	15.576	14.926	12.398	12.435	12.416	13.067	15.093
16.766	18.680	19.796	19.424	17.398	16.877	13.866	12.658
13.048	11.301	12.063	14.164	16.041	18.067	19.108	18.736
17.212	16.394	13.922	11.691	11.970	11.970	12.900	14.888
16.747	17.955	19.201	18.903	17.714	16.097	13.234	12.435
11.952	11.970	12.639	15.204	16.283	18.699	19.851	18.959
18.030	15.316	14.126	12.323	12.230	12.230	13.234	14.647
16.729	18.494	19.312	19.164	18.011	16.338	14.368	12.751
11.022	12.658	12.844	15.037	16.190	18.253	19.182	18.513
17.621	16.506	13.717	11.840	10.855	11.636	12.825	15.019
16.952	17.937	20.000	19.535	17.770	16.283	13.327	10.818
11.264	12.026	12.230	14.126	15.818	18.420	18.699	18.903
17.175	16.041	13.866	12.844	10.613	12.082	12.955	15.000
16.747	18.792	19.312	18.680	18.067	15.632	14.349	12.732
11.152	10.372	12.639	14.981	16.097	17.602	19.331	18.978
17.695	15.539	14.182	10.781	41.059	11.152	12.621	14.981
17.100	18.160	19.015	19.331	18.346	16.115	13.532	12.398

NYC Monthly Births 1946–1959

26.663	23.598	26.931	24.740	15.806	24.364	24.477	23.901
23.175	23.227	21.672	21.870	21.439	21.089	23.709	21.669
21.752	20.761	23.479	23.824	23.105	23.110	21.759	22.073
21.937	20.035	23.590	21.672	22.222	22.123	23.950	23.504
22.238	23.142	21.059	21.573	21.548	20.000	22.424	20.615
21.761	22.874	24.104	23.748	23.262	22.907	21.519	22.025
22.604	20.894	24.677	23.673	25.320	23.583	24.671	24.454
24.122	24.252	22.084	22.991	23.287	23.049	25.076	24.037
24.430	24.667	26.451	25.618	25.014	25.110	22.964	23.981
23.798	22.270	24.775	22.646	23.988	24.737	26.276	25.816
25.210	25.199	23.162	24.707	24.364	22.644	25.565	24.062
25.431	24.635	27.009	26.606	26.268	26.462	25.246	25.180
24.657	23.304	26.982	26.199	27.210	26.122	26.706	26.878
26.152	26.379	24.712	25.688	24.990	24.239	26.721	23.475
24.767	26.219	28.361	28.599	27.914	27.784	25.693	26.881
26.217	24.218	27.914	26.975	28.527	27.139	28.982	28.169
28.056	29.136	26.291	26.987	26.589	24.848	27.543	26.896
28.878	27.390	28.065	28.141	29.048	28.484	26.634	27.735
27.132	24.924	28.963	26.589	27.931	28.009	29.229	28.759
28.405	27.945	25.912	26.619	26.076	25.286	27.660	25.951
26.398	25.565	28.865	30.000	29.261	29.012	26.992	27.897

*Total 168 observations (read row-wise).

Index

aliasing 74–75
applications
  acoustics 411, 416–418
  bioelectric signals 2, 4, 7–9, 48, 70, 101, 182, 187, 267, 309, 343, 358, 430–431, 436
  counting process 129
  detection 414
  electrical 8–9, 103, 127, 158, 323, 378, 380
  imaging 5–6
  industrial process 200, 252, 309, 400–401, 408
  natural sciences 1, 10, 32, 193, 217, 269, 361, 420
  physical sciences 21–22, 33, 41, 50, 69, 128, 160, 179
  physiology 19, 20, 21, 38–40, 162, 342, 344, 439–440
  point process 6–7
  speech 1–3, 182, 253, 257, 383, 385
  vibration and noise 69, 102, 267, 429–430
autocorrelation function
  definition 184, 194, 235, 244
  estimation 192
  signal modeling 244–249, 258–261
  system relationship 239–241
autocorrelation matrix 354–355
autocovariance function
  definition 183, 194
  estimation 192–198, 228–230
  properties 188
autoregressive moving average process 238, 252–253, 343–344
autoregressive process 216, 238, 243–252
autoregressive signal modeling
  See parametric signal modeling
averages
  frequency domain 288–303
  time domain 14–15
axioms of probability 128–130

bandwidth 305, 318–320
bias 162–163, 189, 195
bivariate distribution
  moments 141–145
  properties 139–141
Blackman–Tukey method
  bias and variance 316–320
  confidence limits 320–326

Blackman–Tukey method (*cont'd.*)
estimation 270–271, 311–316
variance 336–338
Burg algorithm 369–373

coherence function
definition 421–423
estimation
alignment 434
confidence limits 434–435
estimator and statistical properties 431–434
conditional probability 140–143
consistency 163, 189, 195
convolution 81–82, 236
correlation
definition 141–142
estimation 151–154
random process 210–217
cospectrum 423–425
covariance 141, 183
cross correlation function 239, 400–403
cross covariance function
definition 400
estimation 403–405
cross magnitude spectrum 423
cross phase spectrum
definition 423, 426–427
estimation
confidence limits 436–438
estimator 432–434
cross spectral density function
definition 240, 421
estimation 424–427
curve fitting
error, definition 23–24, 29
error, minimization 26–28
model order 28–30
normal equations 27

degrees of freedom 150
dependence 142, 184, 188, 199
deterministic signal 11–12
detrending 98–99, 303, 351
discrete Fourier transform 75–100
detrending 98–99
frequency resolution 95–98
frequency spacing 75, 83, 86
leakage error 89–93, 98
periodic repetition 74–80
process loss 94
properties 77–79
theorems 79–82
windows 88–94
zero padding 86–87

energy 15, 233
ensemble 178–180, 183, 188–191
ergodicity 188–192, 196
error function 137–138, 172
expectation 134–136
extrapolation 41

fast Fourier transform 82, 87
Fejer's kernel 273–274
Fourier series 10, 72, 115–120
Fourier transform
continuous time 71–73, 120–124
discrete 75–100
implementation 82–100
properties 77–79
theorem 79–82
discrete time 73–75, 233–235
frequency resolution 95–98
frequency spacing 75, 83, 86

histogram 128, 157–162, 192

independence 141–142
interpolation 20, 40–55, 86

Lagrange polynomials 43–48
leakage error 89–93, 98, 289
Levinson–Durbin algorithm 363–369, 392–395
linear prediction
definition 345, 348
parameter estimation 369–372, 391–392, 395–397
linearity 79
linearization, polynomial 30–34

maximum entropy method 384
model order, polynomial 28–30
model order, signal 356–362
 Akaike's information criteria 359–362
 final prediction error 359–362
moments of random variable
 mean and variance 134–136
  confidence limits 148–151
  estimators 146–148
 correlation 142
  confidence limits 153–155
  estimator 152
moving average process 210–217, 238, 241–243

normal equations 27
normalized autocovariance function
 definition 185, 194
 estimation 198–201
 signal modeling 246–251
normalized cross covariance function
 definition 401
 estimation 403, 405–410, 420–421

orthogonal functions 34–40, 65–70, 115–120

parametric signal modeling
 definition 343–344
 first order model 345–361
 general model 353–363, 373–374
 See linear prediction
 See model order, signal
Parseval's theorem 37, 234
partial correlation 363
periodic repetition 74–80
periodicity 8, 80, 115
periodogram
 averaging 288–303
 confidence limits 292–302
 estimator 272, 276
 variance 332–336
phase spectrum 11, 422
picket fence effect 86, 375
Poisson sum formula 71–73, 107

polynomial modeling
 See curve fitting
polynomial sets
 gram 36–37
 Lagrange 43–48
 spline 48–55
power 15, 233
power spectral density
 definition and properties 234–235, 254–257
 signal model 258–261
power spectral density estimation
 bias and variance 272–278
 See Blackman–Tukey
 See spectral modeling
 See periodogram
 See spectral smoothing
power spectral density, sample properties
 white noise 278–282
 random signal 283–288
prewhitening 407–410
probability density function
 estimation 155–162, 192
 properties 131
 types 131–134, 136–139, 170–171
probability distribution function
 properties 130
 types 132–134
probability distributions, sampling
 chi square 155–156, 174–175, 282, 426
 F 219–220, 230–231, 434
 Gaussian 172, 199, 382, 435–436
 Student's *t* 148–150, 173–174, 219, 230
 uniform 434
probability transformation 205–210
process loss 94, 290

quadrature spectrum 423–425

random number generation 201–205, 225
random process/signal
 definition 12, 178–180

random process/signal (*cont'd.*)
moments 183–186
system models 238–254
random variable
definition 128–130
Gaussian 132, 136–139
moments 134–136
relative frequency 130, 146

sampling frequency/interval 4, 74–75
scalloping loss 89
scatter diagram 18, 21
signal data
airline passengers 115
distillation process yields 226
gas furnace 445–448
grinding wheel roughness 397
heartbeat intervals 176–177
hospital census 340
rainfall 398
river flow 173
temperature and birth 449
variable star brightness 110–114
vibration 398
Wolfer sunspot numbers 227
simulation
correlation 210–216
probability density function 206–210
random process 205–217
spectral averaging 288–303
spectral modeling
confidence 382–384
estimator 374–377
line splitting 378–381
statistical properties 381
spectral smoothing 304–308
spectrum
energy 124
line 119
magnitude 11, 85, 118–119
phase 11, 85, 118–119
spline functions 48–55
squared error
definition 23, 163, 345
minimization 26–27, 35–37, 66, 349–355
standardization 137
stationarity
definition 12, 180–182, 188
tests 217–220, 230–231
system, discrete time 236–251, 342
correlation function relationships 239–240
power spectral density relationships 240–241
transfer function 236, 241

time shift 80
Toeplitz matrix 347
transfer function 236, 241, 251, 375
truncation 87–88

Welch method 303–304, 432
white noise 188, 199, 202–204, 207–209, 278–282, 375
windows
characteristics 338
data 88, 94, 124–125, 273
lag 273–274, 311–312, 338–339
spectral 89, 92–94, 124–125, 274, 331, 338–339
tables 124–125, 338–339

Yule–Walker equations
first order 346–347
general order 354, 365
matrix form 392–395

Wold decomposition 347

zero padding 86–87